AF361724

Challenges and Successes in Identifying the Transfer and Transformation of Phosphorus from Soils to Open Waters and Sediments

Challenges and Successes in Identifying the Transfer and Transformation of Phosphorus from Soils to Open Waters and Sediments

Editors

Donald S. Ross
Eric O. Young
Deb P. Jaisi

MDPI • Basel • Beijing • Wuhan • Barcelona • Belgrade • Manchester • Tokyo • Cluj • Tianjin

Editors
Donald S. Ross   Eric O. Young   Deb P. Jaisi
University of Vermont   USDA—Agricultural Research Service   University of Delaware
USA   USA   USA

Editorial Office
MDPI
St. Alban-Anlage 66
4052 Basel, Switzerland

This is a reprint of articles from the Special Issue published online in the open access journal *Soil Systems* (ISSN 2571-8789) (available at: https://www.mdpi.com/journal/soilsystems/special_issues/transfer_transformation_phosphorus).

For citation purposes, cite each article independently as indicated on the article page online and as indicated below:

LastName, A.A.; LastName, B.B.; LastName, C.C. Article Title. *Journal Name* **Year**, *Volume Number*, Page Range.

ISBN 978-3-0365-2432-0 (Hbk)
ISBN 978-3-0365-2433-7 (PDF)

Cover image courtesy of Donald S. Ross

Contents

About the Editors . . . . **vii**

Donald S. Ross, Eric O. Young and Deb P. Jaisi
Challenges and Successes in Identifying the Transfer and Transformation of Phosphorus from
Soils to Open Waters and Sediments
Reprinted from: *Soil Syst.* **2021**, *5*, 65, doi:10.3390/soilsystems5040065 . . . . **1**

Collin J. Weber and Christoph Weihrauch
Autogenous Eutrophication, Anthropogenic Eutrophication, and Climate Change: Insights
from the Antrift Reservoir (Hesse, Germany)
Reprinted from: *Soil Syst.* **2020**, *4*, 29, doi:10.3390/soilsystems4020029 . . . . **3**

Shreeram Inamdar, Nathan Sienkiewicz, Alyssa Lutgen, Grant Jiang and Jinjun Kan
Streambank Legacy Sediments in Surface Waters: Phosphorus Sources or Sinks?
Reprinted from: *Soil Syst.* **2020**, *4*, 30, doi:10.3390/soilsystems4020030 . . . . **33**

**Qingxin Zhang, Mackenzie Wieler, David O'Connell, Laurence Gill, Qunfeng Xiao and
Yongfeng Hu**
Speciation of Phosphorus from Suspended Sediment Studied by Bulk and Micro-XANES
Reprinted from: *Soil Syst.* **2020**, *4*, 51, doi:10.3390/soilsystems4030051 . . . . **53**

Anish Subedi, Dorcas Franklin, Miguel Cabrera, Amanda McPherson and Subash Dahal
Grazing Systems to Retain and Redistribute Soil Phosphorus and to Reduce Phosphorus Losses
in Runoff
Reprinted from: *Soil Syst.* **2020**, *4*, 66, doi:10.3390/soilsystems4040066 . . . . **65**

William Osterholz, Kevin King, Mark Williams, Brittany Hanrahan and Emily Duncan
Stratified Soil Sampling Improves Predictions of P Concentration in Surface Runoff and
Tile Discharge
Reprinted from: *Soil Syst.* **2020**, *4*, 67, doi:10.3390/soilsystems4040067 . . . . **79**

Rebecca M. Dzombak and Nathan D. Sheldon
Weathering Intensity and Presence of Vegetation Are Key Controls on Soil Phosphorus
Concentrations: Implications for Past and Future Terrestrial Ecosystems
Reprinted from: *Soil Syst.* **2020**, *4*, 73, doi:10.3390/soilsystems4040073 . . . . **95**

**Sarah Doydora, Luciano Gatiboni, Khara Grieger, Dean Hesterberg, Jacob L. Jones,
Eric S. McLamore, Rachel Peters, Rosangela Sozzani, Lisa Van den Broeck and
Owen W. Duckworth** Accessing Legacy Phosphorus in Soils
Reprinted from: *Soil Syst.* **2020**, *4*, 74, doi:10.3390/soilsystems4040074 . . . . **131**

**Jessica F. Sherman, Eric O. Young, William E. Jokela, Michael D. Casler, Wayne K. Coblentz
and Jason Cavadini**
Influence of Soil and Manure Management Practices on Surface Runoff Phosphorus and
Nitrogen Loss in a Corn Silage Production System: A Paired Watershed Approach
Reprinted from: *Soil Syst.* **2021**, *5*, 1, doi:10.3390/soilsystems5010001 . . . . **153**

Eric O. Young, Donald S. Ross, Deb P. Jaisi and Philippe G. Vidon
Phosphorus Transport along the Cropland–Riparian–Stream Continuum in Cold Climate
Agroecosystems: A Review
Reprinted from: *Soil Syst.* **2021**, *5*, 15, doi:10.3390/soilsystems5010015 . . . . **173**

Stella Gypser, Elisabeth Schütze and Dirk Freese
Single and Binary Fe- and Al-hydroxides Affect Potential Phosphorus Mobilization and Transfer
from Pools of Different Availability
Reprinted from: *Soil Syst.* **2021**, *5*, 33, doi:10.3390/soilsystems5020033 . . . . **193**

About the Editors

Donald S. Ross (Emeritus Research Professor) worked for 40 years at the University of Vermont as a technician, laboratory director and faculty member. He directed the Environmental Sciences Program for the College of Agriculture and Life Sciences and oversaw Vermont's soil fertility testing program. In addition to teaching in both Environmental Sciences and Soils, Ross conducted research on a wide variety of soil problems including riparian corridor phosphorus dynamics, hydropedology, forest soil monitoring, cation exchange in acidic forest soils, nitrification, earthworm effects on forest soil carbon, mercury dynamics at high elevation, and manganese biogeochemistry.

Eric O. Young grew up on a dairy farm in central New York and has over twenty years of experience in nutrient management research and outreach. A main research theme is evaluating agronomic and environmental practice effects on water (overland and subsurface flows) and soil quality in cold climate agroecosystems. Prior to joining USDA-ARS in 2018 as a Research Soil Scientist, Eric was a Research Agronomist for nine years at the William H. Miner Agricultural Research Institute focusing mainly on nitrogen and phosphorus loss in tile-drained systems.

Deb P. Jaisi has been an associate professor of Environmental Biogeochemistry at the University of Delaware since 2011. He was an Interdepartmental Batemen Scholar at Yale University and NSF Fellow at California Institute of Technology. He is a founding co-director of the Environmental Isotope Science (EIS) Center at UD. His research focuses on the biogeochemical processes revolving around phosphorus under three axes- phosphorus as an essential nutrient for all living beings, phosphorus as a contaminant to open waters, and phosphorus as a biocidal agent, at various scales, from molecular to ecosystem-scale processes.

 *soil systems*

Editorial

Challenges and Successes in Identifying the Transfer and Transformation of Phosphorus from Soils to Open Waters and Sediments

Donald S. Ross [1],*, Eric O. Young [2] and Deb P. Jaisi [3]

1   Department of Plant and Soil Sciences, University of Vermont, Burlington, VT 05405, USA
2   USDA—Agricultural Research Service, Marshfield, WI 54449, USA; eric.young@ars.usda.gov
3   Department of Plant and Soil Sciences, University of Delaware, Newark, DE 19808, USA; jaisi@udel.edu
*   Correspondence: dross@uvm.edu

check for
updates

Citation: Ross, D.S.; Young, E.O.; Jaisi, D.P. Challenges and Successes in Identifying the Transfer and Transformation of Phosphorus from Soils to Open Waters and Sediments. *Soil Syst.* **2021**, *5*, 65. https:// doi.org/10.3390/soilsystems5040065

Received: 9 October 2021
Accepted: 12 October 2021
Published: 19 October 2021

Publisher's Note: MDPI stays neutral with regard to jurisdictional claims in published maps and institutional affiliations.

The anthropogenic loading of phosphorus (P) to water bodies continues to increase worldwide, in many cases leading to increased eutrophication and harmful algal blooms [1]. Determining the sources of P and the biogeochemical processes responsible for this increase is often difficult because of the complexity of inputs and pathways, which vary both in spatial and temporal scales [2]. In order to effectively develop strategies to improve water quality, it is essential to develop a comprehensive understanding of the relationship of P pools with biological uptake and cycling under varied soil and water conditions. A wide variety of processes, including changes in P speciation; transformations between organic and inorganic species; and the transfer between biotic and abiotic forms occur along the route from soils to open waters and to sediments until ultimate burial, and together increase the complexity of quantifying processes, cycling, or tracing sources [3–5]. In addition, climate-change-related effects and feedback thereof often exacerbate a number of processes, including the redox-mediated release of legacy P in sediments.

In this special issue, we invited research and review articles that address the topic of soil P processes involving transfer and transformation across the landscape, either presenting novel research methods or synergy among non-traditional research fields; a review of existing successes and failures with underlying causes; or data-driven recommendations on the various approaches necessary to mitigate P loss and achieve the tangible goal of improving water quality. This volume contains eight original research articles [6–13] and two review articles [14,15].

General contribution papers covered the various aspects of basic–applied research on mineral–P interaction and how these reactions impact P mobilization, bioavailability, transfer [7], and speciation of P in different soil matrices using advanced analytical methods. Some of these methods included the application of XANES [13] and field-based research related to stream bank legacy nutrients [8]; natural and anthropogenic eutrophication, and its relationship to climate change [12]; and the evaluation of the impact of P due to (i) grazing systems [11], (ii) weathering and vegetation [6], and soil and manure management practices [10]. Together, these contributions improved our current understanding of the reactions and processes that impact P concentration, speciation, cycling, loss, and transfer from agroecosystems.

The two review papers took a holistic approach to cover an expansive area of P transformation processes along the cropland–riparian–stream continuum [15] and the assessment of legacy P [14]. The first review paper provided a broader assessment of P transformation and highlighted various approaches to improve and assess the effectiveness of riparian buffer zones in cold climate agroecosystems and highlighted the need of connecting hydro-biogeochemical and hydro-climatic data for the risk assessment on P loss to open waters. The chronic issue of legacy P was highlighted [14] by synthesizing the current knowledge of the bioaccessibility of different P forms, the transformations of

1

legacy P, and by proposing research and management approaches for potentially tapping legacy P for crop production.

We would like to thank all contributing authors in this special issue on 'Challenges and Successes in Identifying the Transfer and Transformation of Phosphorus from Soils to Open Waters and Sediments' and all reviewers for their constructive criticisms to improve the quality of science and delivery during the review process.

Funding: This research received no external funding.

Conflicts of Interest: The authors declare no conflict of interest.

References

1. Heisler, J.; Glibert, P.M.; Burkholder, J.M.; Anderson, D.M.; Cochlan, W.; Dennison, W.C.; Dortch, Q.; Gobler, C.J.; Heil, C.A.; Humphries, E.; et al. Eutrophication and harmful algal blooms: A scientific consensus. *Harmful Algae* **2008**, *8*, 3–13. [CrossRef] [PubMed]
2. Ator, S.W.; Denver, J.M. *Understanding Nutrients in the Chesapeake Bay Watershed and Implications for Management and Restoration—The Eastern Shore*; US Geological Survey Circular: Reston, VA, USA, 2015; Circular 1406; p. 72.
3. Jaisi, D.P.; Mingus, K.A.; Joshi, S.R.; Upreti, K.; Sun, M.; McGrath, J.; Massudieh, A. Linking sources, transformation, and loss of phosphorus in the soil-water continuum in a coastal environment. In *Multi-Scale Biogeochemical Processes in Soil Ecosystems: Critical Reactions and Resilience to Climate Changes*; Yu, Y., Marco, K., Nicola, S., Baoshan, X., Eds.; Wiley: Hoboken, NJ, USA, 2021.
4. Young, E.O.; Ross, D.S. Phosphate release from seasonally flooded soils: A laboratory microcosm study. *J. Environ. Qual.* **2001**, *30*, 91–101. [CrossRef] [PubMed]
5. Young, E.O.; Ross, D.S.; Alves, C.; Villars, T. Soil and landscape influences on native riparian phosphorus availability in three Lake Champlain Basin stream corridors. *J. Soil Water Conserv.* **2012**, *67*, 1–7. [CrossRef]
6. Dzombak, R.M.; Sheldon, N.D. Weathering Intensity and Presence of Vegetation Are Key Controls on Soil Phosphorus Concentrations: Implications for Past and Future Terrestrial Ecosystems. *Soil Syst.* **2020**, *4*, 73. [CrossRef]
7. Gypser, S.; Schutze, E.; Freese, D. Single and Binary Fe- and Al-hydroxides Affect Potential Phosphorus Mobilization and Transfer from Pools of Different Availability. *Soil Syst.* **2021**, *5*, 33. [CrossRef]
8. Inamdar, S.; Sienkiewicz, N.; Lutgen, A.; Jiang, G.; Kan, J.J. Streambank Legacy Sediments in Surface Waters: Phosphorus Sources or Sinks. *Soil Syst.* **2020**, *4*, 30. [CrossRef]
9. Osterholz, W.; King, K.; Williams, M.; Hanrahan, B.; Duncan, E. Stratified Soil Sampling Improves Predictions of P Concentration in Surface Runoff and Tile Discharge. *Soil Syst.* **2020**, *4*, 67. [CrossRef]
10. Sherman, J.F.; Young, E.O.; Jokela, W.E.; Casler, M.D.; Coblentz, W.K.; Cavadini, J. Influence of Soil and Manure Management Practices on Surface Runoff Phosphorus and Nitrogen Loss in a Corn Silage Production System: A Paired Watershed Approach. *Soil Syst.* **2021**, *5*, 1. [CrossRef]
11. Subedi, A.; Franklin, D.; Cabrera, M.; McPherson, A.; Dahal, S. Grazing Systems to Retain and Redistribute Soil Phosphorus and to Reduce Phosphorus Losses in Runoff. *Soil Syst.* **2020**, *4*, 66. [CrossRef]
12. Weber, C.J.; Weihrauch, C. Autogenous Eutrophication, Anthropogenic Eutrophication, and Climate Change: Insights from the Antrift Reservoir (Hesse, Germany). *Soil Syst.* **2020**, *4*, 29. [CrossRef]
13. Zhang, Q.X.; Wieler, M.; O'Connell, D.; Gill, L.; Xiao, Q.F.; Hu, Y.F. Speciation of Phosphorus from Suspended Sediment Studied by Bulk and Micro-XANES. *Soil Syst.* **2020**, *4*, 51. [CrossRef]
14. Doydora, S.; Gatiboni, L.; Grieger, K.; Hesterberg, D.; Jones, J.L.; McLamore, E.S.; Peters, R.; Sozzani, R.; van den Broeck, L.; Duckworth, O.W. Accessing Legacy Phosphorus in Soils. *Soil Syst.* **2020**, *4*, 74. [CrossRef]
15. Young, E.O.; Ross, D.S.; Jaisi, D.P.; Vidon, P.G. Phosphorus Transport along the Cropland-Riparian-Stream Continuum in Cold Climate Agroecosystems: A Review. *Soil Syst.* **2021**, *5*, 15. [CrossRef]

 *soil systems*

Article

Autogenous Eutrophication, Anthropogenic Eutrophication, and Climate Change: Insights from the Antrift Reservoir (Hesse, Germany)

Collin J. Weber and Christoph Weihrauch *

Department of Geography, Philipps University of Marburg, 35037 Marburg, Germany;
collin.weber@geo.uni-marburg.de
* Correspondence: christoph.weihrauch@geo.uni-marburg.de

Received: 10 February 2020; Accepted: 3 May 2020; Published: 7 May 2020

Abstract: Climate change is projected to aggravate water quality impairment and to endanger drinking water supply. The effects of global warming on water quality must be understood better to develop targeted mitigation strategies. We conducted water and sediment analyses in the eutrophicated Antrift catchment (Hesse, Germany) in the uncommonly warm years 2018/2019 to take an empirical look into the future under climate change conditions. In our study, algae blooms persisted long into autumn 2018 (November), and started early in spring 2019 (April). We found excessive phosphorus (P) concentrations throughout the year. At high flow in winter, P desorption from sediments fostered high P concentrations in the surface waters. We lead this back to the natural catchment-specific geochemical constraints of sediment P reactions (dilution- and pH-driven). Under natural conditions, the temporal dynamics of these constraints most likely led to high P concentrations, but probably did not cause algae blooms. Since the construction of a dammed reservoir, frequent algae blooms with sporadic fish kills have been occurring. Thus, management should focus less on reducing catchment P concentrations, but on counteracting summerly dissolved oxygen (DO) depletion in the reservoir. Particular attention should be paid to the monitoring and control of sediment P concentrations, especially under climate change.

Keywords: eutrophication; phosphorus; water quality; sediment; dissolved oxygen; phosphorus mobilization; climate change; algae bloom

---

1. Introduction

In the environmental sciences, eutrophication is considered problematic under excessive anthropogenic nitrogen (N) and phosphorus (P) enrichment in aquatic ecosystems [1–4]. The eutrophication of coastal marine waters is driven by N as the nutrient that limits biomass production. Instead, P is the major driver of primary production in rivers, lakes, and reservoirs [5,6]. A surplus of N and P triggers plant growth, mainly of algae [7–9]. Such algae blooms can drastically decrease water quality in terms of turbidity, foul odor/taste, and (brown-) green color [4]. Some species (e.g., cyanobacteria) even produce toxic substances that make water consumption harmful to human and animal health [1,2,10]. Moreover, dead algae sink to the ground and are decayed by microorganisms under the consumption of oxygen. Due to the large number of dead algae during algae blooms, decay increasingly depletes dissolved oxygen (DO) in the water until the ecosystem finally becomes hypoxic (low concentrations of DO) or even anoxic (depletion of DO) [1,11]. In the latter case, higher aquatic organisms depending on DO (e.g., fish) die and mass "fish kills" may result [3,6,12]. Such extermination of certain species undermines the structure and functioning of the aquatic ecosystem until it finally collapses [2,4].

Due to global warming, the frequency and geographical extent of harmful algae blooms has already increased in recent decades [1]. Climate change is expected to further intensify eutrophication [5,13]. Increasing air temperatures will lead to increasing water temperatures of surface waters [2,14,15]. Warmth fosters most biochemical processes [5,16]. It will likely trigger algae growth, especially of cyanobacteria at temperatures ≥25 °C [13,14]. Due to global warming, algae blooms are expected to start earlier and persist longer in the year [16,17]. Higher water temperatures also enhance the decay of dead organic matter at the bottom of water bodies, thus exacerbating the depletion of DO, and, in turn, P dissolution from sediment and fish kills [2,13]. Eutrophication risk is assumed to increase substantially in the summer, when high temperatures and low precipitation lead to low water levels. As a consequence, the dilution of nutrients is low [3,5,14]. Moreover, long residence times and a slow velocity of water are expected in warmer summers [16,18]. Such conditions foster the build-up of algae blooms. Additionally, climate change is predicted to increase the frequency and magnitude of extreme summer storms [2,13,18]. Such events trigger soil erosion and overland flow, transporting P-containing sediment into water bodies [5,14,19].

Climate change is projected to decrease surface water quality significantly [1,2,16]. This might become a challenge for the global community, because freshwater ecosystems are suppliers of drinking water [2,4]. As eutrophication reduces water quality, the production of drinking water from affected water bodies becomes more cost-intensive and insecure [3,5]. Moreover, due to lacking aesthetics or health risks, affected water bodies often cannot be used for tourism and recreational activities anymore [3,9]. There might even be considerable loss of value of waterfront real estate [4]. Furthermore, harmful algae blooms and fish kills result in lower profits for fisheries [9,12].

To assess the effects of climate change on eutrophication, a case study was conducted in the catchment of the Antrift reservoir (Hesse, Germany). Because 2018 and 2019 were uncommonly warm and dry years in Germany, our water quality assessment gives a preview of eutrophication under climate change conditions. We consider our data empirically illustrative of the assumptions made in modelling-based studies on climate change impacts on water quality. The goals of our study were (1) to empirically depict the temporal development of eutrophication in an uncommonly warm year, (2) to elucidate the effects of sediment P loads on eutrophication, and (3) to seek an indication of diffuse P sources beside erosion and surface runoff. A better understanding of eutrophication drivers and their relations with climate change is a prerequisite for targeted management to secure the availability of water in a warmer world [2,5].

2. Study Area

In Hesse (Germany), 79% of surface waters were classified as of poor ecological status in 2015, because they overstepped the total P (TP) threshold value of 0.1 mg/l [20]. According to modelling estimations, 1100 t of P are anthropogenically introduced into the Hessian surface waters every year; 65% resulting from communal point sources (wastewater treatment plants), 15% from erosion and surface runoff, 17% from other diffuse sources (e.g., rainwater channels), and 3% from industrial point sources [20]. On the basis of these results, the target values for effluents from wastewater treatment plants were drastically lowered from 1.0 to 0.2 mg TP/l in 2015–2021, which required the technical improvement of many sewage plants [21].

Improved wastewater treatment is estimated to reduce P inputs by only 43% [21]. Further technical improvement is currently considered technically impossible or not economically feasible. The Hessian State Ministry is skeptical that point-source-management would suffice to transfer the Hessian surface waters to a good ecological state [20]. When the proportion of point sources decreases, the relative contribution of diffuse sources to P inputs in water bodies increases. Hence, improved knowledge on diffuse P sources is required to develop targeted and efficient management strategies, to prevent or mitigate future P transfer to water bodies [3,5,14].

We conducted a case study in the catchment of the Antrift reservoir 5 km to the northwest of Alsfeld (Hesse, Germany) (Figure 1) [22]. The climate is warm and temperate [22,23], with an average annual temperature of 8.4 °C and with annual mean precipitation of 714 mm [24]. For 2035, regional climate projections depict a significant increase in the annual mean temperature (+0.8 °C; high confidence), an increase in heat extremes and droughts, an increase in events with intensive precipitation (>20 mm precipitation), and a slight increase in total annual precipitation [25,26].

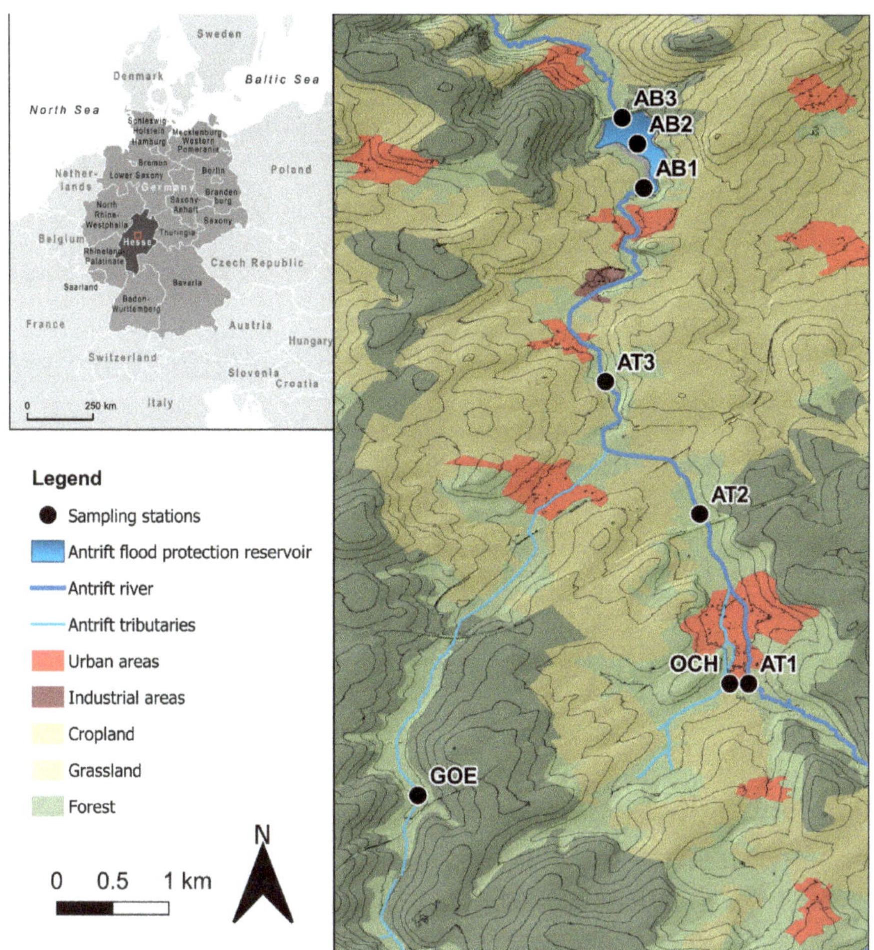

Figure 1. Location of the Antrift catchment with main water courses and sampling/measurement stations. (Data source: CORINE land use classification by Umweltbundesamt, 2019, DEM 1 m by Hessian Administration for Soil Management and Geoinformation, and overview map according to Weihrauch et al. 2020).

The catchment of the Antrift reservoir has a total area of 61.7 km^2 and an altitudinal difference of 152 m from the higher southern parts to the lower northern area. The population ranges between 49–69 persons/km^2 (Hessian average: 269 persons/km^2). The catchment is a rural region [27] composed of ca. 7.9% settlement area, 45.6% forest, and 45.2% agricultural areas (53.0% cropland and 47.0% grassland) [27,28]. Its landscape is shaped by geological and pedological features (i.e., suitability for land use; Table 1). Forests are mainly found on wooded hills and ridges, where Cambisols and Leptosols (WRB classification) developed from Tertiary alkali basalts or red Mesozoic sandstones.

Agricultural lands are largely situated in moderately sloping terrain on Luvisols, Stagnic Luvisols, and Anthrosols developed from Pleistocene loess. Grassland can largely be found in the floodplain areas and valley floors on Gleysols, Gleyic Fluvisols, and Anthrosols developed from Quaternary alluvial sediments [29–33]. Wooded areas are used for forestry, and grassland for grazing [28]. The farms mainly produce wheat, winter wheat, and barley. Out of 67 agricultural holdings, only five practice organic farming [27].

The Antrift catchment comprises the main river Antrift, with a length of 16.5 km, and the headwaters Göringer Bach (11.0 km) and Ocherbach (10.0 km) [34]. Both headwaters join with the main river before it flows into the Antrift reservoir (Figure 1). North of the reservoir, the Antrift river continues to drain towards the Weser River, which finally flows into the North Sea.

The Antrift reservoir has an average water surface of 0.31 km^2 and a maximum depth of 10.0 m [35]. It has been characterized as carbonatic and polymictic, with an unstratified water column throughout the year (except for May 2015) [35]. While construction began in 1971, the reservoir was first used for water retention in 1981 [34]. Besides flood protection, it was initially planned to be used for recreational activities and swimming. However, this has never been realized due to strong eutrophication, frequent algae blooms, and fish kills since the 1980s [34]. In 2015, the Antrift reservoir was classified as being in a poor ecological state (polytrophic) according to the European Water Framework Directive [35,36]. High TP concentrations of 0.14 mg TP/l were already measured at the reservoir's inlet [35], indicating that the local eutrophication issue was catchment-based instead of solely reservoir-based.

Table 1. Environmental characterization of the Antrift catchment.

Water Body Features		Riparian Features		Features of Surrounding Landscape	
Water Body Type	Sampling Stations	Bedrock/Sub-strata and Soils	Land Use	Bedrocks/Substrata and Soils	Land Use
Tributaries	GOE OCH	Alluvial sediments, Gleysols and Anthrosols	Grassland (meadow), strip of trees (alder) next to water courses	Basalt ridge with some loess accumulation along slopes, Leptosols, Cambisols, Luvisols	Forest (low percentage of cropland)
Main river	AT-1 AT-2 AT-3	Alluvial sediments, Gleyic Fluvisols and Anthrosols	Grassland (meadow), strip of trees (alder) next to water courses	Dominantly loess, Luvisols, Stagnic Luvisols	Cropland (low percentage of grassland)
Flood protection reservoir	AB-1 AB-2 AB-3	Alluvial sediments, Gleyic Fluvisols and Anthrosols	Strip of trees (alder, willow), dam structures, nature reserve	Oligocene clays (S, SE), Mesozoic sandstones (direct reservoir surroundings), Pleistocene loess (surrounding slopes), Luvisols, Stagnic Luvisols	Cropland (low percentage of forest)

The poor ecological status of the Antrift reservoir has remained unexplained to the present day. Several official assessments were carried out to develop effective strategies for water quality improvement. In line with the research history, the eutrophication of the Antrift reservoir was first attributed to P inputs from point sources [1,34,37,38]. However, eutrophication recurred after the only wastewater treatment facility of the catchment had been closed in 2002 (personal communication, F. Reissig, Regional Administrative Council Giessen). Since then, erosion and surface runoff have been considered the major sources of P inputs [37,39–41], even though Grimm et al. (2000) concluded in their assessment that only few agricultural lands in the Antrift catchment were severely threatened by erosion [42]. Moreover, erosion has been alleviated further due to financial support and far-reaching advice to farmers, with regard to conservation tillage, intercrop cultivation (especially in winter), conversion of arable lands into permanent grassland, as well as the establishment of vegetated buffer strips along surface waters [20,21]. Since the implementation of erosion conservation measures, the ongoing eutrophication of the Antrift reservoir has been increasingly attributed to P-containing sediment in the catchment water bodies, from which P can be desorbed [37,43,44]. As the impacts of erosion are rather small in this catchment [42], there might not be much sediment in the water courses from which P could be desorbed continuously in an extent large enough to lead to poor ecological state. With regard to legacy P [43], we presume a relatively quick equilibration between

P-containing sediments and the water column took place after sediment deposition. This is suggested by Cornel et al. (2015), who showed in a desorption experiment that 20% of the mobilizable P from solids within the effluents of waste-water treatment plants (with Fe and Al precipitation to eliminate P) were set free within a few minutes, and 50% were mobilized within two days [45]. Hence, as sediment deposition has been largely controlled in the Antrift catchment since the 1990s, we do not assume P release from prior sediments to be still relevant today. Thus, theoretically, it is difficult to attribute the present eutrophication of the Antrift reservoir to the P sources commonly discussed in the scientific literature.

3. Material and Methods

3.1. Water Sampling and Water Quality Analyses in the Field

To depict the temporal development of water quality in the Antrift catchment, we determined water quality parameters via field measurements and laboratory analyses of water samples. Our evaluation is based on threshold values and quantitative recommendations for a good ecological status of water bodies given in the "Oberflächengewässerverordnung", a national regulation implementing the guidelines of the EU Water Framework Directive (WFD) [46]. Water sampling and field measurements were performed in a three-week interval from 5 November 2018 until 6 August 2019 (i.e., 14 sampling/ measurement dates). Samples were taken from eight sampling stations in the Antrift catchment (Figure 1, Table 1). GOE and OCH represent the two main headwaters of the Antrift river and pass through forest areas. AT-1 represents the upper reaches of the Antrift river, and AT-2 was chosen because of its location downstream from the major settlement. AT-3 is situated downstream of the tributary Göringer Bach (GOE) and represents the central agricultural area of the catchment. Finally, we chose three sampling stations inside the Antrift reservoir, located at the inlet (AB-1), in the shallow water near the bank of the main reservoir (AB-2), and in the deepest section of the reservoir at the overflow construction of the dam (AB-3).

On each sampling date, three water samples were taken per sampling station from 10–30 cm below the water surface. For GOE, OCH, AT-1, AT2, and AT-3, samples were taken along the channel line. The stations within the Antrift reservoir (AB-1, AB-2, AB-3) were sampled at 3 m distance from the shoreline. Water samples were filled into PE-flasks and kept frozen until further analyses were performed [47].

The field measurements of water quality parameters were performed with an automatic water quality logging system during water sampling (portable logging multiparameter system, HI98290, Hanna Instruments Inc., Woonsocket, RI, USA). The system was operated with the following sensors and accuracies: temperature sensor (°C, ± 0.15), HI7609829-1 sensor for pH (pH, ± 0.02), and HI7609829-2 sensor for DO (%, saturation and concentration, ± 1.5% and ± 0.10 ppm, respectively). The sensor setup was calibrated through autocalibration with standard solutions (Hanna Instruments Inc.) before each measurement date [48]. Temperature, pH, and DO were measured during the whole investigation period.

3.2. Determination of Total Phosphorus in Water Samples

In the laboratory, total phosphorus (TP) was determined according to DIN EN ISO 6878: 2004-09 (2019). Prior to analysis, water samples were unfrozen, brought to an ambient temperature, and mixed with 1 ml 17.97 M sulfuric acid (H_2SO_4) to adjust pH to ~1. Samples were then oxidized with potassium persulfate ($K_2S_2O_8$), while heating to 90–100 °C for 90 minutes in closed vessels [49]. Afterwards, extracts were filtered with ashless blue ribbon filters (2 μm pore size). The extracts' concentration of phosphate was determined on a spectrophotometer (Genesys 10S; Thermo Fisher Scientific; Bremen, Germany) via the molybdenum-blue method at 700 nm [50,51]. Samples were measured three times and averaged. Phosphate concentrations were arithmetically converted into mg TP/l.

We calculated the relative standard deviations of the method (RSDM) and the detection limits for colorimetric TP measurement [52–54]. Data below the detection limit were excluded from data evaluation. In our 14 measurements, the maximum RSDM was 4.07%. Hence, we interpret our TP concentrations with a rounded uncertainty range of ± 5.0%.

3.3. Sediment Analyses

We took sediment samples from the water courses in the Antrift catchment to assess potential correlations between water quality parameters and sediment features, especially with regard to P mobilization from sediments. The sediment sampling sites correspond to the water sampling sites (Figure 1). Sediments were sampled about 5–10 cm deep from the bottom of the rivers and the reservoir with a plastic receptacle attached to an extensible stick. The first sampling date (5 November 2018) was in the end phase of a summerly algae bloom after a prolonged period with warm, dry weather and low flow conditions (average in the week before sampling: 0.22 m^3/s). The second sampling date (15 March 2019) was after an expected regeneration of water quality during winter. Even though the winter of 2018/2019 was relatively mild and dry, some precipitation occurred prior to our second sampling. Thus, the second date marks relatively moist weather and higher flow conditions (average in the week before sampling: 1.03 m^3/s). Sediment samples were stored airtight in plastic bags for about one day until further processing in the laboratory. The sediments were then dried at 70 °C in a drying furnace for eight days. Afterwards, they were ground in a mortar and sieved (2 mm mesh).

We determined the content of organic matter (OM; via loss on ignition; DIN 19684–3:2000–08) and pH (with 0.01 M $CaCl_2$; m:V = 1:2.5; DIN ISO 10390:1997-05) of the prepared samples. Furthermore, we assessed the sediments' texture with the integral suspension pressure method [55] after samples were prepared according to DIN ISO 11277:2002–08. We used three aliquots of each sediment sample for a P fractionation. We determined the following: (1) easily soluble P with 0.1 M hydrochloric acid (P_{dHCl}) [56]); (2) pedogenic oxide-bound P soluble in an ammonium oxalate–oxalic acid solution (P_{ox}; DIN 19684–6:1997–12 [56]); and (3) pseudo-total P (P_{AR}) after extraction with *aqua regia* (12.1 M HCl and 14.4 M HNO_3 in a ratio of 1:3; [56]). P_{dHCl} was measured on the spectrophotometer according to Murphy and Riley (1962) [48]. P_{ox} and P_{AR} were quantified with an ICP-MS (X Series 2; Thermo Fisher Scientific; Bremen, Germany), as well as Fe_{ox}, Al_{ox}, Mn_{ox}, Fe_{AR}, Al_{AR}, Mn_{AR}, Na_{AR}, Mg_{AR}, K_{AR}, and Ca_{AR}. All data were converted into the unit mg element/kg. The results of the three aliquots were averaged for further evaluation.

We calculated RSDM and the detection limits for ICP measurement of each element [52–54]. Data below the detection limits were excluded from evaluation. Data measured with a relative standard deviation (RSD) of ≥20% were also excluded [57,58]. To quantify measurement uncertainty, we added RSDM (as a standard parameter of calibration-based measurement) with RSD (as a parameter depicting data reproduction, and reflecting effects of heterogeneous matrixes typical for environmental samples [59]). As our sediments were measured in one calibration, only one RSDM resulted for each element. From the several RSD calculated during this measurement, we chose the median RSD for each element, because it is insensitive to outliers. In sum, we calculated rounded-up measurement uncertainty ranges of ± 2.0% (P_{ox}, P_{AR}, Al_{ox}, Al_{AR}, Fe_{ox}, Fe_{AR}, Ca_{AR}, K_{AR}, Na_{AR}, Mg_{AR}), ± 3.0% (Mn_{ox}, Mn_{AR}), and ± 9.0% (P_{dHCl}).

3.4. Discharge and Precipitation Datasets

At the inlet of the Antrift reservoir, there is a gauging station operated by the Wasserverband Schwalm e.V. (Schwalmstadt, Germany), which measured daily means of flow (m^3/s) during our investigation period. Because there is no climate station in the Antrift catchment, average daily precipitation and air temperature were calculated based on data from three climate stations close to the catchment (7–16 km off). These stations were Neustadt (Station-ID: DWD3516, 257 m a.s.l.), Alsfeld-Eifa (Station-ID: DWD91, 300 m a.s.l.), and Meiches (Station-ID: HLNUG4288360, 467 m a.s.l.).

The stations are operated by the German Weather Service (DWD) and the Hessian Agency for Nature Conservation, Environment and Geology (HLNUG) [60,61]. All climate data are freely available online.

3.5. Statistical Analyses

Basic statistical operations were performed in Microsoft Excel 2013 (Microsoft; Redmond, WA, USA), in R (R Core Team, 2013) and RStudio (Version 1.1.447; RStudio Inc.; Boston, MA, USA). Data visualization, tests for normal distribution (Shapiro–Wilk test), and Spearman correlation analyses were conducted with the R-packages "graphics", "stats", and "corrplot" [62,63]. Significances were tested on different levels. We interpret significant ($p \leq 0.05$) correlation coefficients as: weak (r_{SP} 0.4–<0.6), clear (r_{SP} 0.6–<0.8), and strong ($r_{SP} \geq 0.8$) [64]. For the sediment data, we performed Spearman correlation analyses for all our data, as well as separately for the data of each sampling date.

Each of our P fractions was determined for a new aliquot (i.e., 1.0 g) of the respective sediment sample. Hence, the resulting P contents are cumulative and contain P forms of different solubility (i.e., P_{ox} includes P_{dHCl}, and P_{AR} includes P_{ox}). To deduce the reactive behavior of sediment P, we calculated differential P fractions, which represent only one class of similar P solubility. Next to easily soluble P_{dHCl}, we calculated moderately labile P_{ml} (P_{ox} minus P_{dHCl}), and recalcitrant P_{rc} (P_{AR} minus P_{ox}).

To depict tendencies in the reactive behavior of sediment P, we determined the degrees of P mobilization (DPM). These ratios between two differently soluble P fractions elucidate whether the sediment has a tendency for P mobilization (large DPM) or P bonding (small DPM [59]). We calculated DPM2 via P_{dHCl}: P_{ox}, and DPM3 via P_{ox}: P_{AR}.

The poor ecological status of the water courses of the Antrift catchment is largely attributed to erosion and overland flow [65–67]. Hence, there should be P inputs into the surface waters shortly (i.e., maximally within a few hours) after precipitation events with sufficient intensity [68–71]. Thus, we used our limited dataset to seek an indication of temporal relationships between TP concentrations in the water and the occurrence of precipitation events. We applied a three-step data evaluation: (1) we tested for similarities between the time series of precipitation and discharge, as a function of the increments of precipitation relative to discharge ("cross correlation"; $p \leq 0.05$) [62], to find out if there is a temporal connection between a precipitation event and a discharge event, and to elucidate how quickly the first affects the later. (2) Next, we performed a Spearman correlation analysis between TP concentrations in the water and the number of days between each precipitation event and the next sampling date. For this, rainfall events were classified according to the daily precipitation sum into heavy (>10 mm/day), and maximal precipitation (maximal daily precipitation sum during each three-week measurement period). Moreover, we determined "higher precipitation" for events with a daily precipitation sum between the third quartile and the maximum of each three-week measurement period. Because we wanted to investigate the (short-term) effects of precipitation on erosion and P loss, we picked the "higher precipitation" event that occurred closest to our measuring date. (3) Finally, we checked the absolute number of days between each precipitation event and the next sampling date to evaluate the correlation results and exclude all events which are backdated more than two days. This two-day threshold was chosen according to the results of step (1) and the normal duration of erosion or overland flow events [68,69].

4. Results

4.1. Evaluation of Climate Data: Discharge and Precipitation

According to data from the climate station Alsfeld-Eifa (ca. 7 km northeast of the Antrift reservoir), the last six years show a positive temperature deviation between 0.29 °C and 1.30 °C from the reference period 1991–2020 (Figure 2) [72]. Between 2009 and 2019, the highest positive temperature increase was documented in 2018. A strong negative deviation in annual precipitation sums was reported.

Precipitation was ca. 19% lower in 2018 (−135.75 mm), and ca. 10.5% lower in 2019 than in the reference period (1991–2020).

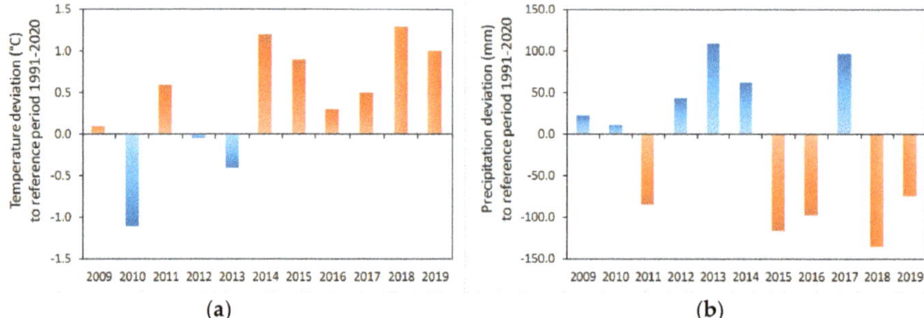

Figure 2. Temperature (**a**) and precipitation trend (**b**) in the Antrift catchment (2009–2019). (Data source: HLNUG (2019); operator: Deutscher Wetterdienst, data for the climate station Alsfeld-Eifa, station-ID: DWD91, 50.7447° N, 9.345° E.).

Between July and November 2018, monthly precipitation sums were lower by between 17.8% (August) and 38.3% (November), compared with the long-term (1979–2020) average monthly precipitation sums (Figure 3). Except for March and May 2019, monthly precipitation sums were below the long-term average precipitation sum all through our field measurement and sampling campaign. The clear increase of the precipitation sum in May 2019 (74.7% higher than in April 2019) resulted from a thunderstorm event with heavy rainfall on 20 May 2019 (Figure 4) [73]. For that day, a 24-h precipitation sum of 45.21 mm was calculated on the basis of data from three regional climate stations (see Materials and Methods). During our investigation period, eleven events with precipitation of >10 mm per day occurred.

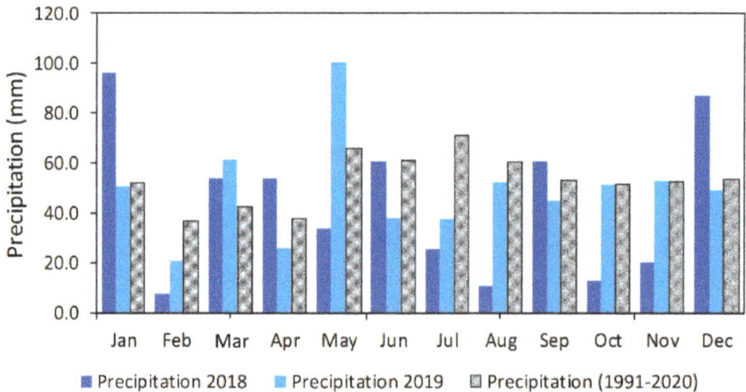

Figure 3. Monthly precipitation sums in the Antrift catchment during the investigation period, compared to long-term average monthly precipitation (reference period 1991–2020). (Data source: HLNUG (2019); operator: Deutscher Wetterdienst, data for the climate station Alsfeld-Eifa, station-ID: DWD91, 50.7447° N, 9.345° E.).

Discharge remained below its long-term average (MQ), with the exception of a few increases in discharge during the winter of 2018/2019, and as a result of the precipitation events previously mentioned (Figure 4). Discharge was low from late spring to summer 2018 and 2019 (data not shown).

Except for the weeks with higher rainfall, daily discharge declined below the long-term average of lowest discharge (MNQ).

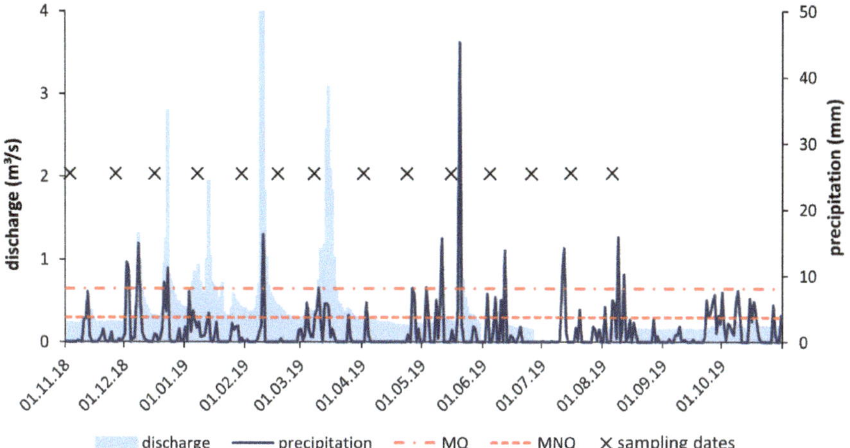

Figure 4. Monthly precipitation and discharge trend of 2018 and 2019 in the Antrift catchment, compared to average discharge in the reference period 1991–2000 (MQ = average discharge; MNQ = average low-flow discharge). Data source for discharge: Wasserverband Schwalm e.V. (2019), water level station 42881009, 50.76245° N, 9.20654° E. Precipitation data triangulated from HLNUG (2019): climate stations DWD3561 (50.8494° N, 9.1253° E), DWD91 (50.7447° N, 9.345° E), and HLNUG4288360 (50.6281° N, 9.26143° E).

4.2. Water Quality Assessment

4.2.1. Water Temperature

During the investigation period, water temperature rose steadily from winter to summer (Figure 5a). The increase in water temperature corresponds to the temporal development of air temperature, which only collapsed shortly in May 2019 due to a thunderstorm (see Supplementary Materials) [72]. During winter (December 2018–March 2019), the maximum water temperatures indicated good ecological status. The threshold for poor ecological status (20 °C) was reached in July and (locally) during the following months.

In the Antrift reservoir, maximum water temperatures ranged between 24.4 °C (26 June 2019) and 21.6 °C (5 June 2019), whereas the temperatures in its tributaries were ca. 6.3 °C lower on these dates (Figure 6a). Since February 2019, the water temperature has been higher in the Antrift reservoir than in the headwaters and the main river. During this time, median water temperatures ranged below the threshold for good ecological quality. Single values, mainly recorded at station AB-2 and AB-3 inside the Antrift reservoir, exceeded the threshold during summer.

4.2.2. pH

Our investigation period started with low surface water pH (6.8–6.9; 5 November 2018; Figures 5b and 6b). All our pH data of the first measurement date, as well as single outliers in the following periods, were within the range for good ecological status [46]. Since April 2019, the pH values inside the Antrift reservoir have been exceeding the threshold for good ecological quality until the end of the investigation period.

With incipient precipitation in late November 2018, pH increased all over the catchment and remained between pH 7.5–8.5 until March 2019. Since March 2019, pH was significantly higher (mean:

+ 1.5 pH) in the Antrift reservoir than in the main river and headwaters. The lowest pH was recorded in the headwaters, ranging by about 1.9 pH units under the reservoir pH.

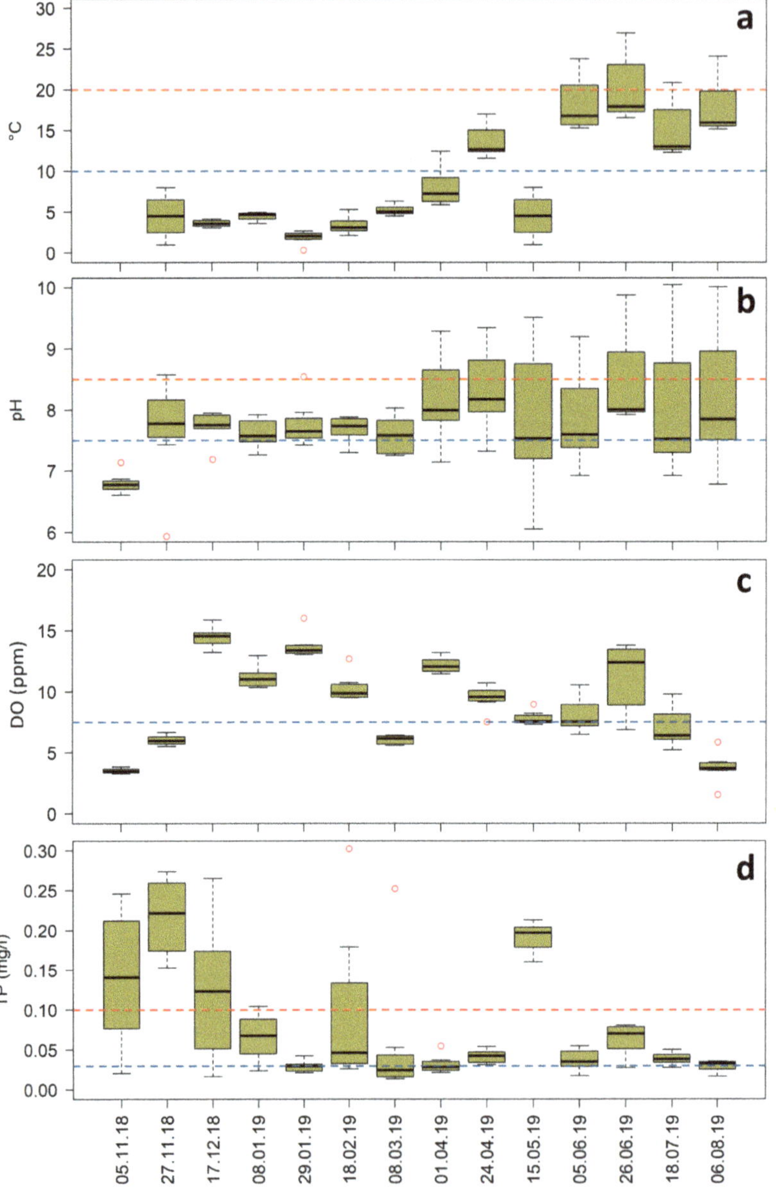

Figure 5. Time series of water quality data (n = 8 per date). Dashed lines represent thresholds for good ecological status according to OGewV (2016): (**a**) water temperature, winter maximum (blue line), and summer maximum (red line); (**b**) pH, minimum (blue line) and maximum (red line); (**c**) dissolved oxygen (DO), minimum; (**d**) total phosphorus (TP), threshold between good (<0.03 mg TP/l) and moderate ecological status (>0.03 mg TP/l; blue line), threshold between moderate and poor ecological status (>0.1 mg TP/l; red line).

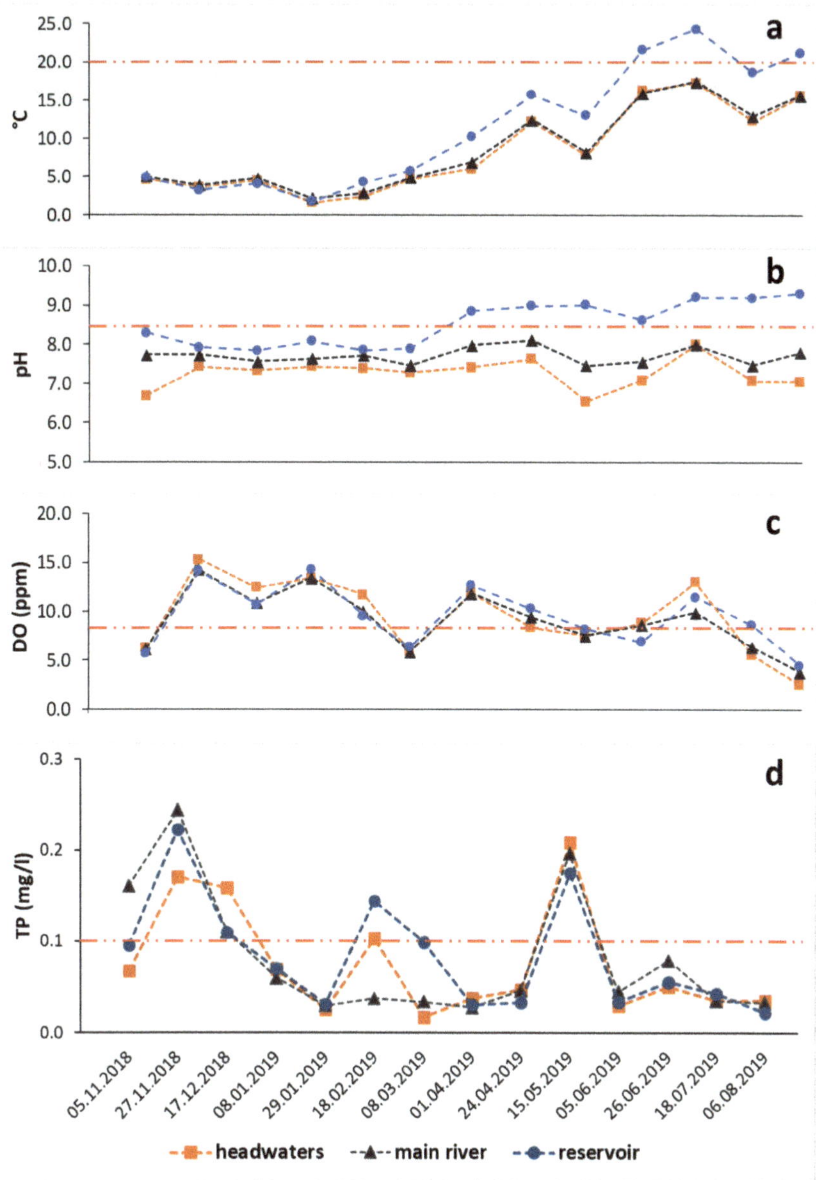

Figure 6. Temporal trends of water quality parameters during the investigation period; means according to types of water bodies (headwaters (n = 2): GOE, OCH; main river (n = 3): AT-1, AT-2, AT-3; reservoir (n = 3): AB-1, AB-2, AB-3). Dashed lines represent thresholds for good ecological status according to Bundesministerium der Justiz und für Verbraucherschutz (2016): (**a**) water temperature, summer maximum; (**b**) pH, maximum; (**c**) dissolved oxygen (DO), minimum; (**d**) total phosphorus (TP), threshold between moderate and poor ecological status (0.1 mg TP/l).

4.2.3. Dissolved Oxygen

Our DO data show no clear trend (Figures 5c and 6c). Generally, DO concentrations increased after a minimum in November 2018 (mean: 4.8 ppm) to stay above the threshold for good ecological status (8 ppm DO). Between July and August 2019, DO concentrations clearly decreased in the headwaters and the main river (80.3% and by 61.0%, respectively), compared to DO concentrations in June. The minimum DO concentrations in August 2019 ranged between 2.57 ppm in the headwaters and 4.52 ppm inside the reservoir. No significant spatial variability of DO concentrations was found for the Antrift catchment (Figure 6c). However, headwaters had slightly higher DO concentrations at single measurement dates.

4.2.4. Total Phosphorus

The TP concentrations were relatively high, varied strongly, and indicated bad ecological status in the investigated water courses (Figure 5d). During November and December 2018, TP levels remained high (>0.10 mg TP/l). From February 2019 to August 2019, lower TP concentrations were recorded, ranging between the lower and upper thresholds of eutrophic nutritional status. However, outliers occurred with >0.25 mg TP/l. The largest outliers were found at station AB-2 (reservoir), with 0.30 mg TP/l (18 February 2019) and 0.25 mg TP/l (3 March 2019). In May 2019, a single increase was detectable to TP concentrations of >0.20 mg TP/l. We found no systematic spatial trend of TP concentrations in the catchment despite slight differences between the hydrological water body types (Figures 6d and 7).

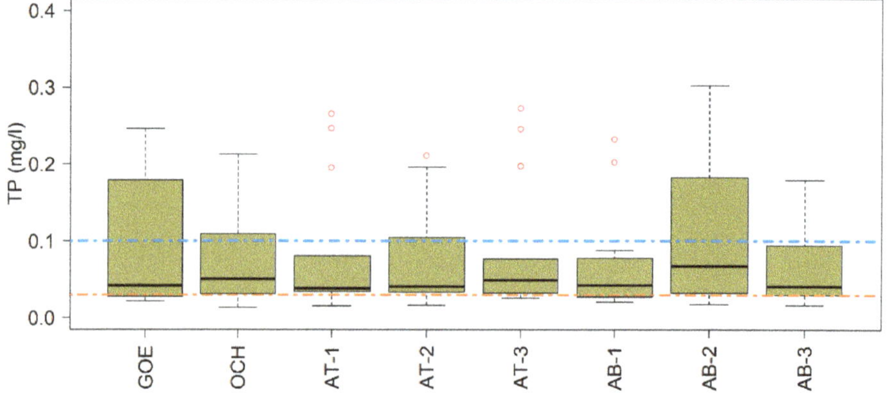

Figure 7. Total phosphorus (TP) concentration during the investigation period, according to sampling stations (n = 14 per station). Dashed lines represent thresholds of nutritional status according to OGewV (2016): orange line = threshold between good (<0.03 mg TP/l) and moderate ecological status (>0.03 mg TP/l); blue line = threshold between moderate and poor ecological status (>0.1 mg TP/l).

4.3. Correlation of Total Phosphorus Concentrations and Precipitation Events

The cross correlation between the time series of precipitation and discharge data shows a significant ($p \leq 0.05$) mutual influence for a period from 0 to ± 2 days (see Supplementary Materials). Within this two-day period, the results of the auto correlation function (ACF) were above the 0.1 ACF threshold. At point zero (i.e., precipitation and discharge event on the same day), the highest ACF value (0.416) is reached, followed by a continuous decrease until point +2 (i.e., discharge event two days after precipitation event).

Based on this result, Spearman correlation analyses (Figure 8) show a strong and significant ($p \leq 0.05$) positive correlation between the number of days since the last high, maximum, or heavy precipitation event (>10 mm/day) and the maximal drainage for each measurement period.

	drain_max	prec_heavy	prec_max	drain_high	prec_high	TP_mean
drain_max	1 (***)	1 (*)	1 (*)		-0.63 (*)	-0.84 (**)
prec_heavy		1 (***)	1 (*)		-0.63 (*)	-0.84 (**)
prec_max			1 (***)		-0.63 (*)	-0.84 (**)
drain_high				1 (***)		
prec_high					1 (***)	0.47 (**)
TP_mean						1 (***)

Abbreviations explanation:

drain_max	Number of days between last maximal drainage event and the next measurement date
prec_heavy	Number of days between last heavy precipitation event (>10 mm rainfall/day) and the next measurement date
prec_max	Number of days between last maximal precipitation event and the next measurement date
drain_high	Number of days between last higher drainage event (> 3rd quartile, see methods) and the next measurement date
prec_high	Number of days between last higher precipitation event (> 3rd quartile, see methods) and the next measurement date
TP_mean	Average total phosphorus (TP) concentration of all our measurement stations per measurement date

Figure 8. Spearman correlation coefficients of TP concentrations in the water and the number of days between each precipitation event and the next sampling date. (Significance levels: $p \leq 0.001$ (***); $p \leq 0.01$ (**); $p \leq 0.05$ (*)).

Comparing the average TP concentrations with the number of days since the last maximal drainage or maximal and heavy precipitation events, a strong significant ($p \leq 0.01$) negative correlation was found. Moreover, a moderate positive correlation was observed between average TP and the number of days since the last higher precipitation event. Instead, no significant correlation was found for TP and the number of days since the last higher drainage event.

Beside these results of the correlation analyses, the absolute number of days since an event was mostly larger than two days for all the precipitation events during our investigation period (see Supplementary Materials). Only in August 2019 did a maximal precipitation event coincide with our sampling date. For higher precipitation, six of 14 measurement periods showed an event within two days. No heavy precipitation event occurred within the two-day interval.

4.4. Sediment Analyses

4.4.1. General Sediment Features

The sediments from the Antrift catchment were largely silty-loamy (average: 63.9 mass-% silt, 18.1 mass-% sand, 18.0 mass-% clay; Table 2). However, the clay content decreased and the sand content increased from the headwaters (25.7 mass-% clay, 5.5 mass-% sand; averages) to the reservoir (16 mass-% clay, 27.3 mass-% sand; averages). The silt content decreased slightly from the headwaters/main river (mean: 68.7 mass-%) to the reservoir (mean: 56.7 mass-%). There was also temporal variation in the sediments' textures from the first to the second sampling (average deviations: +6.2 mass-% silt; +9.3 mass-% sand, −15.5 mass-% clay).

Table 2. General sediment features.

Station	Date	pH (CaCl$_2$)	OM [a] (Mass-%)	Grain Size (Mass-%)			P Fractions (mg/kg)		
				Clay	Silt	Sand	Pd$_{HCl}$	P$_{ox}$	P$_{AR}$
GOE	05.11.2018	6.03	18.79	44.34	47.81	7.84	379.61	720.53	1214.73
	15.03.2019	6.16	7.53	22.98	74.00	3.02	228.44	630.00	919.80
OCH	05.11.2018	6.66	15.30	28.96	65.08	5.96	326.37	634.67	1093.40
	15.03.2019	6.01	9.13	6.53	88.51	4.96	226.64	672.00	1094.80
AT1	05.11.2018	n.d.[b]	n.d.	n.d.	n.d.	n.d.	n.d.	n.d.	n.d.
	15.03.2019	6.26	10.72	10.05	75.83	14.12	429.26	796.13	1255.57
AT2	05.11.2018	7.17	8.85	22.12	54.26	23.61	559.00	952.23	1727.60
	15.03.2019	6.26	11.76	15.59	73.27	11.14	601.05	966.47	1658.77
AT3	05.11.2018	7.16	6.92	18.03	67.61	14.36	465.44	796.13	1311.10
	15.03.2019	6.70	3.92	5.54	72.28	22.18	482.89	767.67	1277.73
AB1	05.11.2018	7.03	7.42	23.72	66.22	10.06	331.99	836.73	1376.90
	15.03.2019	6.71	17.27	7.66	65.52	26.82	666.55	1188.37	1630.53
AB2	05.11.2018	7.61	3.17	21.11	57.07	21.83	84.33	221.20	507.27
	15.03.2019	7.14	2.89	12.59	72.86	14.55	304.29	546.70	1197.70
AB3	05.11.2018	7.48	3.41	25.77	66.19	8.04	3.94	107.33	429.80
	15.03.2019	7.69	2.09	5.38	12.29	82.33	230.10	154.93	402.73

[a] Organic matter; [b] not determined.

With an overall mean pH of 6.8, the sediments depict slightly acidic to neutral conditions. However, over our study period, the catchment pH shifted between slightly alkaline conditions (first sampling) and slightly acidic conditions (second sampling). The average pH increased from the headwaters (slightly acidic), to the main river, to the reservoir (slightly alkaline).

The sediments in the Antrift catchment were relatively high in organic matter (OM; overall mean: 8.6 mass-%). There was a slight tendency for OM contents to decrease from the headwaters to the main river to the reservoir at the first sampling (17.1 vs. 7.9 vs. 4.7 mass-%, respectively). However, OM contents were relatively similar at the second sampling (8.3 vs. 8.8 vs. 7.4 mass-%, respectively).

4.4.2. Sediment Phosphorus Contents

The sediments in the Antrift catchment are relatively high in each of the determined P fractions (Table 3). We found spatial and temporal variability in the sediment P data. The average P$_{dHCl}$ contents were significantly larger at the second sampling (+30.3%). Besides, we found relatively large ranges of the P$_{dHCl}$ means of the hydrological units. Through our study period, the largest mean P$_{dHCl}$ occurred in the main river stations. However, for the first sampling, the smallest mean P$_{dHCl}$ was in the reservoir. On the second date, the minimum average P$_{dHCl}$ was instead observed in the headwaters.

The P$_{ml}$ contents changed insignificantly from the first to the second sampling (+26%) and among the hydrological units. However, there was an increase in P$_{ml}$ in the headwaters and the reservoir at the second sampling. The mean P$_{rc}$ contents also increased in the reservoir at the second sampling, but decreased in the headwaters. The largest mean P$_{rc}$ contents were generally found in the main river.

We found relatively large DPM2 and DPM3 (i.e., >0.4). From the first to the second sampling, DPM2 remained constant in the main river. It was extremely small in the reservoir at the first sampling, but almost doubled to the second sampling. By contrast, DPM2 decreased by about 33% in the headwaters from Sampling 1 to Sampling 2. DPM3 increased slightly to the second sampling (+19%). However, we found no significant changes between the sampling dates and hydrological units.

Table 3. Differential P fractions and degrees of P mobilization of sediments (means).

	P_{dHCl} ^{a} (mg/kg)		P_{ml} ^{a} (mg/kg)		P_{rc} ^{a} (mg/kg)		DPM2 ^{a}		DPM3 ^{a}	
all data	355.8		341.1		484.2		0.47		0.57	
sampling	1	2	1	2	1	2	1	2	1	2
catchment	307.2 ^{a} *	400.2 ^{b}	302.6 ^{a}	381.6 ^{a}	484.6 ^{a}	483.8 ^{a}	0.43 ^{a}	0.51 ^{a}	0.52 ^{a}	0.62 ^{b}
headwaters	353.0 ^{a,A}	227.5 ^{b,A}	324.6 ^{a,A}	423.5 ^{b,A}	476.5 ^{a,A}	356.3 ^{b,A}	0.52 ^{a,A}	0.35 ^{b,A}	0.59 ^{a,A}	0.65 ^{a,A}
main river	512.2 ^{a,A}	504.4 ^{a,B}	362.0 ^{a,A}	339.0 ^{a,A}	645.2 ^{a,B}	553.9 ^{a,B}	0.59 ^{a,A}	0.60 ^{a,B}	0.58 ^{a,A}	0.61 ^{a,A}
reservoir	140.1 ^{a,B}	412.3 ^{b,AB}	248.3 ^{a,A}	407.8 ^{a,A}	382.9 ^{a,A}	498.6 ^{a,AB}	0.27 ^{a,B}	0.56 ^{b,B}	0.44 ^{a,A}	0.59 ^{a,A}

[a] for explanation, see 3.5; * different letters indicate significant differences between means ($p \leq 0.05$); lower-case letters refer to comparison by line (sampling dates per P fraction), upper-case letters refer to comparison by row (hydrological units per P fraction and sampling date).

4.4.3. Correlation Analyses

We conducted Spearman correlation analyses for all our data, as well as separately for the data of each sampling date. Different correlation coefficients result in these three versions of data aggregation. The specific results for the correlation of all data are shown in Figure 9. However, we also found overarching correlations (i.e., present in all three versions of data aggregation).

	Sediment data oxalate extraction			Sediment data aqua regia extraction							Sediment features					Water sample data			
	Al.ox	Fe.ox	Mn.ox	Al.AR	Fe.AR	Mn.AR	Na.AR	K.AR	Mg.AR	Ca.AR	pH	OM	clay	silt	sand	TP	Temp	pH	DO
PdHCl	0.82	0.50	-0.05	0.65	0.67	-0.12	0.75	0.55	0.79	0.81	-0.11	0.49	-0.34	-0.09	0.56	-0.04	0.24	0.06	0.20
P.ox	0.94	0.49	-0.08	0.74	0.69	-0.11	0.89	0.69	0.82	0.86	-0.17	0.55	-0.30	0.00	0.42	-0.31	0.14	-0.10	0.11
P.AR	0.90	0.42	0.03	0.73	0.83	0.06	0.87	0.66	0.84	0.81	-0.02	0.35	-0.30	-0.06	0.53	-0.10	0.12	-0.14	0.09
Al.ox	1.00	0.32	-0.05	0.86	0.71	-0.01	0.87	0.83	0.87	0.91	-0.11	0.53	-0.19	-0.02	0.32	-0.30	0.00	-0.24	0.03
Fe.ox		1.00	-0.07	0.04	0.32	-0.20	0.36	0.07	0.31	0.26	-0.36	0.22	-0.49	0.31	0.13	-0.27	0.40	0.38	0.38
Mn.ox			1.00	0.03	0.09	0.84	0.02	0.16	-0.13	-0.02	0.53	-0.27	0.27	-0.38	0.21	0.58	-0.12	-0.19	-0.15
Al.AR				1.00	0.53	0.03	0.67	0.70	0.81	0.83	-0.05	0.44	0.09	-0.14	0.16	-0.15	-0.24	-0.45	-0.21
Fe.AR					1.00	0.24	0.72	0.63	0.81	0.49	0.31	-0.12	-0.48	0.06	0.64	-0.01	0.04	-0.13	0.07
Mn.AR						1.00	0.00	0.18	-0.04	-0.01	0.72	-0.43	0.23	-0.47	0.37	0.58	-0.30	-0.34	-0.18
Na.AR							1.00	0.74	0.82	0.79	-0.10	0.48	-0.16	-0.09	0.27	-0.19	-0.13	-0.36	-0.18
K.AR								1.00	0.70	0.75	0.12	0.39	-0.14	-0.10	0.31	-0.20	-0.08	-0.23	-0.05
Mg.AR									1.00	0.79	0.07	0.31	-0.20	-0.13	0.42	-0.13	-0.24	-0.37	-0.20
Ca.AR										1.00	-0.15	0.71	0.06	-0.34	0.27	-0.13	-0.14	-0.31	-0.10
pH											1.00	-0.66	0.13	-0.46	0.59	0.46	-0.33	-0.28	-0.24
OM												1.00	0.14	-0.14	-0.23	-0.27	0.02	-0.06	-0.05
clay													1.00	-0.59	-0.47	0.41	-0.68	-0.67	-0.71
silt														1.00	-0.31	-0.57	0.61	0.53	0.54
sand															1.00	0.21	0.21	0.18	0.29
TP																1.00	-0.29	-0.23	-0.33
Temp																	1.00	0.92	0.93
pH																		1.00	0.88
DO																			1.00

Figure 9. Spearman correlation of sediment and water data for both samplings. $p \leq 0.05$ (green), $p \leq 0.01$ (blue) and $p \leq 0.001$ (brown), not significant (no color).

First, there are regular correlations between the moderately labile to recalcitrant P fractions and the acid metal cations, as well as the base cations. P_{dHCl}, P_{ox}, and P_{AR} are correlated positively and clearly to strongly with Al_{ox}, Al_{AR}, Fe_{AR}, Na_{AR}, Mg_{AR}, Ca_{AR}. For all data, there are furthermore positive correlations of P_{dHCl}, P_{ox} and P_{AR} with Fe_{ox} (weak) and K_{AR} (weak to clear; Figure 9).

Second, we mostly found weak positive correlations of P_{dHCl}, P_{ox}, and P_{AR} with the sediments' OM content.

5. Discussion

5.1. Evaluation of Climate Data: Discharge and Precipitation

The trends of air temperature and monthly precipitation showed uncommonly warm and dry conditions in the Antrift catchment in 2018 and 2019 (Figure 2). These years could be termed 'extreme', regardless of the general trend for precipitation to decrease on average by 68 mm per decade in Hesse [22,72]. Despite the long-term drought, increased heat and low flow conditions, individual higher to heavy precipitation events occurred during the investigation period, whose rainfall intensity might have enabled soil erosion [68,74]. However, a high potential for causing soil erosion can only be attributed to the storm event on 20 May 2019, which exceeded a daily precipitation sum of 70 mm, and could hence be termed an extreme rainfall event [68,74,75]. Such extreme weather conditions are assumed to increase in frequency under climate change [16].

Climate forecasts (e.g., IPCC, HLNUG) predict an increase of annual mean temperature, heat extremes, drought periods, and heavy precipitation events [25,26,76]. We documented all these factors during our investigation period. Therefore, our data are likely to draw a picture of water quality impairment under climate change.

5.2. Water Quality Assessment

5.2.1. Water Temperature

In general, water temperature is correlated with air temperature. An increase in air temperature through climate change leads to increasing water temperatures [16]. In addition, water temperature is a main factor controlling chemical and physical properties of aquatic ecosystems (e.g., pH, diffusion rates) [2,14]. The water temperature within the Antrift reservoir and its tributaries ranged between the lower and upper thresholds for good ecological status during our investigation period. Although a moderate increase of air temperature from winter to summer was detected (see Supplementary Materials), the influence of climate change and a long-term increase of water temperature could not be investigated, because this study considers a 10-month period only. However, a long-term increase in water temperature is documented for rivers and lakes by other studies and can be assumed for the Antrift catchment under aggravated climate change conditions [5,77]. Deviations from the moderate annual increase of water temperatures from spring to summer occurred only due to heavy rainfall events (e.g., a thunderstorm in May 2019). Moreover, minor differences were observed due to spatial variation between lower water temperatures at sampling sites shaded by riparian vegetation (e.g., headwaters, the main river) and higher temperatures in the unshaded Antrift reservoir.

Water temperature and sufficient nutrient concentrations stimulate algal growth [1]. To do so, temperatures were too low (<20 °C to 23.0 °C) at the beginning of our measurement period [77,78]. Climate change is projected to increase the persistence of high water temperatures—especially in shallow lakes—and thus lead to extended eutrophication periods and algal blooms [2,16,17]. However, Richardson et al. (2019) observed factors limiting total phytoplankton growth and fostering the decrease of harmful cyanobacteria in nutrient-rich environments. Thus, climate change effects on algae blooms might be more complex than commonly assumed [79].

In our study, temperatures have been favorable for algal growth since June 2019. On 5 June 2019, an incipient bloom of blue-green algae was observed, leading to a widespread green-brown algal carpet on the water surface of the Antrift reservoir in August 2019 (see Supplementary Materials). At the Antrift main river and the headwaters, no such visible water quality impairment was observed, probably due to more riparian shading and higher flow velocity [80,81].

5.2.2. pH

On 5 November 2018, pH in the Antrift catchment was relatively low (Figure 5b), probably as a result of the bacterial degradation of the 2018 algae bloom [1,15]. During the winter period, pH oscillated around the lower threshold (7.5 pH) for good ecological quality. Beginning in April 2019, pH rose to a more alkaline level [2]. The high fluctuation (i.e., range) of pH is typical for increasing eutrophication appearances [1,82,83]. At that time, we observed increased biomass formation in the reservoir. A second decrease of pH in connection with biomass increase was not observed, because our investigation period ended in August 2019.

5.2.3. Dissolved Oxygen

Despite the three-week interval between measurements, two trends in DO were observed. First, both algae blooms (end phase in November 2018, build-up in June 2019) were correlated with low DO concentrations. The minimum DO concentrations were measured during the end of the 2018 algae bloom, probably resulting from oxygen depletion through the intensified bacterial degradation of biomass [1,84,85]. Hence, after the winter increase, DO concentrations decreased again after August 2019, when the 2019 algae bloom had already built up for at least two months. Second, increases in DO coincided with increases in precipitation and discharge after the long drought phase in the second half of 2018. Increasing flow velocity leads to a higher level of DO in the surface waters [13,86]. Hence, during winter and spring 2019, DO stayed above the threshold for good ecological status.

The spatial differentiation of DO concentrations reflects the water temperature differences between the reservoir, the main river, and the headwaters [87]. Higher DO concentrations occurred at the headwater stations on single dates, probably as a result of higher flow velocity and shading by riparian vegetation (see Supplementary Materials). Still, during the 2019 algae bloom, the lowest DO concentrations were measured in the headwaters, most likely because of low flow rates and low velocity [88,89]. This indicates a higher vulnerability of the headwaters for water quality impairment under climate change (e.g., due to prolonged low flow conditions).

5.2.4. Total Phosphorus

Our data confirm that the Antrift reservoir is eutrophicated according to the EU Water Framework Directive. High TP concentrations occurred throughout our investigation period (Figure 5). However, TP concentrations indicating poor ecological quality were measured only from November to December 2018, and sporadically (event-driven) in February and May 2019. On the one hand, the high TP concentrations between November and December 2018 can be explained by low flow rates and a low nutrient dilution in the water column [16]. On the other hand, this period marks the end phase of the 2018 algae bloom with low DO concentrations and pH.

The average TP concentrations decreased continuously until January 2019. This can be attributed to the increase in precipitation and flow rates within the Antrift catchment. Even with the relaxation of the eutrophication situation during winter 2018, the median TP concentrations stayed close to the lower threshold of eutrophic conditions (except for single outliers; Figure 5d). After the autumnal decrease, TP concentrations varied slightly for the rest of the studied period.

Since July 2019, there was a strong negative correlation of TP concentrations with water temperature ($r_{SP} = -0.66$; $p = 0.01$). On 5 June 2019, the water temperature reached the 20 °C-threshold favorable for algae growth. As a consequence, the TP concentrations decreased, suggesting a consumption of P. This decrease of TP is corroborated by our field observations of increasing algal growth, which was also documented in other case studies [67,90].

With regard to our sampling stations, some conclusions can be drawn regarding potential P sources. We found no increase in TP concentrations with increasing percentage of agricultural land use or settlements [65,91,92]. Interestingly, TP concentrations are somewhat lower in the headwater stations than in the main river and reservoir, but they are still within the range of eutrophic nutritional status.

The reservoir and both stream classes are clearly different in hydrology and morphology. However, no significant difference in stream morphology is assumed between the headwaters and the main river [67]. Generally, our TP data mostly depict no clear differences between the three types of surface water bodies. Thus, not only is the Antrift reservoir eutrophicated, but all the investigated water bodies in the catchment, even the headwaters, are. This indicates that agriculture (erosion, surface runoff) and settlements (point sources) should not be termed the major factors for causing eutrophication in the Antrift catchment. Because the headwater station GOE is situated in a forest region, high TP concentrations also occur in areas without erosion risk and point sources (see Supplementary Materials). Hence, there seem to be further, so far unnoticed, sources for P losses from soil to water.

The TP trend, in combination with the trend of water temperature, pH, DO, and field observations, indicates that, under climate change, the relaxation of water quality impairment during winter could be shortened. Our data depict an extended duration of the 2018 algae bloom and a new worsening of water quality in April 2019. P stocks conserved from the prior algae bloom could accelerate the next algal bloom in spring. This is corroborated by Richardson et al. (2019), who showed that an increase in water temperature, sufficient nutrient levels, and hydrological conditions (e.g., intensive precipitation after drought events) might trigger cyanobacteria growth and complex algal bloom dynamics [79].

5.3. Phosphorus Sources in the Antrift Catchment

After the only wastewater treatment plant closed in 2002, no major point sources for P inputs remained in the Antrift catchment [93]. Erosion and surface runoff are thus expected to be the major P sources due to the large percentage of agricultural lands (>45%) [27]. However, the local erosion risk has been judged low in an official assessment [18,68]. Besides, vegetated buffer strips have been established along the water courses in the catchment, even along smaller drainage channels. Moreover, the influence of direct surface runoff and drainage must be considered minor, as a result of the very limited proportion of sealed surfaces.

We tested the influence of possible P inputs by erosion and surface runoff, with a correlation analysis between TP concentrations and their occurrence (i.e., number of days) after the last precipitation event (see Section 3.5. Statistical Analyses). The cross-correlation between the time series data of precipitation and discharge shows that precipitation events influenced the discharge trend significantly (see Supplementary Materials). This correlation is significant for a range between 0 and +2 days after rainfall events. Precipitation events thus had a significant impact on runoff in water bodies for a maximum of two days after the respective event.

The Spearman correlation analyses resulted in a significant negative correlation between average TP concentrations and the number of days since different kinds of precipitation events (Figure 8). Thus, a larger number of days since the last precipitation event coincided with lower TP concentrations in the water bodies. This indicates an influence of precipitation events and intensity on the TP concentrations (e.g., erosion and surface runoff) [71,93]. However, both processes are short-term processes, with a potential direct impact on water quality within a few minutes to hours, maximally in <2 days [68].

During our investigation period, no precipitation event occurred so shortly before a measurement date (i.e., <2 days) that it could be attributed plausibly to erosion or surface runoff [68]. In particular, all the potentially highly erosive higher and heavy precipitation events happened more than two days before our measurements [68–70]. Therefore, neither point sources nor extensive erosion (with surface runoff) can be considered as the major reasons for the all-season high P concentrations in the Antrift catchment. Instead, our results might tentatively point to an effect of underground P translocation (interflow-P), with a delay after precipitation events due to water infiltration and percolation [94,95]. However, our approach is limited by the relatively coarse temporal resolution of our measurements (three weeks) [68,69].

5.4. Sediment Analyses

5.4.1. General Sediment Features

The texture of the sediments results from the typical bedrocks and substrata in the Antrift catchment. The dominance of silt probably results from the large spatial extension of loess [29,96]. The somewhat higher clay content in the headwater area could be explained by the abundance of basalt in this section of the catchment. By contrast, the increasing percentage of sand in the reservoir sediments could be due to the occurrence of sandstones in the reservoir's surroundings. The variation we found between sediment textures of Samplings 1 and 2 generally demonstrates that sediments are dynamic components of the soil-water-interface, which undergo constant changes by transport, selection, and biochemical reactions [97,98]. Those processes could affect the textural composition of the sediments.

Over our investigation period, the sediments oscillated between slightly acidic and slightly alkaline conditions (Table 3). This general pH range can also be attributed to the dominance of alkali basalt and loess in the Antrift catchment [96]. That pH was lower in the headwater area than in the reservoir is probably geochemically determined. The headwaters are surrounded by basalts, which contain relatively many base cations (Ca, Mg, K), but even more Al/Fe (among other elements). Instead, base cations are probably more abundant in the loess areas (main river, reservoir) [96]. Moreover, pH could increase towards the reservoir due to the increasing concentration of leached basic cations, with increasing flow distance through the catchment.

We found the highest variance of pH between our sampling dates in the main river stations. This seems plausible because the main river represents the largest part of the catchment area and hence the most hydrologically diverse conditions [99]. Instead, the headwaters and the reservoir depict relatively small sections of the catchment with rather controlled hydrological conditions [97,99].

That we found large OM contents is not surprising for fluvial and lacustrine sediments, especially with eutrophication-enhanced primary productivity and the increased accumulation of dead OM [1,5]. Our OM contents decrease from the headwaters to the reservoir, which might be due to increased OM mineralization during summerly eutrophication and algae blooms [3]. Moreover, this spatial differentiation of sediment OM might be related to the spatial distribution of land uses in the Antrift catchment. The headwater areas are largely composed of forests, which produce and transfer more OM to the adjacent water courses. The main river and reservoir stations are instead located in agricultural (conventional farming) and grassland areas, where less OM is produced [96].

5.4.2. Sediment Phosphorus Contents

The large sediment P contents might in part result from the bedrocks in the Antrift catchment (e.g., basalt). Besides, they could be the legacy of prior land management (especially in the agricultural areas around the main river stations). However, the build-up of sediment is generally relatively small and also well-controlled in the Antrift catchment since the 1990s [42]. Instead, sediment pH might be a relevant and overarching driver of sediment P in this catchment. The mean pH levels of the sediments were between 6.0 and 6.5. P mobility is large in this pH range [100,101] because the solubility of P-sorbing metal oxides is relatively low (lability increases with pH), while at the same time, P-bearing minerals precipitating with base cations are also relatively easily soluble (stability increases with pH [96,102]). Hence, any shift of pH would foster P mobilization: Under more acidic conditions, the P-bearing minerals of base cations would be dissolved [103,104]; under more alkaline conditions, P would increasingly be desorbed from metal oxides [105,106]. It is a particularity of the Antrift catchment that sediment P retention is governed by sorption to metal oxides and by precipitation with base cations, instead of either of both. Thus, sediment P contents could be large due to the large number of potential reaction partners. However, a decrease in pH might be most favorable for P retention in the Antrift sediments, because metal cations are much more abundant in the sediments than the base cations.

We found larger sediment P_{dHCl} contents at the second sampling. This might result from an increased tendency to P_{ml} mobilization (i.e., conversion to P_{dHCl}) under high flow conditions (high dilution). Such conditions would cause a disequilibrium between bound P in the sediment and dissolved P in the water column [102,107]. As equilibrium would be shifted towards bound P, P mobilization would be enhanced to restore the equilibrium state [108,109].

The differences between the average P_{dHCl} contents at both samplings suggest that the three hydrological units might be subject to different P dynamics, especially under an acute algae bloom. In the reservoir, P_{dHCl} was very low at the first sampling, possibly due to enhanced P mobilization (i.e., loss of P_{dHCl}) during the 2018 algae bloom and severe DO depletion (Figure 5c). Without an acute algae bloom, P_{dHCl} increased significantly at the second sampling. By contrast, P_{dHCl} decreased in the headwaters at the second sampling. This could be the consequence of increased P mobilization under high flow conditions.

The P_{ml} contents increased slightly but non-significantly on the second date. This might suggest a tendency to P mobilization under high flow conditions, which could successively convert recalcitrant P forms into P_{ml} [6]. Even though not statistically significant, P_{ml} increased in the headwaters and in the reservoir. Most likely, in the reservoir, the small P_{ml} content at Sampling 1 resulted from the acute algae bloom and enhanced P mobilization from the sediment. In the headwaters, the increase in P_{ml} might instead have resulted from successive equilibrium-driven P mobilization under high flow conditions [6,107]. This might also explain the decrease in P_{rc} in the headwaters at the second sampling. However, the increase in P_{rc} in the reservoir might also have resulted from a relatively small P_{rc} at Sampling 1 due to the acute algae bloom with severe DO depletion.

The largest P_{rc} contents were generally found in the main river. With regard to the local bedrocks/substrata (loess), this is possibly due to the high abundance of base cations in the water, which can precipitate with dissolved P, especially under low flow conditions (Sampling 1) [96,109]. Moreover, the main river stations are not affected by algae-bloom–enhanced DO depletion and resulting P mobilization from sediments. Hence, more P_{rc} might accumulate

The relatively large DPM2 and DPM3 suggest that the sediments in the Antrift catchment are generally prone for P mobilization due to their large share of readily and moderately soluble P forms. We found slightly larger DPM2 (+19%) and DPM3 (+19%) at the second sampling, which might point to increased P mobilization under high flow conditions. By contrast, the small DPM2 in the reservoir at Sampling 1 might have resulted from algae-bloom-mediated P mobilization.

To the second sampling, DPM2 decreased significantly in the headwaters. Because the lowest pH were observed for the headwaters, there might be more reactions between P and Al/Fe, which are favored under acidic conditions [103,104]. P associated with Al/Fe would raise sediment P_{ml} (based on P_{ox}) [105,106] and thus result in a smaller DPM2. Furthermore, P_{ml} might have increased due to the above-mentioned successive mobilization of P_{rc}.

For DPM3, we found no significant changes between the sampling dates and hydrological units. However, the smallest DPM3 was found in the reservoir on the first date. This might also have resulted from the above-stated tendency for sediment P mobilization under severe DO depletion.

5.5. Spearman Correlation Analyses of Sediments and Water Samples

Our correlation analyses resulted in different relationships between the sediment and water parameters when all our data were combined (Figure 9) or grouped according to sampling dates (Figures A1 and A2). Hence, temporal dynamics of rather short-term processes like acute eutrophication with an algae bloom are probably depicted more adequately as snapshots according to sampling dates instead of sums or averages of longer timespans (e.g., yearly averages [3]).

5.5.1. Correlations with Acid and Base Metal Cations

We found overarching positive correlations of P_{dHCl}, P_{ox}, and P_{AR} with Al_{ox}, Al_{AR}, Fe_{AR}, Na_{AR}, Mg_{AR}, and Ca_{AR}. This indicates that P bonding in the sediments is strongly governed by the acid metal cations (especially Al), probably largely via sorption, which the strong correlations with Al_{ox} point to [102,103]. The correlations of P_{dHCl}, P_{ox}, and P_{AR} with Al_{AR} and Fe_{AR} might furthermore indicate that the precipitation of Al/Fe and P-containing minerals could play a role. However, the correlations of P_{dHCl}, P_{ox}, and P_{AR} with Na_{AR}, Mg_{AR}, and Ca_{AR} indicate that mineral precipitation, either with instantaneous (e.g., Ca-P-minerals) or with subsequent P bonding (e.g., adsorption and surface precipitation [110,111]), might also be relevant for sediment P retention.

5.5.2. Correlations with Sediment Organic Matter

We found weak positive correlations of P_{dHCl}, P_{ox}, and P_{AR} with the sediments' OM content. Because OM contains P, an increasing amount of OM also means an increase of the bound P fractions. Furthermore, the mineralization of organic matter leads to P mobilization to the water column and might result in an increase of TP. This is indicated by the weak negative correlation between sediment OM and TP in the water (Figures 9, A1 and A2). However, this correlation is visible only at the second sampling. On the first date, other eutrophication-related processes might instead have controlled TP.

Generally, the relatively large sediment OM contents (average: 8.61 mass-%) might hamper precipitation reactions and favor the formation of more weakly crystalline mineral phases [104,112]. This might explain why we found correlations of the P fractions with sand and silt but not with clay.

Generally, P bonding increases with clay content because most of the major P sorbing particles belong to the clay fraction (e.g., Fe/Al oxyhydroxides, OM [100,113]). However, Weihrauch (2018) found correlations of soil P fractions with sand and silt instead of clay in a study area characterized by weakly acidic to weakly alkaline soils [59]. The author explained this finding with the precipitation of carbonates adsorbing P weakly in a monolayer (at low P concentration) and with the precipitation of rather amorphous Ca-P-minerals (at higher P concentration) of larger diameter, due to the interference with OM [112]. Moreover, P might become occluded in a recalcitrant form within Mn and Fe concretions, which could grow with time [96,114,115]. This might explain the correlations between P_{AR} and sand (Figures 9, A1 and A2). The correlations between sand and P_{dHCl} might instead result from P sorption and surface precipitation on calcite, the formation of easily soluble Ca-P-minerals, and/or amorphous metal oxides [102,116]. Such a P retention in larger particles might also explain the clear negative correlation between the sediments' silt content and TP in the water (e.g., apatite often forms crystals in the silt fraction [96]). Hence, sediment OM is indicated to affect sediment P contents directly as well as indirectly by influencing the potential P bonding sites.

The above-mentioned correlations show for only the second sampling, probably due to strong effects of eutrophication on P dynamics on the first date. However, we found a strong negative correlation of sediment OM and pH in the water on the first date, which could have resulted from decreasing pH with increasing OM abundance and related mineralization [88,117].

6. Conclusions

We investigated the eutrophication of the Antrift reservoir in Hesse (Germany) in the uncommonly warm and dry years 2018/2019. Our results give an empirical preview on the development of local water quality under climate change. They furthermore enabled us to answer crucial questions regarding the poor ecological status of the Antrift reservoir.

Our results clarify that not only the Antrift reservoir is affected by eutrophication due to high TP concentrations, but so is the entire catchment. The catchment TP concentrations are high throughout the year, but apparently not due to the P sources commonly stated in eutrophication literature (e.g., agriculture). Possibly, natural pH-driven P mobilization from catchment soils and sediments (depending on soil moisture and flow conditions) might foster the eutrophication of the local surface waters. This hypothesis of an "autogenous eutrophication" should be investigated further.

Conceptually, we differentiate between two kinds of constraints on TP concentrations (graphical abstract): biological and geochemical constraints. Biological constraints are largely determined by water temperature, pH, and nutrient availability (especially P). In the Antrift catchment, pH and TP concentrations are favorable for algal growth throughout the year. Hence, temperature might be the major driver of algal growth, DO depletion, and resulting sediment P mobilization in summer. During the winter, water temperatures <20 °C prevent algal growth and DO depletion. Thus, only in summer can biological constraints lead to algae blooms with acute DO depletion. Under such conditions, they outweigh the geochemical constraints.

The geochemical constraints are determined by flow conditions (elemental dilution), water and sediment pH, redox status, as well as, to a lesser degree, by the local bedrocks/substrata (supply of P, Al, Fe, Mn, Ca, Mg, K). At low flow conditions and low dilution (summer), P might increasingly be retained in sediments. By contrast, P is likely to be increasingly mobilized from sediments at high flow conditions and high dilution (winter). This should be studied further.

In the literature, P dissolution/remobilization from sediment mostly describes P desorption/dissolution from sediment that has been transferred to water courses due to land use (especially agriculture) [43,118]. For our argument, it is not relevant where the sediments come from. Sediment P equilibrates relatively rapidly with P in the water column [6]. Hence, any P desorption/dissolution from P-enriched sediment from agricultural lands would likely happen quickly after deposition [45]. Then, the sediments would be subject to the geochemical constraints in the Antrift catchment, which generally foster a high abundance of labile sediment P and temporary P mobilization. This would apply to any sediment, regardless of its original P load and origin regarding land use. Hence, we are skeptical about the scientific theories of a "hysteresis" of P dissolution from sediment, which is sometimes evoked to explain why best-management to improve water quality practices do not succeed right away, but might need decades [43,119].

With regard to climate change, our data from the recent uncommonly warm years corroborate the assumptions of others (e.g., LeMoal et al. 2019; Moss et al. 2011; Whitehead et al. 2009) that global warming will likely foster eutrophication, as well as prolonged and spatially extended algae blooms. Climate change might hence aggravate biological constraints in summer. In winter, however, lowered flow conditions might instead lead to reduced P mobilization, or, in other words, to increased P retention in the catchment sediments. However, combined effects between cyanobacteria and phytoplankton, short-term hydrological changes, and nutrient availability in freshwater ecosystems and sediments might complicate a clear prediction of the interaction of constraints on P mobilization/retention under climate change [79,120,121].

To mitigate the eutrophication of the Antrift catchment, it is important to register that TP concentrations are naturally high throughout the year. Hence, any artificial reduction of TP concentrations (e.g., by common best-management practices) would foster geochemical P mobilization from the sediment. As mentioned above, the Antrift catchment could theoretically self-regulate its ecological status under natural conditions (e.g., by flushing winterly nutrient excesses downstream). The Antrift reservoir undermines this self-regulation and creates an environment of all-year P accumulation in the water column, which artificially fosters algae blooms in summer.

We conclude that the eutrophication (i.e., TP concentrations of 0.03–<0.1 mg TP/l) of the Antrift catchment cannot be prevented. Still, best-management practices should be kept active (e.g., erosion conservation) to prevent further P inputs that could aggravate eutrophication and lead to a poor ecological state (TP concentrations $\geq$0.1 mg/l). To restrict the summertime P mobilization from the sediment, DO concentrations should be started to be managed besides TP concentrations. Hence, algae blooms (and resulting DO depletion) might be controlled with algicides or the introduction of certain key species into the water bodies [1], bearing in mind the possible ecological consequences. Furthermore, artificial aeration of the Antrift reservoir could significantly relax summerly hypoxia and anoxia. Without hypoxia, the geochemical constraints would probably be dominant in summer and trigger P retention in the sediments. A certain monitoring and the eventual removal of the catchment sediments should also be considered, especially regarding sediments loaded with P in summer. In addition, the respective material would have beneficial features for application as fertilizer (e.g., high in carbonate and OM, pH 6–7, much labile P), if it was free of toxins, etc.

Our study indicates that—beside the known sources of P inputs—there might be currently unknown and unregulated diffuse P sources which contribute to the high P concentrations of the surface waters in the Antrift catchment. These sources seem to depend on the flow conditions, but are activated >2 days after precipitation events. Hence, we hypothesize an underground P translocation with the soil water (interflow-P) [94]. Such P sources should be investigated further so that they could be effectively managed and restricted. Hence, our study demonstrates that freshwater eutrophication is not yet conclusively understood despite the successes in research and practice. Challenges remain for science and should possibly be tackled apart from the disciplinary and conceptual mainstream.

Supplementary Materials: The following are available online at http://www.mdpi.com/2571-8789/4/2/29/s1, Figure S1: Average monthly air temperature for the climate station Alsfeld-Eifa, Figure S2: Pictures of algal bloom in the Antrift reservoir, Figure S3: Stations of water and sediment sampling in the Antrift catchment, Figure S4: Cross-correlation matrix of precipitation and discharge data, Table S5: Days since precipitation/discharge events according to measurement periods.

Soil Syst. **2020**, *4*, 29

Author Contributions: Conceptualization, C.J.W., C.W.; Methodology, C.J.W., C.W.; Software, C.J.W., C.W.; Validation, C.J.W., C.W.; Formal Analysis, C.J.W., C.W.; Investigation, C.J.W., C.W.; Resources, C.J.W., C.W.; Data Curation, C.J.W., C.W.; Writing—Original Draft Preparation, C.J.W., C.W.; Writing—Review & Editing, C.J.W., C.W.; Visualization, C.J.W., C.W.; Supervision, C.W.; Project Administration, C.W.; Funding Acquisition, C.W. All authors have read and agreed to the published version of the manuscript.

Funding: This research was funded by the Regional Administrative Council Gießen (Regierungspräsidium Gießen).

Acknowledgments: The authors gratefully acknowledge supply with discharge data by the Wasserverband Schwalm e.V. We furthermore thank Alexander Santowski for support in statistical data evaluation and Christin Wedra for support in laboratory analyses. Finally, we thank our students who partook in the research-based seminar and laboratory courses on the eutrophication of the Antrift reservoir.

Conflicts of Interest: The authors declare no conflict of interest.

Appendix A

	Sediment data oxalate extraction			Sediment data aqua regia extraction							Sediment features					Water sample data		
	Al.ox	Fe.ox	Mn.ox	Al.AR	Fe.AR	Mn.AR	Na.AR	K.AR	Mg.AR	Ca.AR	pH	OM	clay	silt	sand	TP	pH	DO
PdHCl	0.82	0.39	0.18	0.68	0.54	0.43	0.79	0.39	0.89	0.64	-0.36	0.50	-0.25	-0.14	0.39	0.32	-0.79	0.21
P.ox	0.96	0.39	0.21	0.86	0.64	0.43	0.89	0.57	0.93	0.75	-0.32	0.43	-0.25	0.00	0.43	-0.07	-0.89	0.21
P.AR	0.96	0.39	0.21	0.86	0.64	0.43	0.89	0.57	0.93	0.75	-0.32	0.43	-0.25	0.00	0.43	-0.07	-0.89	0.21
Al.ox	1.00	0.32	0.25	0.93	0.68	0.46	0.82	0.75	0.96	0.82	-0.29	0.39	-0.29	0.11	0.39	-0.21	-0.82	0.32
Fe.ox		1.00	0.14	0.07	0.14	0.25	0.21	-0.21	0.36	0.00	-0.39	0.07	-0.32	0.39	0.04	0.04	-0.21	0.00
Mn.ox			1.00	0.29	0.46	0.43	0.50	0.32	0.32	-0.14	0.07	-0.21	-0.32	0.79	0.07	0.21	-0.50	-0.46
Al.AR				1.00	0.50	0.21	0.79	0.86	0.86	0.89	-0.39	0.54	0.00	0.07	0.14	-0.21	-0.79	0.14
Fe.AR					1.00	0.93	0.64	0.57	0.71	0.25	0.46	-0.36	-0.82	0.36	0.86	-0.25	-0.64	0.43
Mn.AR						1.00	0.39	0.36	0.57	0.00	0.57	-0.54	-0.96	0.43	0.86	-0.14	-0.39	0.50
Na.AR							1.00	0.50	0.79	0.54	-0.21	0.32	-0.18	0.11	0.39	0.11	-1.00	-0.14
K.AR								1.00	0.71	0.75	-0.04	0.21	-0.18	0.21	0.21	-0.39	-0.50	0.36
Mg.AR									1.00	0.75	-0.25	0.36	-0.39	0.14	0.43	-0.04	-0.79	0.36
Ca.AR										1.00	-0.50	0.71	0.18	-0.29	0.04	-0.21	-0.54	0.36
pH											1.00	-0.89	-0.64	0.14	0.68	-0.29	0.21	0.36
OM												1.00	0.68	-0.46	-0.50	0.32	-0.32	-0.18
clay													1.00	-0.46	-0.82	0.21	0.18	-0.54
silt														1.00	0.00	-0.21	-0.11	-0.21
sand															1.00	-0.21	-0.39	0.57
TP																1.00	-0.11	-0.46
pH																	1.00	0.14
DO																		1.00

Figure A1. Spearman correlation of sediment and water data for Sampling 1 (5 November 2018). $p \leq 0.05$ (green), $p \leq 0.01$ (blue) and $p \leq 0.001$ (brown), not significant (no color).

	Sediment data oxalate extraction			Sediment data aqua regia extraction							Sediment features					Water sample data			
	Al.ox	Fe.ox	Mn.ox	Al.AR	Fe.AR	Mn.AR	Na.AR	K.AR	Mg.AR	Ca.AR	pH	OM	clay	silt	sand	TP	Temp	pH	DO
PdHCl	0.64	0.68	-0.14	0.61	0.71	-0.14	0.57	0.54	0.75	0.82	0.58	0.57	-0.11	-0.75	0.71	0.43	0.25	0.31	0.00
P.ox	0.93	0.61	-0.43	0.86	0.50	-0.32	0.82	0.75	0.82	1.00	0.09	0.89	-0.21	-0.32	0.43	-0.14	-0.11	-0.13	-0.32
P.AR	0.75	0.61	-0.32	0.68	0.86	-0.21	0.71	0.64	0.86	0.82	0.45	0.57	-0.18	-0.57	0.61	0.39	0.00	0.09	-0.29
Al.ox	1.00	0.50	-0.54	0.86	0.46	-0.32	0.89	0.82	0.79	0.93	-0.14	0.93	-0.07	-0.04	0.14	-0.29	-0.32	-0.41	-0.54
Fe.ox		1.00	0.14	0.18	0.64	0.14	0.71	0.43	0.64	0.61	0.16	0.50	-0.07	-0.64	0.32	0.25	-0.18	-0.05	-0.18
Mn.ox			1.00	-0.79	-0.11	0.93	-0.36	-0.07	-0.61	-0.43	0.45	-0.32	0.11	-0.43	0.21	0.36	0.54	0.58	0.75
Al.AR				1.00	0.29	-0.68	0.61	0.54	0.75	0.86	-0.09	0.75	-0.14	0.04	0.21	-0.29	-0.18	-0.27	-0.46
Fe.AR					1.00	0.00	0.64	0.46	0.75	0.50	0.45	0.21	-0.29	-0.61	0.54	0.61	-0.18	0.07	-0.36
Mn.AR						1.00	-0.14	0.21	-0.54	-0.32	0.36	-0.14	0.11	-0.29	0.14	0.25	0.39	0.40	0.57
Na.AR							1.00	0.79	0.82	0.82	-0.18	0.79	-0.25	-0.14	0.14	-0.18	-0.54	-0.47	-0.64
K.AR								1.00	0.46	0.75	0.11	0.82	-0.18	-0.14	0.32	-0.18	-0.07	-0.13	-0.21
Mg.AR									1.00	0.82	0.02	0.57	-0.29	-0.32	0.32	0.11	-0.43	-0.29	-0.64
Ca.AR										1.00	0.09	0.89	-0.21	-0.32	0.43	-0.14	-0.11	-0.13	-0.32
pH											1.00	-0.16	-0.11	-0.83	0.85	0.81	0.77	0.90	0.63
OM												1.00	0.04	-0.04	0.11	-0.43	-0.14	-0.32	-0.29
clay													1.00	0.18	-0.54	0.07	0.21	-0.18	0.18
silt														1.00	-0.82	-0.75	-0.50	-0.70	-0.43
sand															1.00	0.54	0.54	0.77	0.39
TP																1.00	0.43	0.61	0.32
Temp																	1.00	0.88	0.93
pH																		1.00	0.85
DO																			1.00

Figure A2. Spearman correlation of sediment and water data for Sampling 2 (15 March 2019). $p \leq 0.05$ (green), $p \leq 0.01$ (blue) and $p \leq 0.001$ (brown), not significant (no color).

References

1. Le Moal, M.; Gascuel-Odoux, C.; Ménesguen, A.; Souchon, Y.; Étrillard, C.; Levain, A.; Moatar, F.; Pannard, A.; Souchu, P.; Lefebvre, A.; et al. Eutrophication: A new wine in an old bottle? *Sci. Total Environ.* **2019**, *651*, 1–11. [CrossRef]
2. Nazari-Sharabian, M.; Ahmad, S.; Karakouzian, M. Climate Change and Eutrophication: A Short Review. Engineering. *Technol. Appl. Sci. Res.* **2018**, *8*, 3668–3672.
3. Bowes, M.; Davison, P.; Hutchins, M.; McCall, S.; Prudhomme, C.; Sadowsko, J.; Soley, R.; Wells, R.; Willets, S. *Climate Change and Eutrophication Risk in English Rivers*; Technical Report No. SC140013/R; Environment Agency: Bristol, UK, 2016.
4. Dodds, W.K.; Bouska, W.W.; Eitzmann, J.L.; Pilger, T.J.; Pitts, K.L.; Riley, A.J.; Schloesser, J.T.; Thornbrugh, D.J. Eutrophication of US Freshwaters: Analysis of Potential Economic Damages. *Environ. Sci. Technol.* **2009**, *43*, 12–19. [CrossRef]
5. Charlton, M.B.; Bowes, M.J.; Hutchins, M.G.; Orr, H.G.; Soley, R.; Davison, P. Mapping eutrophication risk from climate change: Future phosphorus concentrations in English rivers. *Sci. Total Environ.* **2018**, *613*, 1510–1526. [CrossRef]
6. Correll, D.L. The role of phosphorus in the eutrophication of receiving waters: A review. *J. Environ. Qual.* **1998**, *27*, 261–266. [CrossRef]
7. Fisher, M. Subsoil Phosphorus Loss: A complex problem with no easy solutions. *CSA News* **2015**, 4–9. [CrossRef]
8. Breuning-Madsen, H.; Bloch Ehlers, C.; Borggaard, O.K. The impact of perennial cormorant colonies on soil phosphorus state. *Geoderma* **2008**, *148*, 51–54. [CrossRef]
9. Sharpley, A.; Rekolainen, S. Phosphorus in agriculture and its environmental implications. In *Phosphorus Loss from Soil to Water*; Tunney, H., Carton, O.T., Brookes, P.C., Johnston, A.E., Eds.; CAB International: Wallingford, Ireland, 1997; pp. 1–53. ISBN 0851991564.
10. Lal, R. Phosphorus and the environment. In *Soil Phosphorus*; Lal, R., Stewart, B.A., Eds.; CRC Press: Boca Raton, FL, USA, 2017; pp. 209–223. ISBN 9781482257847.
11. Rockström, J.; Steffen, W.; Noone, K.; Persson, A.; Chapin, F.S.; Lambin, E.F.; Lenton, T.M.; Scheffer, M.; Folke, C.; Schellnhuber, H.J.; et al. A safe operating space for humanity. *Nature* **2009**, *461*, 472–475. [CrossRef]
12. Baker, L.A. Can urban P conservation help to prevent the brown devolution? *Chemosphere* **2011**, *84*, 779–784. [CrossRef]
13. Jeppesen, E.; Moss, B.; Bennion, H.; Carvalho, L.; DeMeester, L.; Feuchtmayr, H.; Friberg, N.; Gessner, M.O.; Hefting, M.; Lauridsen, T.L.; et al. Interaction of climate change and eutrophication. In *Climate Change Impacts on Freshwater Ecosystems*; Kernan, M.R., Battarbee, R.W., Moss, B., Eds.; Wiley-Blackwell: Chichester, UK, 2010; pp. 119–151. ISBN 9781405179133.
14. Moss, B.; Kosten, S.; Meerhoff, M.; Battarbee, R.W.; Jeppesen, E.; Mazzeo, N.; Havens, K.; Lacerot, G.; Liu, Z.; de Meester, L.; et al. Allied attack: Climate change and eutrophication. *IW* **2011**, *1*, 101–105. [CrossRef]
15. Nickus, U.; Bishop, K.; Erlandsson, M.; Evans, C.D.; Forsius, M.; Laudon, H.; Livingstone, D.M.; Monteith, D.; Thies, H. Direct impacts of climate change on freshwater ecosystems. In *Climate Change Impacts on Freshwater Ecosystems*; Kernan, M.R., Battarbee, R.W., Moss, B., Eds.; Wiley-Blackwell: Chichester, UK, 2010; pp. 38–64. ISBN 9781405179133.
16. Whitehead, P.G.; Wilby, R.L.; Battarbee, R.W.; Kernan, M.; Wade, A.J. A review of the potential impacts of climate change on surface water quality. *Hydrol. Sci. J.* **2009**, 101–123. [CrossRef]
17. Hering, D.; Haidekker, A.; Schmidt-Kloiber, A.; Barker, T.; Buisson, L.; Graf, W.; Grenouillet, G.; Lorenz, A.; Sandin, L.; Stendera, S. Monitoring the responses of freshwater ecosystems to climate change. In *Climate Change Impacts on Freshwater Ecosystems*; Kernan, M.R., Battarbee, R.W., Moss, B., Eds.; Wiley-Blackwell: Chichester, UK, 2010; pp. 84–118. ISBN 9781405179133.
18. Verdonschot, P.F.M.; Hering, D.; Murphy, J.; Jähnig, S.C.; Rose, N.L.; Graf, W.; Brabec, K.; Sandin, L. Climate change and the hydrology and morphology of freshwater ecosystems. In *Climate Change Impacts on Freshwater Ecosystems*; Kernan, M.R., Battarbee, R.W., Moss, B., Eds.; Wiley-Blackwell: Chichester, UK, 2010; pp. 65–83. ISBN 9781405179133.
19. Quinton, J.N.; Catt, J.A.; Hess, T.M. The selective removal of phosphorus from soil: Is event size important? *J. Environ. Qual.* **2001**, *30*, 538–545. [CrossRef]

20. Hessisches Ministerium für Umwelt, Klimaschutz, Landwirtschaft und Verbraucherschutz. *Bewirtschaftungsplan 2015–2021*; HMUKLV: Wiesbaden, Germany, 2015.
21. Hessisches Ministerium für Umwelt, Klimaschutz, Landwirtschaft und Verbraucherschutz. *Maßnahmenprogramm Hessen 2015–2021*; HMUKLV: Wiesbaden, Germany, 2015.
22. Hessisches Landesamt für Naturschutz, Umwelt und Geologie. Umweltatlas Hessen—Die Naturräume Hessens. Available online: http://atlas.umwelt.hessen.de/atlas/ (accessed on 10 October 2019).
23. Strahler, A.H. *Introducing Physical Geography*, 5th ed.; Wiley: Hoboken, NJ, USA, 2011; ISBN 9780470418116.
24. Climate-Data. Klima Romrod. Available online: https://de.climate-data.org/europa/deutschland/hessen/romrod-22186/ (accessed on 13 December 2019).
25. Intergovernmental Panel on Climate Change (Ed.) Global Warming of 1.5 °C. In *An IPCC Special Report on the Impacts of Global Warming of 1.5 °C above Preindustrial Levels and Related Global Greenhouse Gas Emission Pathways, in the Context of Strengthening the Global Response to the Threat of Climate Change, Sustainable Development, and Efforts to Eradicate Poverty*; Intergovernmental Panel on Climate Change: Geneva, Switzerland, 2018.
26. Masson-Delmotte, V.; Zhai, P.; Pörtner, H.O.; Roberts, D.E.A. IPCC, 2018: Summary for Policymakers. In *Global Warming of 1.5 °C. An IPCC Special Report on the Impacts of Global Warming of 1.5 °C above Preindustrial Levels and Related Global Greenhouse Gas Emission Pathways, in the Context of Strengthening the Global Response to the Threat of Climate Change, Sustainable Development, and Efforts to Eradicate Poverty*; Intergovernmental Panel on Climate Change, Ed.; Intergovernmental Panel on Climate Change: Geneva, Switzerland, 2018.
27. Hessisches Statistisches Landesamt. *Hessische Gemeindestatistik 2019 (Hessian Municipal Statistics 2019)*; Hessisches Statistisches Landesamt (Hessian State Statistical Office): Wiesbaden, Germany, 2019.
28. Friedrich, K.; Sabel, K.-J. *Standorttypisierung für Biotopenentwicklung, Maßstab 1:50.000, Blatt L 5320 Alsfeld*; Hessisches Landesamt für Naturschutz, Umwelt und Geologie: Wiesbaden, Germany, 2002.
29. Rosenberger, W.; Sabel, K.-J. *Geologische Karte von Hessen, Maßstab 1:50.000, Blatt L 5320 Alsfeld*; Hessisches Landesamt für Naturschutz, Umwelt und Geologie (HLNUG): Wiesbaden, Germany, 2002.
30. Lotz, K. *Einführung in die Geologie des Landes Hessen*; Hitzeroth: Marburg, Germany, 1995.
31. Meschede, M.; Warr, L.N. *The Geology of Germany: A Process-Oriented Approach*; Springer: Cham, Switzerland, 2019.
32. Diehl, O. *Geologische Karte von Hessen, Maßstab 1:25.000, Blatt 5221 Alsfeld*; Hessisches Landesamt für Naturschutz, Umwelt und Geologie (HLNUG): Darmstadt, Germany, 1926.
33. Nesbor, H.-D. *Geologische Karte von Hessen, Maßstab 1:25.000, Blatt 5321 Storndorf*; Hessisches Landesamt für Naturschutz, Umwelt und Geologie (HLNUG): Wiesbaden, Germany, 2009.
34. Frede, H.-G. *Nährstoffeinträge im Einzugsgebiet der Antrifttalsperre—Abschätzung der Anteile aus Diffusen und Punktuellen Quellen. Gutachten im Auftrag des Wasserverbandes Antrifttal*; unpublished assessment on behalf of the Wasserverband Antrifttal: Giessen, Germany, 1994.
35. Hessisches Landesamt für Naturschutz, Umwelt und Geologie. *Gütebewertung Seen 2015 Bereich Regierungspräsidium Kassel*; Hessisches Landesamt für Naturschutz, Umwelt und Geologie (HLNUG): Wiesbaden, Germany, 2015.
36. Hessisches Landesamt für Naturschutz, Umwelt und Geologie. *Umsetzung der Wasserrahmenrichtlinie in Hessen—Bewirtschaftungsplan Hessen 2009–2015, 1. Aufl.*; Hessisches Ministerium für Umwelt, Energie, Landwirtschaft und Verbraucherschutz: Wiesbaden, Germany, 2009; ISBN 9783892742890.
37. Landesbetrieb Landwirtschaft Hessen. *Pilotprojekt—Umstellung der Landwirtschaftlichen Bewirtschaftung zur Verminderung des Erosiven Nährstoffeintrages in den Antrift-Stausee*; Landesbetrieb Landwirtschaft Hessen: Alsfeld, Germany, 2007.
38. Tunney, H.; Coulter, B.; Daly, K.; Kurz, I.; Coxon, C.; Jeffrey, D.; Mills, P.; Kiely, G.; Morgan, G. *Quantification of Phosphorus Loss from Soil to Water: End of Project Report, ARMIS 4365*; Teagasc (Agriculture and Food Development Authority): Dublin, Ireland, 2000.
39. Bergström, L.; Kirchmann, H.; Djodjic, F.; Kyllmar, K.; Ulén, B.; Liu, J.; Andersson, H.; Aronsson, H.; Börjesson, G.; Kynkäänniemi, P.; et al. Turnover and losses of phosphorus in Swedish agricultural soils: Long-term changes, leaching trends, and mitigation measures. *J. Environ. Qual.* **2015**, *44*, 512–523. [CrossRef]
40. Andersson, H.; Bergström, L.; Djodjic, F.; Ulén, B.; Kirchmann, H. Topsoil and subsoil properties influence phosphorus leaching from agricultural soils. *J. Environ. Qual.* **2013**, 455–463. [CrossRef]
41. Delgado, A.; Scalenghe, R. Aspects of phosphorus transfer from soils in Europe. *J. Plant Nutr. Soil Sci.* **2008**, *171*, 552–575. [CrossRef]

42. Grimm, M. *Ermittlung des Erosiven P-Eintrages in den Antrift-Stausee und Maßnahmen zu ihrer Reduzierung*; Gesellschaft für Boden- und Gewässerschutz: Giessen, Germany, 2000.

43. Sharpley, A.; Jarvie, H.P.; Buda, A.; May, L.; Spears, B.; Kleinman, P. Phosphorus legacy: Overcoming the effects of past management practices to mitigate future water quality impairment. *J. Environ. Qual.* **2013**, *42*, 1308–1326. [CrossRef]

44. Borchardt, D.; Fischer, J. *Untersuchungen zum P-Rücklösungspotential der Sedimente der Antrifttalsperre*; Institut für Gewässerforschung und Gewässerschutz (IAG) und Universität Kassel: Kassel, Germany, 2001.

45. Cornel, P.; Schaum, C.; Lutze, R. *Gutachten zur Rücklösung von Phosphor im Gewässer aus Feststoffen von Kläranlagenabläufen*; unpublished Assessment on behalf of the Hessian Agency of Nature Conservation, Environment and Geology (HLNUG); Hessisches Landesamt für Naturschutz, Umwelt und Geologie (HLNUG): Wiesbaden, Germany, 2015.

46. Bundesministeriums der Justiz und für Verbraucherschutz. *Oberflächengewässerverordnung (National Implementation of the EU Water Framework Directive)*; OGewV: Berlin, Germany, 2016.

47. Kianpoor Kalkhajeh, Y.; Jabbarian Amiri, B.; Huang, B.; Henareh Khalyani, A.; Hu, W.; Gao, H.; Thompson, M.L. Methods for Sample Collection, Storage, and Analysis of Freshwater Phosphorus. *Water* **2019**, *11*, 1889. [CrossRef]

48. Hanna Instruments Inc. *HI9829 Unstruction Manual*; Hanna Instruments Inc.: Woonsocket, RI, USA, 2016.

49. DIN EN ISO 6878: 2004-09. Water quality—Determination of phosphorus—Ammonium molybdate spectrometric method (ISO 6878:2004); German version EN ISO 6878:2004. In *Handbuch der Bodenuntersuchung. Terminologie, Verfahrensvorschriften und Datenblätter; Physikalische, Chemische, Biologische Untersuchungsverfahren; Gesetzliche Regelwerke*; Deutsches Institut für Normung, Ed.; Wiley: Berlin, Germany, 2004.

50. Murphy, J.; Riley, J.P. A modified single solution method for the determination of phosphate in natural waters. *Anal. Chim. Acta* **1962**, *27*, 31–36. [CrossRef]

51. Tiessen, H.; Moir, J.O. Characterization of available P by sequential extraction. In *Soil Sampling and Methods of Analysis*, 2nd ed.; Carter, M.R., Gregorich, E.G., Eds.; CRC Press: Boca Raton, FL, USA, 2006; pp. 293–306.

52. Telgheder, U. Leitfaden zur Auswertung Analytischer Ergebnisse. Available online: https://www.uni-due.de/imperia/md/content/water-science/ws1314/1351_2661_ws1314_leitfaden_ausertung.pdf (accessed on 13 April 2017).

53. Molt, K.; Telgheder, U. Berechnung der Verfahrensstandardabweichung und Nachweis-, Erfassungs- und Bestimmungsgrenze Einer Kalibrierung Gemäß DIN 32645. Available online: https://www.uni-due.de/imperia/md/content/iac/git_erw_1.pdf (accessed on 13 April 2017).

54. JCGM = Joint Committee for Guides in Metrology. Evaluation of Measurement Data. Guide to the Expression of Uncertainty in Measurement. Available online: http://www.bipm.org/en/publication/guides/gum.html (accessed on 12 January 2020).

55. Durner, W.; Iden, S.C.; von Unold, G. The integral suspension pressure method (ISP) for precise particle-size analysis by gravitational sedimentation. *Water Resour. Res.* **2017**, *53*, 33–48. [CrossRef]

56. Weihrauch, C.; Schupp, A.; Söder, U.; Opp, C. Could oxalate-extractable phosphorus replace phosphorus fractionation schemes in soil phosphorus prospections? A case study in the prehistoric Milseburg hillfort (Germany). *Geoarchaeology* **2020**, *35*, 98–111. [CrossRef]

57. Thomas, R. A beginner's guide to ICP-MS—Part VII: Mass separation devices—Double-focusing magnetic-sector technology. *Spectroscopy* **2001**, *16*, 22.

58. Thomas, R. A beginner's guide to ICP-MS—Part VIII—Mass analyzers: Time-of-flight technology. *Spectroscopy* **2002**, *17*, 36.

59. Weihrauch, C. *Phosphor-Dynamiken in Böden: Grundlagen, Konzepte und Untersuchungen zur Räumlichen Verteilung des Nährstoffs*; Springer: Wiesbaden, Germany, 2018; ISBN 9783658223489.

60. Deutscher Wetterdienst. Precipitation and temperature data 2018–2019 climate station Alsfeld-Eifa (DWD91). In *Wetterextreme Hessen*; Hessian Agency for Nature Conservation, Environment and Geology: Wiesbaden, Germany, 2019.

61. Deutscher Wetterdienst. Precipitation and temperature data 2018–2019 climate station Neustadt (DWD 3561). In *Wetterextreme Hessen*; Hessian Agency for Nature Conservation, Environment and Geology: Wiesbaden, Germany, 2019.

62. R Core Team. *R: A Language and Environment for Statistical Computing*; R Foundation: Vienna, Austria, 2018.

63. Wei, T.; Simko, V. R Package "corrplot": Visualization of a Correlation. 2017. Available online: https://github.com/taiyun/corrplot (accessed on 21 December 2019).

64. Zimmermann-Janschitz, S. *Statistik in der Geographie. Eine Exkursion durch die Deskriptive Statistik*; Springer: Berlin, Germany, 2014; ISBN 9783827426116.

65. Sharpley, A.N.; Chapra, S.C.; Wedepohl, R.; Sims, J.T.; Daniel, T.C.; Reddy, K.R. Managing Agricultural Phosphorus for Protection of Surface Waters: Issues and Options. *J. Environ. Qual.* **1994**, *23*, 437–451. [CrossRef]

66. Bol, R.; Gruau, G.; Mellander, P.-E.; Dupas, R.; Bechmann, M.; Skarbøvik, E.; Bieroza, M.; Djodjic, F.; Glendell, M.; Jordan, P.; et al. Challenges of Reducing Phosphorus Based Water Eutrophication in the Agricultural Landscapes of Northwest Europe. *Front. Mar. Sci.* **2018**, *5*, 296. [CrossRef]

67. Withers, P.J.A.; Jarvie, H.P. Delivery and cycling of phosphorus in rivers: A review. *Sci. Total Environ.* **2008**, *400*, 379–395. [CrossRef]

68. Dunkerley, D.L. Rainfall intensity bursts and the erosion of soils: An analysis highlighting the need for high temporal resolution rainfall data for research under current and future climates. *Earth Surf. Dynam.* **2019**, *7*, 345–360. [CrossRef]

69. Lane, L.J.; Shirley, E.D.; Singh, V.P. Modeling erosion on hillslopes. In *Modeling Geomorphological Systems*; Abderson, M.G., Ed.; Wiley: New York, NY, USA, 1988; pp. 287–308.

70. Piacentini, T.; Galli, A.; Marsala, V.; Miccadei, E. Analysis of Soil Erosion Induced by Heavy Rainfall: A Case Study from the NE Abruzzo Hills Area in Central Italy. *Water* **2018**, *10*, 1314. [CrossRef]

71. Reid, K.; Schneider, K.; McConkey, B. Components of Phosphorus Loss from Agricultural Landscapes, and How to Incorporate Them into Risk Assessment Tools. *Front. Earth Sci.* **2018**, *6*, 179. [CrossRef]

72. Wetterextreme Hessen; Hessian Agency for Nature Conservation, Environment and Geology, Ed.; Wiesbaden. 2019. Available online: https://www.hlnug.de/?id=11522 (accessed on 17 December 2019).

73. Hessian Agency for Nature Conservation, Environment and Geology. Hochwasservorwarnung 20.05.2019. (Flood Advance Warning). Available online: http://www2.hochwasser-hessen.de/hochwasserportal-hessen/aktuelle-hochwasserlage/meldung.html?tx_pghochwasser_pi1%5Baction%5D=show&tx_pghochwasser_pi1%5Bmeldung%5D=1157&tx_pghochwasser_pi1%5Bcontroller%5D=Meldung&cHash=1cd845bb5cc04584adace543515f5498 (accessed on 6 July 2019).

74. Umweltbundesamt. Erosion. Available online: https://www.umweltbundesamt.de/en/topics/soil-agriculture/land-a-precious-resource/erosion#textpart-1 (accessed on 17 December 2019).

75. Schwertmann, U.; Vogel, W.; Kainz, M. *Bodenerosion durch Wasser—Vorhersage des Abtrags und Bewertung von Gegenmaßnahmen (Soil Erosion by Water—Prediction of the Removal and Evaluation of Countermeasures)*; Ulmer: Stuttgart, Germany, 1987.

76. Hessisches Ministerium für Umwelt, Klimaschutz, Landwirtschaft und Verbraucherschutz. *Integrierter Klimaschutzplan Hessen 2025*; Hessisches Ministerium für Umwelt, Energie, Landwirtschaft und Verbraucherschutz: Wiesbaden, Germany, 2019.

77. Foley, B.; Jones, I.D.A.N.; Maberly, S.C.; Rippey, B. Long-term changes in oxygen depletion in a small temperate lake: Effects of climate change and eutrophication. *Freshwater Biol.* **2012**, *57*, 278–289. [CrossRef]

78. O'Neil, J.M.; Davis, T.W.; Burford, M.A.; Gobler, C.J. The rise of harmful cyanobacteria blooms: The potential roles of eutrophication and climate change. *Harmful Algae* **2012**, *14*, 313–334. [CrossRef]

79. Richardson, J.; Feuchtmayr, H.; Miller, C.; Hunter, P.D.; Maberly, S.C.; Carvalho, L. Response of cyanobacteria and phytoplankton abundance to warming, extreme rainfall events and nutrient enrichment. *Glob. Change Biol.* **2019**, *25*, 3365–3380. [CrossRef]

80. Neufeld, J.D. Can Shade Structures Help Riparian Areas? A look at using constructed shades to pull cattle off riparian areas in northeastern Nevada. *Rangelands* **2005**, *27*, 24–30. [CrossRef]

81. Rutherford, J.C. Some approaches for measuring and modelling riparian shade. *Ecol. Eng.* **2005**, *24*, 525–530. [CrossRef]

82. Smith, V.H. Eutrophication of freshwater and coastal marine ecosystems a global problem. *Environ. Sci. Pollut. Res.* **2003**, *10*, 126–139. [CrossRef]

83. Liu, W.-C.; Chen, W.-B.; Kimura, N. Impact of phosphorus load reduction on water quality in a stratified reservoir-eutrophication modeling study. *Environ. Monit. Assess.* **2009**, *159*, 393–406. [CrossRef]

84. Ansari, A.A.; Gill, S.S. (Eds.) *Eutrophication: Causes, Consequences and Control*; Springer: Dordrecht, The Netherlands, 2014.

85. Hudnell, H.K. *Cyanobacterial Harmful Algal Blooms: State of the Science and Research Needs*; Springer: New York, NY, USA, 2008.

86. Kernan, M.R.; Battarbee, R.W.; Moss, B. (Eds.) *Climate Change Impacts on Freshwater Ecosystems*; Wiley: Chichester, UK, 2010; ISBN 9781405179133.

87. Williams, R.J.; White, C.; Harrow, M.L.; Neal, C. Temporal and small-scale spatial variations of dissolved oxygen in the Rivers Thames, Pang and Kennet, UK. *Sci. Total Environ.* **2000**, 497–510. [CrossRef]

88. Hilton, J.; O'Hare, M.; Bowes, M.J.; Jones, J.I. How green is my river? A new paradigm of eutrophication in rivers. *Sci. Total Environ.* **2006**, *365*, 66–83. [CrossRef]

89. Dorgham, M.M. Effects of Eutrophication. In *Eutrophication: Causes, Consequences and Control*; Ansari, A.A., Gill, S.S., Eds.; Springer: Dordrecht, The Netherlands, 2014; pp. 29–44.

90. Young, K.; Morse, G.K.; Scrimshaw, M.D.; Kinniburgh, J.H.; MacLoad, C.L.; Lester, J.N. The relation between phosphorus and eutrophication in the Thames catchment, UK. *Sci. Total Environ.* **1999**, *228*, 157–183. [CrossRef]

91. Djodjic, F.; Markensten, H. From single fields to river basins: Identification of critical source areas for erosion and phosphorus losses at high resolution. *Ambio* **2019**, *48*, 1129–1142. [CrossRef]

92. Cassidy, R.; Thomas, I.A.; Higgins, A.; Bailey, J.S.; Jordan, P. A carrying capacity framework for soil phosphorus and hydrological sensitivity from farm to catchment scales. *Sci. Total Environ.* **2019**, *687*, 277–286. [CrossRef]

93. Kleinman, P.J.A.; Church, C.; Saporito, L.S.; McGrath, J.M.; Reiter, M.S.; Allen, A.L.; Tingle, S.; Binford, G.D.; Han, K.; Joern, B.C. Phosphorus leaching from agricultural soils of the delmarva peninsula, USA. *J. Environ. Qual.* **2015**, *44*, 524–534. [CrossRef]

94. Weihrauch, C. Dynamics need space—A geospatial approach to soil phosphorus' reactions and migration. *Geoderma* **2019**. [CrossRef]

95. Christianson, L.E.; Harmel, R.D.; Smith, D.; Williams, M.R.; King, K. Assessment and Synthesis of 50 Years of Published Drainage Phosphorus Losses. *J. Environ. Qual.* **2016**, *45*, 1467–1477. [CrossRef]

96. Blume, H.-P.; Brümmer, G.W.; Fleige, H.; Horn, R.; Kandeler, E.; Kögel-Knabner, I.; Kretzschmar, R.; Stahr, K.; Wilke, B.-M. *Scheffer/Schachtschabel Soil Science*, 1st ed.; Springer: Berlin/Heidelberg, Germany, 2016; ISBN 9783642309410.

97. Miall, A. *Fluvial Depositional Systems*; Springer: Cham, Switzerland, 2014.

98. Robert, A. *River Processes: An Introduction to Fluvial Dynamics*; Oxford University Press: Oxford, UK, 2003.

99. Fryirs, K.A.; Brierley, G.J. *Geomoprhic Analysis of River Systems: An Approach to Reading the Landscape*; Wiley-Blackwell: Oxford, UK, 2013.

100. Holliday, V.T.; Gartner, W.G. Methods of soil P analysis in archaeology. *J. Archaeol. Sci.* **2007**, *34*, 301–333. [CrossRef]

101. Amberger, A. *Pflanzenernährung. Ökologische und Physiologische Grundlagen, Dynamik und Stoffwechsel der Nährelemente*, 3rd ed.; 74 Tabellen; Ulmer: Stuttgart, Germany, 1988; ISBN 3800125536.

102. Weihrauch, C.; Opp, C. Ecologically relevant phosphorus pools in soils and their dynamics: The story so far. *Geoderma* **2018**, *325*, 183–194. [CrossRef]

103. Kruse, J.; Abraham, M.; Amelung, W.; Baum, C.; Bol, R.; Kuehn, O.; Lewandowski, H.; Niederberger, J.; Oelmann, Y.; Rueger, C.; et al. Innovative methods in soil phosphorus research: A review. *J. Plant Nutr. Soil Sci.* **2015**, *178*, 43–88. [CrossRef]

104. Oburger, E.; Jones, D.L.; Wenzel, W.W. Phosphorus saturation and pH differentially regulate the efficiency of organic acid anion-mediated P solubilization mechanisms in soil. *Plant Soil* **2011**, *341*, 363–382. [CrossRef]

105. Prietzel, J.; Klysubun, W.; Werner, F. Speciation of phosphorus in temperate zone forest soils as assessed by combined wet-chemical fractionation and XANES spectroscopy. *J. Plant Nutr. Soil Sci.* **2016**, *179*, 168–185. [CrossRef]

106. Binner, I.; Dultz, S.; Schenk, M.K. Phosphorus buffering capacity of substrate clays and its significance for plant cultivation. *J. Plant Nutr. Soil Sci.* **2015**, *178*, 155–164. [CrossRef]

107. Sharpley, A.N. Soil phosphorus dynamics: Agronomic and environmental impacts. *Ecol. Eng.* **1995**, *5*, 261–279. [CrossRef]

108. Gerke, J. The acquisition of phosphate by higher plants: Effect of carboxylate release by the roots. A critical review. *J. Plant Nutr. Soil Sci.* **2015**, *178*, 351–364. [CrossRef]

109. Reddy, K.R.; Wetzel, R.G.; Kadlec, R.H. Biogeochemistry of phosphorus in wetlands. In *Phosphorus: Agriculture and the Environment*; Sims, J.T., Sharpley, A.N., Eds.; American Society of Agronomy: Madison, WI, USA, 2005; pp. 263–316. ISBN 9780891181576.

110. Samadi, A. Phosphorus sorption characteristics in relation to soil properties in some calcareous soils of Western Azarbaijan Province. *J. Agric. Sci. Technol.* **2006**, *8*, 251–264.

111. Holford, O.C.R.; Mattingly, G.E.G. phosphate sorption by jurassic oolitic limestones. *Geoderma* **1975**, *13*, 257–264. [CrossRef]

112. Alvarez, R.; Evans, L.A.; Milham, P.J.; Wilson, M.A. Effects of humic material on the precipitation of calcium phosphate. *Geoderma* **2004**, *118*, 245–260. [CrossRef]

113. Holzmann, S.; Missong, A.; Puhlmann, H.; Siemens, J.; Bol, R.; Klumpp, E.; von Wilpert, K. Impact of anthropogenic induced nitrogen input and liming on phosphorus leaching in forest soils. *J. Plant Nutr. Soil Sci.* **2016**, *179*, 443–453. [CrossRef]

114. Cosette Galvan-Tejada, N.; Pena-Ramirez, V.; Mora-Palomino, L.; Siebe, C. Soil P fractions in a volcanic soil chronosequence of Central Mexico and their relationship to foliar P in pine trees. *J. Plant Nutr. Soil Sci.* **2014**, *177*, 792–802. [CrossRef]

115. Smeck, N.E. Phosphorus dynamics in soils and landscapes. *Geoderma* **1985**, *36*, 185–199. [CrossRef]

116. Wang, M.K.; Tzou, Y.M. Phosphate sorption by calcite, and iron-rich calcareous soils. *Geoderma* **1995**, *65*, 249–261. [CrossRef]

117. Yang, X.-E.; Wu, X.; Hao, H.-L.; He, Z.-L. Mechanisms and assessment of water eutrophication. *J. Zhejiang Univ. Sci. B* **2008**, *9*, 197–209. [CrossRef]

118. Trimble, S.W. Streams, valleys and floodplains in the sediment cascade. In *Sediment Cascades: An Integrated Approach*; Burt, T.P., Allison, R.J., Eds.; Wiley-Blackwell: Chichester, UK, 2010; ISBN 9780470849620.

119. Meals, D.W.; Dressing, S.A.; Davenport, T.E. Lag Time in Water Quality Response to Best Management Practices: A Review. *J. Environ. Qual.* **2010**, *39*, 85–96. [CrossRef]

120. Paerl, H.W.; Gardner, W.S.; Havens, K.E.; Joyner, A.R.; McCarthy, M.J.; Newell, S.E.; Qin, B.; Scott, J.T. Mitigating cyanobacterial harmful algal blooms in aquatic ecosystems impacted by climate change and anthropogenic nutrients. *Harmful Algae* **2016**, *54*, 213–222. [CrossRef] [PubMed]

121. Paerl, H.W.; Paul, V.J. Climate change: Links to global expansion of harmful cyanobacteria. *Water Res.* **2012**, *46*, 1349–1363. [CrossRef] [PubMed]

 *soil systems*

Article

Streambank Legacy Sediments in Surface Waters: Phosphorus Sources or Sinks?

Shreeram Inamdar [1,*], Nathan Sienkiewicz [2], Alyssa Lutgen [2], Grant Jiang [2] and Jinjun Kan [3]

[1] Plant and Soil Sciences, University of Delaware, Newark, DE 19716, USA
[2] Water Science & Policy Graduate Program, University of Delaware, Newark, DE 19716, USA; nsien@udel.edu (N.S.); alutgen@udel.edu (A.L.); gjiang@udel.edu (G.J.)
[3] Stroud Water Research Center, Avondale, PA 19311, USA; jkan@stroudcenter.org
* Correspondence: inamdar@udel.edu

Received: 31 March 2020; Accepted: 9 May 2020; Published: 11 May 2020

Abstract: Streambank legacy sediments can contribute substantial amounts of sediments to Mid-Atlantic waterways. However, there is uncertainty about the sediment-bound P inputs and the fate of legacy sediment P in surface waters. We compared legacy sediment P concentrations against other streambank sediments and upland soils and evaluated a variety of P indices to determine if legacy sediments are a source or sink of P to surface waters. Legacy sediments were collected from 15 streambanks in the mid-Atlantic USA. Total P and M3P concentrations and % degree of phosphorus saturation (DPS) values for legacy sediments were lower than those for upland soils. % DPS values for legacy sediments were below the water quality threshold for P leaching. Phosphorus sorption index (PSI) values for legacy sediments indicated a large capacity for P sorption. On the other hand, equilibrium phosphorus concentration (EPC_0) for legacy sediments suggested that they could be a source or a sink depending on stream water P concentrations. Anoxic conditions resulted in a greater release of P from legacy sediments compared to oxic conditions. These results suggest that legacy sediment P behavior could be highly variable and watershed models will need to account for this variability to reliably quantify the source-sink behavior of legacy sediments in surface waters.

Keywords: legacy sediments; phosphorus; equilibrium phosphorus concentration; sorption; desorption; anoxic; water quality

---

1. Introduction

Sediment and sediment-bound nutrients such as phosphorus (P) can be detrimental to the health of aquatic ecosystems and are an important concern for watershed managers and natural resource agencies [1–3]. Fine sediments can decrease light penetration and reduce the primary productivity of aquatic vegetation [4]. Inputs of excess nitrogen (N) and/or P to aquatic systems can cause eutrophication and enhance the production of harmful algal blooms [5]. Subsequently, algal decomposition can reduce dissolved oxygen levels in the water column resulting in fish kills [6,7]. The Chesapeake Bay Program ranks nutrients (N and P) and sediment as the top two polluters for the Bay, and the Chesapeake Bay commission seeks to reduce P inputs by 24% and sediment loads by 20% per year [8]. While billions of dollars are being invested in agricultural management practices (e.g., $3.6 billion by 2025 in 2010 dollars in the Chesapeake Bay; [9]) such as riparian buffers, cover crops, nutrient management, etc., water quality improvements have not achieved their targets, especially in agricultural watersheds [2,10]. In 2017, the Chesapeake Bay Model indicated that an additional 0.46 million kg of P reductions will be required from agricultural sources per year to meet the watershed management plan goals for the endpoint year 2025 [2]. While much of the current management and regulatory focus is on upland agricultural and urban sources that constitute 56% and 18% of the total bay P loads [2], other potential sources such as streambanks [11,12] and associated legacy sediments [13,14] have also been gaining increasing attention [15].

Legacy sediments have been defined as deposits of historic sediments that have accumulated in the valley-bottoms of eastern US following European settlement [13,14]. These sediments have resulted from extensive erosion from land clearing and agriculture and the simultaneous construction of mill dams and other structures on streams [13,14,16,17]. Many of the milldams, which were particularly ubiquitous across the mid-Atlantic region (every few kilometers), have now breached or have been removed, leaving incised streams vulnerable to bank erosion [14,17,18] (Figure 1). Not surprisingly, recent studies reveal that streambank erosion of legacy sediments could constitute as much as 50 to 100% of the watershed fine sediment exports [19–21] and need to be accounted for in watershed sediment budgets and management plans. Beyond sediments, however, very few studies have characterized the concentrations of P in legacy sediments and their contributions to sediment-bound and total P budgets [15]. We know of only three studies, including two of our own that have determined P concentrations in legacy sediments and estimated their contributions to watershed P loadings [21–23]. One other study [24] investigated anthropogenic signatures in legacy sediments through the use of elemental ratios. Our work showed that while P concentrations in legacy sediments were on the lower side (25-1293 mg/kg; [22]), streambank legacy sediment contributions to watershed fluxes could be as much as 50% and 32% for sediments and sediment-bound P, respectively [21].

Figure 1. Examples of two streambank legacy sediment sites of the 15 sampled in this study which were upstream of now breached mill dams. Left: Nate Sienkiewicz at the streambank at Nature Center Beach (NCB; bank height 2.59 m). Red flags indicate locations for bank sampling. Right: Alyssa Lutgen in front of the eroding/slumping streambank at Scotts Mill Dam (SM2; bank height 2.74 m).

Beyond total amounts, we know even less about the fate of legacy sediment P as it is eroded from streambanks and deposited in the stream channel or transported downstream to receiving water bodies. While there is substantial information on streambank sediments and associated nutrients [11,25], we do not know if legacy sediments, deposited in streams, become sources or sinks of P and how this behavior varies with sediment and stream water P concentrations and redox conditions in stream sediments. A recent report by the Chesapeake Bay Science and Technology Advisory Committee (STAC) [15] highlighted the lack of information on N and P concentrations in legacy sediments and their fate as a major knowledge gap that urgently needs to be addressed. Without consideration of legacy sediments and sediment-bound P, there is concern that we may have an inaccurate assessment of the watershed loadings and may not be appropriately allocating sediment and nutrient reductions and remedial/management resources [26].

Our interest here was to address these knowledge gaps for legacy sediment P and gain a better understanding of the fate of legacy sediment P in aquatic ecosystems. Key questions we addressed were: (1) What are the concentrations of P in streambank legacy sediments and how do they compare against concentrations for upland soils, stream bed sediments, and water quality P thresholds? (2) What is the fate and net source-sink behavior of sediment-bound P under varying stream water P concentrations and redox conditions? and (3) What are the broader implications of legacy sediment P for water quality and watershed management? To address the first question we collated our recent legacy sediment P concentrations reported by [21,22], compared them to P values reported for other streambank legacy and non-legacy sediment sites, upland soil concentrations, stream bed sediments, and water quality thresholds. Comparisons were performed for total P, bioavailable fraction of P quantified by Mehlich-3 extracts [27], and the % degree of phosphorus saturation (%DPS; [27]). The %DPS has been used to estimate the amount of P sorbed on soils and therefore the potential for P loss via desorption from soils [27].

To address the second question, we analyzed selected legacy sediment samples that were collected [22] using standard analytical methods and indices that have been used to determine the potential for P leaching and its source-sink behavior [28]. These indices included the phosphorus sorption index (PSI, [27,29]) and equilibrium phosphorus concentration (EPC$_0$; [30]). The PSI provides a measure of the maximum phosphorus sorption capacity and can be used to assess the upper limit for P sorption of legacy sediments. In contrast, EPC$_0$ represents the lower end and is the concentration at which there is no net sorption or desorption from the sediment [30,31]. If EPC$_0$ values of sediments are lower than that of the stream water concentrations, the sediments will sorb nutrients from stream water and considered net sinks for P. Inversely, if the sediment EPC$_0$ value is greater than the stream water concentration, sediment will desorb P and thus act as net P source. Thus, the higher the value of EPC$_0$, the greater the vulnerability of P leaching from sediments. In addition to these metrics, we also determined how legacy sediments with known P concentrations responded to oxic and anoxic redox conditions through laboratory incubation assays. The intent here was to investigate if reductive dissolution of P [32] occurred from legacy sediments under anoxic conditions and its extent. These observations were taken together to address the broader implications and water quality challenges for question 3.

2. Materials and Methods

2.1. Site Description, Streambank Sampling for Legacy Sediments, and Sediment Analysis

Detailed description of the streambank sampling sites, coordinates, and methods is provided in [21,22,33]. Briefly, legacy sediment sampling was performed at 15 streambank sites across five streams in northern Delaware (DE), eastern Maryland (MD), and southeast Pennsylvania (PA) (Figure 2). These included Big Elk Creek and its first order tributary Gramies Run and Christina River and its two major tributaries the White Clay Creek and the Brandywine Creek. Big Elk Creek with a drainage area of 205 km^2 (empties into the Chesapeake Bay) and Christina River with a drainage area of 1463 km^2 (drains into Delaware Bay) straddle the fall line with upper portions of the watersheds extending into the Piedmont and Appalachian regions and the lower portions in the Coastal Plain. The drainage areas for Gramies Run, White Clay Creek, and the Brandywine Creek were 8 km^2, 277 km^2, and 854 km^2, respectively [22].

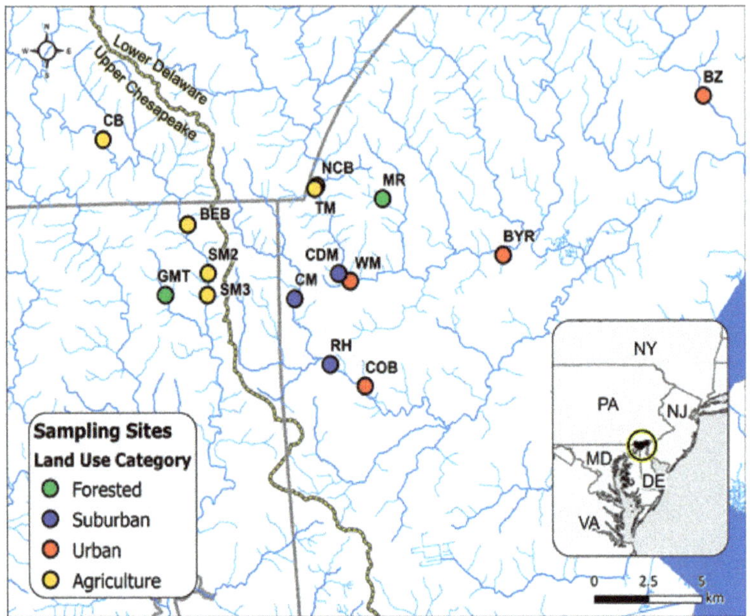

Figure 2. Map showing locations of 15 stream bank sampling sites for legacy sediments across Delaware, Maryland, and Pennsylvania. These sites included: BEB = Big Elk Bridge (Agriculture), CB = Camp Bonsul (Agriculture), NCB = Nature Center Beach (Agriculture), SM2 = Scott's Mill 2 (Agriculture), SM3 = Scott's Mill 3 (Agriculture), TM = Tweed's Mill (Agriculture), RH = Cottage Mill (Suburban), CM = Casho Mill (Suburban), CDM = Cider Mill (Suburban), BZ = Brandywine Zoo (Urban), BYR = Byrnes Mill (Urban), COB = Cooch's Bridge (Urban), GMT = Gramies Run (Forested) and MR = Middle Run (Forested). Inset: Location of study sites in the mid-Atlantic tristate area.

All of the streambank legacy sediment sampling sites were located upstream of formerly breached or existing milldam locations. These sites spanned four different contemporary land use and land covers (LULC) – urban, suburban, forested, and agriculture (see Figure 1 and Table 1 in [22]). Streambank samples were collected for multiple depths (recorded from the top; Table 1 in [22]) by scraping off the surface sediment and collecting a sample using an auger or a hand trowel (where the sediments were too hard to auger). All samples were placed in sterile Ziploc bags and put on ice until they were brought back to the lab. A total of 67 sediment samples were collected across all 15 sites and all sediment sampling was performed in October–November, 2017. Sediments were ground with a mortar and pestle and sieved through a 2-mm mesh to remove small rocks and organic matter. Sediments were then sieved into coarse (>63 μm) and fine (<63 μm) fractions using an RX-29 RoTap®sieve shaker. Percent sand, silt, and clay for the samples were also determined using Beckmann Coulter LS 13 320 Particle Size Analyzer ®(Indianapolis, IN) [22].

Sediment samples were analyzed for Mehlich-3 [27] extractable elements (M3P, M3Fe, and M3Al) and microwave digestion (EPA method 3051) for total P. Using M3P, M3Al, and M3Fe values, the %DPS was computed using the equation:

$$\%\mathrm{DPS} = \left(\frac{\left[\left\{ \frac{\mathrm{M3P}}{(\mathrm{M3Al+M3Fe})} \right\} + 0.019 \right]}{0.0042} \right) \times 100$$

where all the M3 extractable values are in molar concentrations [27]. Representative stream water samples were also collected at the time of bank sediment sampling at all 15 sites to compare stream water PO_4^{3-} against sediment EPC_0 values (to assess source-sink behavior). In addition, stream water PO_4^{3-} data at bi-monthly intervals was also available for Big Elk Creek sampling site (BEB) for a temporal comparison with EPC_0 [28].

2.2. Phosphorus Sorption Index (PSI)

The intent of this experiment was to determine the maximum sorption capacity of streambank legacy sediments and compare them against literature values for other sediments. Sediments from all 15 streambank sites and depths (total samples = 67) were used. Fine and coarse fractions were replicated twice for a total of 268 samples (67 samples × 2 size fractions × 2 replicates). Soils were treated with a 75 mg P L^{-1} solution created by dissolving 0.3295 g of monobasic potassium phosphate (KH_2PO_4) in 1L of deionized water following the protocol of [34] (pages 20-21) based on [35]. About one gram of sediment (the exact amount was recorded) was placed in a 50 mL centrifuge tube along with 20 mL of the 75 mg P L^{-1} sorption solution. This provides a ratio of added P to soil of 1.5 g P kg^{-1} soil. Two drops of chloroform were added to each solution to kill and inhibit microbial activity. Tubes were placed in an end-over-end shaker and shaken for 18 h. After 18 h, samples were centrifuged at 2000 rpm for 30 min and then filtered through 0.45 μm filters using a Millipore filtration unit into 40 mL amber glass vials. The filtered solution samples were analyzed for PO_4^{3-} (mg P L^{-1}) using EPA-118-A Rev 5 method on an AQ2 Discrete Analyzer (Seal Analytical, Mequon, Wisconsin). The PSI (mg kg^{-1}) value was determined using the equation:

$$\mathrm{PSI} = \frac{\left(75\mathrm{mg}\frac{P}{L} - C\right) \times (0.020L)}{(0.001 \text{ kg soil})}$$

where C is the solution equilibrium P concentration after 18 h (mg L^{-1}). Differences in PSI values with particle size class (fine versus coarse fraction) were determined using t tests. Pearson correlations (r) were determined to investigate relationships between PSI values and Mehlich-3 extracted Al and Fe contents of legacy sediments (M3Fe and M3Al). All statistical analyses were performed in JMP (Version 14.0).

2.3. Equilibrium Phosphorus Concentration (EPC_0)

Since EPC_0 analysis required sorption assays with multiple PO_4^{3-} solution concentrations for each sediment sample, to keep things manageable, we limited the sediment analysis to only one randomly-selected depth across the 15 streambank sites and the fine sediment fraction. The intent here was to survey the EPC_0 behavior across sites. With the inclusion of two replicates, 30 sediment samples were analyzed in total (15 sites × 1 depth × 2 replicates). In addition, to assess EPC_0 variability with streambank depth, we determined the EPC_0 values for all depths at two sites—BEB and SM3 (four depths each). These samples were also replicated twice.

Four PO_4^{3-}-P solution concentrations, representing the likely range of stream water P concentrations, were used for the assay and included n.d., 0.25, 0.50, and 2.0 mg P L^{-1}. All solutions were made by dissolving KH_2PO_4 in filtered stream water collected from a forested headwater tributary of Big Elk Creek, Maryland, with undetectable P concentration. Stream water was used to simulate natural ionic strength conditions. About one gram of sediment (exact weight was noted) was added to a 50 mL centrifuge tube to which 20 mL of PO_4^{3-} solution was added; this mixture was created for each of the four P solution concentrations. Two drops of chloroform were added to the solution to inhibit microbial activity. Samples were placed on an end-over-end shaker and incubated for 24 h at 25 ± 2 °C. Once incubated, samples were centrifuged at 2000 rpm for 30 min. Using Sterlitech Glass Microfiber 0.7-μm filters, the centrifugate was filtered into 40 mL amber vials. The solution PO_4^{3-}

concentrations were measured colorimetrically using EPA-118-A Rev 5 method on the AQ2 Discrete Analyzer (Seal Analytical, Mequon, Wisconsin).

The EPC$_0$ was computed using the procedures described by [36] based on [30]. The P sorbed on the sediment phase (S) (μg g^{-1}) after a 24 hour period was calculated using the following equation:

$$S = \frac{v}{m} \left(C_0 - C_{24}\right)$$

where v is the total volume of the solution (0.02 L in this case), m is the mass of dry sediment in (g), C_0 (μg P L^{-1}) is the concentration of solution prior to incubation and C_{24} (μg P L^{-1}) is the concentration of solution after the 24-hour incubation. S was plotted (on the y-axis) against C_{24} (x-axis) and the data points were fitted with a linear regression. At low solution P concentrations (as in this case) the relationship between C and S can be estimated using linear regression [30,37]. The EPC$_0$ (μg P L^{-1}) value is the coordinate at which the linear fit line intercepts the x-axis and is computed by substituting y as 0 and solving the linear equation for x. EPC$_0$ values were then expressed in mg L^{-1}. Additionally, Pearson correlations (r) were determined to investigate the relationships between the EPC$_0$ value, particle size fractions, and M3Fe and M3Al concentrations using JMP software (version 14.0).

2.4. Sorption and Desorption Under Anoxic and Oxic Conditions

The intent of this experiment was to investigate how legacy sediment P sorption or desorption could vary if streambank legacy sediments were eroded from the bank and deposited in the stream under anoxic and oxic conditions [32]. Legacy sediment samples for four sites, one depth each, and only the fine sediment fraction were selected for this analysis. Three replicates were used for each sediment sample. Since the legacy sediments had low initial inorganic P, the sediments were exposed to an elevated P solution (10 mg L^{-1}) prepared using KH$_2$PO$_4$ prior to the experiment. For this, thirty grams of the dry sediment sample was placed in an acid-washed, ethanol cleaned, glass tray and 600 mL of 10 mg L^{-1} PO$_4$$^{3-}$ solution was added to saturate the sediment with P. The sediment was placed on a shaker table for 24 h at 100 rpm, drained, and then placed under a dry hood until any excess liquid was evaporated. The sediment was then removed and placed in a sterile Ziploc bag until the second part of the experiment.

For the second part of the experiment, two grams of P-sorbed legacy sediments were added to 40 mL of filtered stream water in an amber vial and placed on a shaker table for 24 h at 112 rpm. A control treatment was also created where no sediment was added to the stream water. Unlike that for EPC$_0$, stream water PO$_4$$^{3-}$ values were above detection (0.04 mg P L^{-1}) for this experiment. This was likely because of stormflow conditions prior to sampling; this however was not a problem for this experiment since our intent was to investigate the differences between oxic and anoxic conditions. The vials were subject to both oxic and anoxic treatments. To maintain oxic conditions, the cap was left off the amber vial to ensure oxygen would not be depleted in the water. Anoxic conditions were created by adding one gram of sodium sulfite (Na$_2$SO$_3$; an oxygen scavenger) to the solution and the vial was sealed by closing the cap. Anoxic conditions were verified using a dissolved oxygen meter. After 24 h, samples were filtered using Sterlitech glass microfiber 0.7 μm filters. The sample solutions were measured for their solution PO$_4$$^{3-}$ concentrations (EPA-118-A Rev 5) on an AQ2 Discrete Analyzer (Seal Analytical, Mequon, Wisconsin). A t-test was used to determine the significant differences between the oxic and anoxic treatment groups.

3. Results and Discussion

3.1. Concentrations of P in Streambank Legacy Sediments and Comparisons Against Upland Soils, Stream Bed Sediments, and Water Quality Thresholds

P concentrations measured by [21,22] are presented in Table 1 and are compared against other studies with data on streambanks, bed sediments, and upland soils. It should be noted that while P concentrations for both the coarse and fine fractions of the soils/sediments were measured by [38], only the fine fraction values were reported in [21]. The comparisons here provide important insights. Total P concentrations for streambank legacy sediments were the lowest of all sediment sources (Table 1) with mean P concentrations for the banks nearly half of those reported for cropland and developed soils [21]. The difference in mean concentrations between legacy sediments and other sediment categories was less for M3P, but nonetheless, M3P values for streambanks were the lowest among all sources (Table 1). Mean concentrations for total and M3P for streambanks [21,22] were comparable to streambank values from other studies (Table 1). Percent DPS values for streambanks were also low and particularly lower than values reported for cropland and developed soils (Table 1).

Table 1. Comparison of legacy sediment P concentrations against other soil/sediment sources. Comparisons are made for particle size class (<63 µm: fine), coarse, and bulk sediments. Concentrations include a range of values and mean in brackets ( ). Where available, sample numbers are indicated in square brackets [ ]. Table modified from [22].

Reference	Sample Description	Grain Size (µm)	P (mg/kg)	M3-P (mg/kg)	% DPS
[22]	streambank legacy sediment	<63 [67]	80–1293 (551)	0.25–52.8 (11.8)	4.6–16 (6.8)
		>63 [67]	25–668 (255)	0.51–48.8 (10.3)	4.7–19.7 (7.7)
[23]	streambank legacy sediment	Bulk	340-958 (556)	-	-
[36]	streambanks bed sediments	Bulk	209–306 (269)	26–68 (39)	3–8 (5)
			177–454 (315)	17–37 (29)	5–8 (6)
[39]	streambanks bed sediment	Bulk bulk	417 ± 28.7 281 ± 37	14 ± 2.4 22 ± 2.7	- -
[40]	streambanks bed sediment	Bulk Bulk	370–847 558–1134	5–55 13–85	- -
[41]	streambanks	bulk	710 ± 203	-	15–21
[42]	streambanks	bulk	138–1140 (621)	-	15.7–17.3
[21]	forest	<63 [7]	368-1229 (850)	6.4-26.8 (15.4)	5.6–7.9 (6.7)
		>63 [7]	136-620 (372)	5.6–36.3 (17.4)	6.2–10 (8)
	cropland	<63 [7]	924-1780 (1260)	30–237 (149.6)	10.3–60 (40.4)
		>63 [7]	280–1142 (781)	23–223 (126)	10–92 (49)
	developed	<63 [6]	231–2594 (548)	17.5–871 (27.7)	7.1–275.3 (12.9)
		>63 [6]	66–911 (192)	6.4–380 (14.5)	8.4–293 (14.2)
	streambank legacy sediment	<63 [23]	79–719 (549)	0.25–28 (9.8)	4.6–8.9 (6.1)
		>63 [23]	34–526 (248)	0.5–16.9 (8.8)	4.7–10.4 (7)
	bed sediment	<63 [32]	252–921 (668)	14–36 (25)	0–12 (9.1)
		>63 [23]	99–611 (199)	7.6–23 (15)	10–16 (13)

When P concentrations are assessed for particle size class (Table 1), total P concentrations for fine sediments were twice or more than those for coarse fractions across upland soils, banks and bed sediments. The same level of separation, however, was not observed for M3P and %DPS values. Mean %DPS was generally greater for the coarse versus the fine fractions but a similar consistent trend was not observed for M3P. Similarly, when mean %DPS values were compared for bed sediments and streambanks, bed sediment values were slightly greater than the streambank values, with no consistent pattern for total P and M3P.

While M3P was originally developed for agronomic needs, i.e., determining crop nutrient requirements and associated fertilizer application, it has been used for determining environmental risk associated with P leaching [27]. In Delaware, M3P and %DPS values less than 50 mg kg^{-1} and 15%, respectively, are considered "below optimum" and do not pose any risk to water quality [27]. In contrast, M3P and %DPS values in excess of 100 mg kg^{-1} and 58%, respectively, are considered a threat to water quality [27].

In Arkansas, environmental threshold for water quality concern for M3P is higher at 150 mg kg^{-1}. When compared against these water quality thresholds, studies listed in Table 1 arrived at the same conclusion that streambank sediments (legacy and non-legacy) likely posed a low risk for P leaching under well oxygenated conditions and served as a net sink for P [36,41–43]. All of the studies, however, did recognize that while P concentrations were low and most of the P was likely bound to metal hydro-oxides, this P could be released into solution due to the reductive dissolution of the oxides under anoxic conditions [44,45]. The low P concentrations in streambank legacy sediments should not be very surprising considering that much of these sediments were likely deposited prior to the 1950s [13,15,17], before the significant increase in use of synthetic N and P fertilizers on agricultural lands [3]. One way these buried legacy sediments could have acquired elevated P concentrations would be through contact with P-rich streamflow (along the banks or during overbank flooding) and/or upland runoff carrying fertilizer nutrients that infiltrated through the soil profile.

Elevated concentrations of total P in bank versus stream bed sediments (Table 1) were attributed to a greater fraction of fine sediments in the banks, which includes P-sorbing iron oxides [39]. The same pattern, however, has not been reported by other studies in Table 1 (e.g., [21,40]). This could be because bed sediments typically represent a mixture of various sediment sources including P-rich upland sediments. Bed sediments could also acquire elevated P concentrations from P-rich stream runoff. However, broadly, most studies do report that bed sediments are more coarse grained than bank sediments [39] and this would likely result in less P sorption capacity for bed sediments as reflected by the higher %DPS values for bed sediments in Table 1.

3.2. Phosphorus Sorption Index (PSI) for Legacy Sediments

Solution concentrations of PO_4^{3-} after 18 h of legacy sediment incubation were lowest for the fine fraction (<63 µm) (43.7 ± 8.5 mg P L^{-1}; Figure 3) down from the starting concentration of 75 mg P L^{-1}. While these solution concentrations are much greater than what one would typically observe in streams, this experiment indicates that fine legacy sediments have a fairly high capacity for P sorption. In comparison, solution concentrations for the coarse fraction (>63 µm) were higher (61.5 ± 10.9 mg P L^{-1}; Figure 3), indicating lower sorption for this sediment class. Three samples within the coarse fraction had solution concentrations greater than the starting solution of 75 mg P L^{-1} (Figure 3) indicating some release of P from sediments. The mean PSI value for coarse and fine legacy sediment fractions taken together was 472.3 ± 270.4 mg kg^{-1}, while that for the coarse fraction was 292.6 ± 224.4 mg kg^{-1}, and that for the fine fraction was 652.0 ± 177.3 mg kg^{-1} (Table 2). There was a significant difference in PSI values between the coarse and fine size fractions (p < 0.001). There was a strong positive correlation between PSI values for the coarse fraction and M3Al (r = 0.77; p < 0.001) and a weak and insignificant correlation with M3Fe (r = 0.17; p = 0.15). For the fine fraction, the relationship between M3Al was slightly weaker, but still positive (r = 0.69; p < 0.0001) and there was a positive correlation with M3Fe (r = 0.26; p = 0.029).

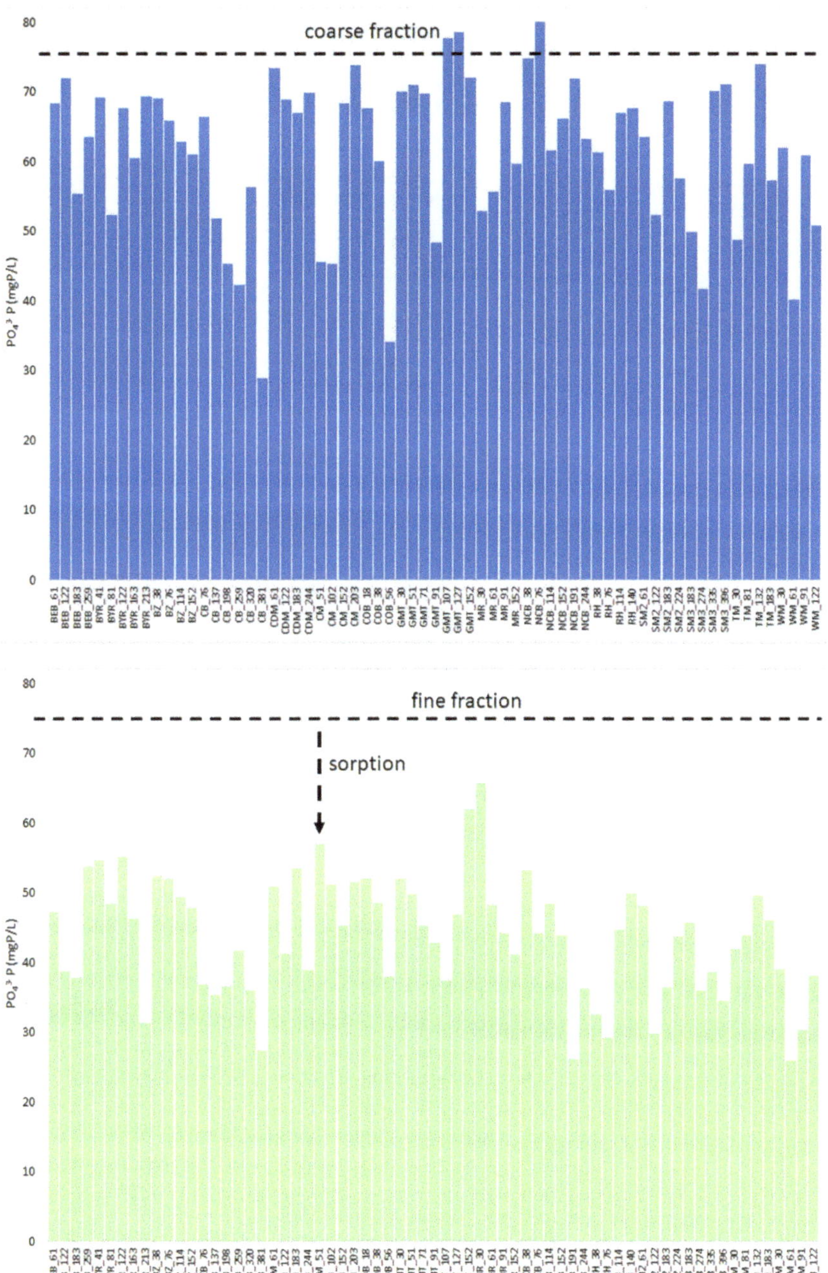

Figure 3. Solution PO_4^{3-}-P (mg P L^{-1}) concentrations after 24-hour incubation for coarse fraction and fine fraction sediments. The dashed line represents the starting solution concentration (75 mg PO_4^{3-}-P L^{-1}). Larger decreases below this value indicate greater sorption by sediments. The labels on the x-axis include site name and bank sampling depth in cm.

Table 2. PSI (mg/kg) values for agricultural and streambank soils in the Mid-Atlantic Region.

Location	Soil Depth	Soil Type	PSI (mg/kg)	Reference
Mid-Atlantic watersheds		Stream bank legacy sediments (Coarse)	293	This study
		Stream bank legacy sediments (Fine)	652	
Delaware Inland Bays Watershed	0–20 cm	Evesboro loamy-sand *	149	[46]
	20–40 cm	Evesboro loamy-sand *	136	
	40–60 cm	Evesboro loamy-sand *	217	
	60–80 cm	Evesboro loamy-sand *	263	
	0–20 cm	Matawan sandy loam *	588	
	20–40 cm	Matawan sandy loam *	2083	
	40–60 cm	Matawan sandy loam *	2564	
	60–80 cm	Matawan sandy loam *	1886	
	0–20 cm	Matawan sandy loam **	434	
	20–40 cm	Matawan sandy loam **	1562	
	40–60 cm	Matawan sandy loam **	2000	
	60–80 cm	Matawan sandy loam **	1923	
	0–20 cm	Pocomoke sandy clay loam *	95	
	20–40 cm	Pocomoke sandy clay loam *	714	
	40–60 cm	Pocomoke sandy clay loam *	212	
	60–80 cm	Pocomoke sandy clay loam *	303	
Mahantango Creek Catchment (Central PA)		Agricultural catchment exposed stream bank	259	[39]
		Agricultural catchment submerged bank sediment	214	

* Agricultural soils; ** Field Border areas separate crop fields from drainage ditches.

The PSI experiment confirmed that fine legacy sediments have a greater sorption capacity than the coarser fractions. The PSI value for the fine legacy sediments was also greater than most of the agricultural soils (e.g., Table 2, Evesboro loamy-sand (136–263 mg kg^{-1}) and Pocomoke sandy clay loam (95–714 mg kg^{-1}), with the exception of the Matawan sandy loam, which had a higher sorption capacity (588–2564 mg kg^{-1}) [46]. In another sorption study [39], maximum sorption values for exposed and submerged streambank sediments in an agricultural catchment in central Pennsylvania were 259 mg kg^{-1} and 214 mg kg^{-1}, respectively (Table 2). These values were much lower than values for our fine legacy sediment fraction, but comparable to the coarse fraction PSI value.

3.3. Equilibrium Phosphorus Concentration (EPC$_0$)

EPC$_0$ values across the 15 legacy sediment sites ranged from 0–0.24 (mean: 0.044) mg L^{-1} (Table 3). Ten of fifteen sites had EPC$_0$ values greater than the baseflow stream water concentration during sediment sampling indicating that the sediment could be a potential P source if deposited into the channel (Table 3). Five of the fifteen sites had EPC$_0$ values that were less than the stream water P concentrations indicating that the sediment would act as a potential sink for P (Table 3). While there was no consistent relationship between EPC$_0$ values and stream water PO$_4$$^{3-}$ concentrations, two of the highest EPC$_0$ values were associated with Brandywine Zoo and Cooch's Bridge locations, both urban stream locations (Table 3).

Table 3. Sediment EPC_0 concentrations for the 15 streambank legacy sediment sites (one selected depth) and comparisons against baseflow stream water PO_4^{3-} concentrations to determine if sediments would act as a source or sink for P.

Site Name	Depth (cm)	EPC_0 (mg/L)	Stream PO_4^{3-} Concentration (mg/L)	Land Use	Sink or Source
Gramies Run (GMT)	107	0.028	0	Forest	Source
Middle Run (MR)	91	0.044	0.036	Forest	Source
Cider Mill (CDM)	183	0.024	0.01	Suburban	Source
Casho Mill (CM)	102	0.020	0.005	Suburban	Source
Cottage Mill (RH)	76	0.000	0.001	Suburban	Sink
Byrnes Mill (BYR)	163	0.035	0.023	Urban	Source
Brandywine Zoo (BZ)	76	0.240	0.064	Urban	Source
Cooch's Bridge (COB)	38	0.136	0.004	Urban	Source
Woolen Mill (WM)	61	0.001	0.205	Urban	Sink
Big Elk Bridge (BEB)	122	0.027	0.004	Agriculture	Source
Camp Bonsul Road (CB)	137	0.010	0.066	Agriculture	Sink
Nature Center Beach (NCB)	114	0.027	0.032	Agriculture	Sink
Scotts Mill 2 (SM2)	122	0.011	0.002	Agriculture	Source
Scotts Mill 3 (SM3)	183	0.033	0.002	Agriculture	Source
Tweeds Mill (TM)	81	0.026	0.039	Agriculture	Sink
10 sources and 5 sinks					

Our EPC_0 values were within the range of values reported by other studies (Table 4). A study in Maryland (Kimages creek) that investigated legacy sediments reported an EPC_0 value of 0.010 mg L^{-1} [47], while that from till bank material near Lake Pepin in Minnesota had an EPC_0 value of less than 0.1 mg L^{-1} [48]. Studies have shown that there is a strong correlation between the EPC_0, soluble reactive phosphorus (SRP), and dissolved reactive phosphorus (DRP) in stream water [43,49]. The EPC_0 in sediments may change depending upon the SRP concentration in the sediment and stream waters [49] as, potentially, in the case of our urban, P-rich, Brandywine zoo site (Table 3). One study [49] looked at the influence of sewage treatment works (STWs) on riverbed sediment to determine if sediments could act as a buffer for the increase in stream water P concentrations. They found that regardless of STW influence, the riverbed sediments always acted as sinks in the water column. Despite higher concentrations of SRP in the rivers downstream of STWs, sediment near STWs had a higher capacity to absorb SRP [49]. Similarly, [48] showed that fine bank sediments transported in stream waters sorb elevated P from sewage and industrial waste and then deliver it to Lake Pepin. Given that controlling bank erosion could be expensive, they suggest reducing P inputs to waterways to control this loading to Lake Pepin.

Table 4. Comparison of EPC_0 values from various sites and sediments types reported in the literature.

Location	EPC_0 (mg/L)	Reference
Delaware, Maryland, & Pennsylvania Legacy Sediment (mean)	0.044	**This study**
Lake Pepin stream bank till sediment	<0.1	[48]
River Wensum Catchment (UK)	0.085	[49]
Mahantango Creek Catchment (Central PA)		[39]
Bed sediment	0.043	
Bank sediment	0.02	
Courthouse Creek Sediment VA	0.090	[47]
Kimages Creek Sediment VA (Legacy sediment)	0.010	[47]
Rathburn Lake Watershed (Iowa):		[36]
Bed sediments (mean)	0.09	
Bank sediments (mean)	0.06	

For our sites, we did not find any significant correlations between EPC_0 concentrations and particle size classes %fine, %sand, %silt, and %clay. In contrast, EPC_0 was significantly negatively correlated with M3Al (r = −0.70; p = 0.0033) and positively correlated with M3Fe (r = 0.54; p = 0.03). Contrary to our observations, a significant inverse correlation between M3Fe and EPC_0 was reported by [36], indicating that as the amount of M3Fe increased, the potential for sorption increased resulting in lower EPC_0 values. In the same study, EPC_0 values were positively correlated with sand content but negatively correlated with silt and clay content [31,36], indicating that with finer fractions greater retention of P occurred.

EPC_0 values also varied with bank depth for both sites that were evaluated with depth. At BEB, for depths 60, 122, 183, and 260 cm from the top, the EPC_0 values were 0.032, 0.028, 0.002, and 0.032, respectively. At SM3, for depths 132, 173, 231, 267 cm from the top, the values were 0.031, 0.024, 0.047, and 0.031, respectively. This variation in EPC_0 values with bank depth was likely due to variation in sediment composition and characteristics with depth [22]. When the BEB EPC_0 values are compared against seasonally varying stream water PO_4^{3-} concentrations (Figure 4), we note that the source-sink behavior of sediments varies temporally. The sediments behave as a sink when stream water P concentrations are elevated, particularly during stormflows (Figure 4), and as a source during low-P baseflow conditions. For example, for the sediment at 60 cm depth (EPC_0 = 0.032; Figure 4), the sediment would serve as a sink for P for 21 out of the 57 sampling points and as a source for the remainder 35 points (assuming stream waters are in contact with sediment at this depth). In comparison, sediments at 183 cm depth behaved as a sink for 49 out of the 57 sampled stream water concentrations. A similar temporal variation in source-sink behavior of sediments was also reported by [36] by valuating sediment EPC_0 values against stream water P concentrations through the year. While we did not measure EPC_0 at multiple times of the year, other studies suggest that sediment EPC_0 could also be temporally variable, driven by sediment conditions and stream water concentrations. For example, EPC_0 can increase under reducing conditions due to the loss of crystalline forms of Fe oxides resulting in a release or reduced retention of P [44,50]. EPC_0 values could also fluctuate as a result of stream water concentrations with an increase in EPC_0 with increasing stream water concentrations, resulting in a reduction of the sediment buffering capacity [51]. On the other hand, hydrologic dilution after storms could result in a release of P from sediments [52]. This dynamic behavior could make determining source-sink behavior of legacy sediments more complicated since such a behavior would now be dictated by stream water concentrations as well as time-variable sediment EPC_0 values.

While we were not able to assess the EPC_0 values for bed sediments, others have made such assessments and evaluated them against streambank sediment values. EPC_0 values for bed sediments at all of the 10 sites in river Wensum in Norfolk, UK [49], were below the stream water SRP concentrations suggesting that bed sediments were always acting as a sink for P. Two studies [36,39] found that bed sediment EPC_0 values were greater than those for bank sediments (Table 4). They suggested that, in general, banks had higher proportion of fine grained material, including P-sorbing clay and metal hydro-oxides (and thus lower EPC_0), as opposed to bed sediments [36,39]. Additionally, erosion and downstream transport or loss of finer fractions from bed sediments could further increase EPC_0 values for bed sediments and thus reduce the sorption potential for bed sediments compared to the original bank sources [39].

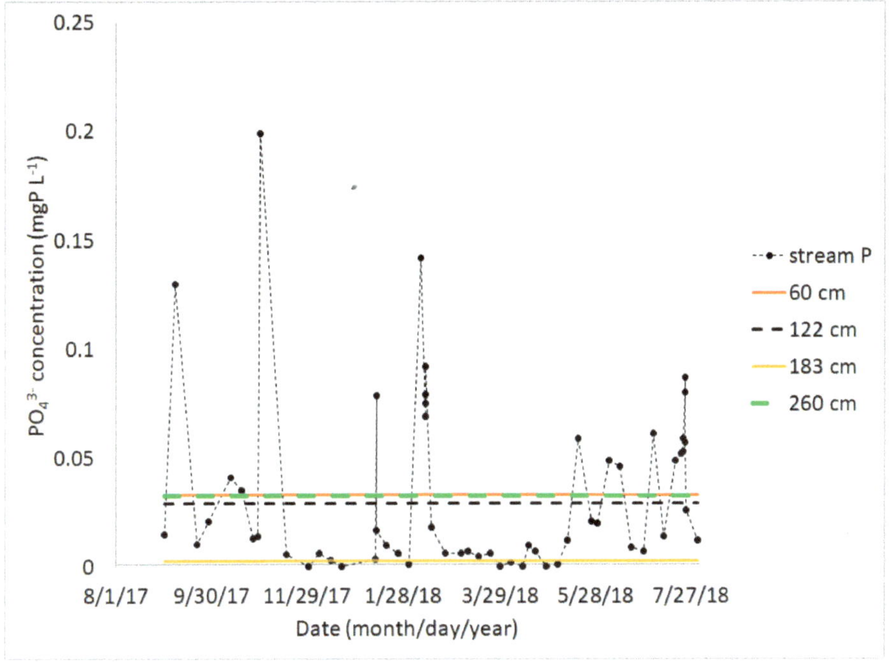

Figure 4. Comparison of sediment EPC$_0$ values for four depths for stream bank site BEB against stream water PO$_4$$^{3-}$ concentrations over the period August 2017 to July 2018. Sediment sampling was done only one time in October–November 2017. EPC$_0$ values for four depths 60, 122, 183, and 260 cm from the top were 0.032, 0.028, 0.002, and 0.032, respectively.

3.4. Legacy Sediment Sorption Under Anoxic and Oxic Conditions:

Our experiment with P-sorbed legacy sediments revealed that the solution PO$_4$$^{3-}$ concentrations under all sediment treatments under anoxic conditions (mean 1.40 ± 0.32 mg P L^{-1}) were significantly greater (t-test, $p < 0.001$) than those measured under oxic conditions (mean 0.26 ± 0.19 mg P L^{-1}) (Figure 5). Similar to studies for other sediments and soils [32,44,53], these results showed that the P-sorbed legacy sediments released more PO$_4$$^{3-}$ under anoxic conditions than under oxic conditions. Thus, legacy sediments with elevated P could release P in greater amounts under anoxic conditions, effectively acting as a source. Anoxic conditions result in the reduction of Fe (III) to Fe (II) [54,55] releasing the tightly bound P from the iron oxide surfaces [32,44]. Reduced conditions also allow Fe (II) to preferentially bind to sulfide, releasing the PO$_4$$^{3-}$ ions [29]. Similar work [40] with aerobic and anaerobic treatments however suggested that the amount of inorganic P release could vary with different soils and would depend on the various P fractions associated with oxides. They found that anaerobic conditions released P associated with Fe, but the same did not extend to slowly-cycling P that was associated with Ca, stable P or residual P [40]. Temperature could also be a factor influencing the release of P under anoxic conditions. Soils under warm flooded and anaerobic conditions released more P than similar anoxic soils under cold or unfrozen/frozen conditions [56]. These responses would likely extend to legacy sediments too with some sediments releasing relatively more P than others under varying anoxic conditions.

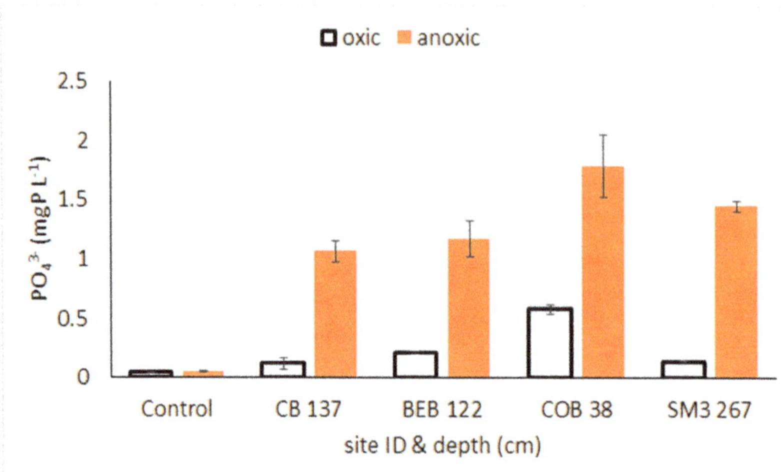

Figure 5. Solution PO_4^{3-} concentrations after a 24-hour incubation of P-sorbed legacy sediments in oxic and anoxic conditions. Site IDs along with their depths (in cm) are listed. Solution PO_4^{3-} concentrations under anoxic conditions (mean = 1.40 ± 0.32 mg P L^{-1}) for sediment treatments were significantly greater (t-test, $p < 0.001$) than those measured under oxic conditions (mean = 0.26 ± 0.19 mg P L^{-1}). Control treatment did not contain any sediments.

3.5. Conceptual Model for Source/Sink Behavior of Legacy Sediment P and Broader Implications for Water Quality and Watershed Management

Evaluation of legacy sediments through a variety of P indices provided important insights into the potential source-sink behavior of legacy sediments under contrasting redox conditions and sediment and water P concentrations. Overall, our results suggest that legacy sediments, which are particularly rich in silts and clays [15,17,22], have a large capacity for P sorption and can act as a sink for P under well-oxygenated conditions and in stream waters with moderate to high levels of dissolved P (e.g., as in the case of agricultural and urban streams). Floodplains could be important reservoirs of such P (e.g., conceptual model in Figure 6). In contrast, if legacy sediments are introduced into stream waters with low dissolved P concentrations (e.g., in forested streams with P concentrations below sediment EPC_0), or are deposited in anoxic conditions, e.g., sediment interfaces in lowland streams or at the bottom of ponds and lakes (Figure 6), these locations could become potential "hotspots" [57] for release P to overlying waters. Because of the greater amount of fine fractions, streambank legacy sediments are likely greater sinks of P than the stream bed sediments.

At the full drainage network scale, legacy sediments in headwater streams, with steeper slope gradients, shallower and faster flows, and likely more oxygenated conditions, could potentially serve as net sinks/repositories of P as opposed to lower positions in the drainage network where deeper and slower flows with anoxic conditions may encourage P-release. Increasing stream water P concentrations lower in the drainage network due to urban/agricultural land use and/or inputs from wastewater and other sources could also increase P sorption on legacy sediment surfaces and thus reduce the buffering potential [48]. Another factor that could affect legacy sediment P dynamics lower in the drainage basin, particularly in estuaries, is the effect of salinity on sediment P. Studies report that increasing salinity could result in release of P from sediments, but to varying extents [58,59]. It should also be noted that beyond abiotic sorption-desorption processes, P release or sequestration in sediments could also be influenced by biotic decomposition of P associated with sediment organic matter [44,60]. Typically, however, organic matter or organic carbon (OC) content of legacy sediments are low (OC < 1%; [22]), hence, such decomposition-related P release/source is likely to be small.

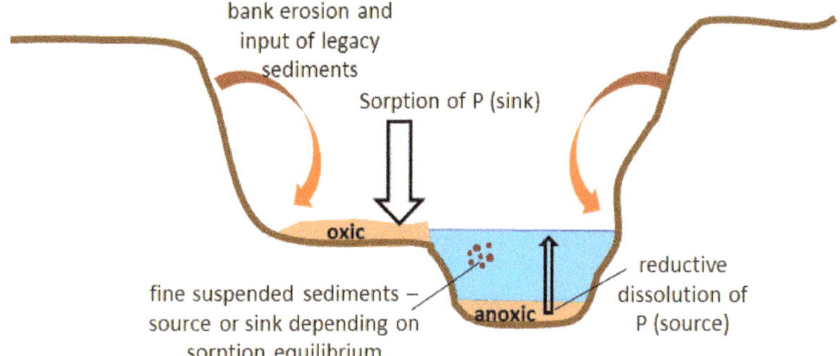

Figure 6. Conceptual model illustrating how stream bank legacy sediments could act as either a source or sink for P at various fluvial positions. Sorption (P sink) could occur on oxygenated floodplains if sediment EPC_0 values are below stream water P concentrations. In contrast, sediments could become a source of P due to reductive dissolution of iron oxides under anoxic conditions or when EPC_0 values are greater than stream water P concentrations. Deposition of legacy sediments in the channel could occur due to fluvial and/or subaerial bank erosion or mass wasting of the bank material.

We recognize here that we investigated legacy sediment behavior through controlled laboratory experiments. Field or in-situ conditions could be much more complicated with multiple factors and conditions (e.g., pH, temperature, salinity, oxygen, water flow and diffusion gradients, dry-wet cycles, etc.) simultaneously affecting P processes and dynamics with one factor reinforcing or cancelling out the effects of the other [11,25,61,62]. Because of this, measuring and understanding legacy sediment responses in-situ could be both, important and challenging [63]. Our observations were also based on only 15 sites and additional legacy sediment sites across the mid-Atlantic and elsewhere, particularly across a gradient of land uses and stream water P concentrations extending from the headwaters to the estuaries, could further help generalize and validate our conceptual model.

Given this conceptual model and the potential source-sink effects of legacy sediments on stream water P, one wonders how the presence of large, valley-bottom stores of P-poor legacy sediments have impacted stream water quality over the past century. Use of synthetic fertilizers started increasing around the 1950s and since then sources such as wastewater, sewage, septic, and fertilizer and manure applications on agricultural lands have been contributing elevated P to our surface waters [2,3,64]. Simultaneously, over the same period or even earlier, streambank legacy sediments have been eroding and contributing sediments to the fluvial system as a result of milldam breaches and removals [17]. Key questions then are: how have the interactions of P-poor legacy sediments and P-rich upland waters affected the trajectory of water quality of our surface waters over the past few decades? Have P-poor legacy sediments buffered the full impacts of upland sources on downstream aquatic ecosystems? On the other hand, have fine legacy sediments enriched with upland-P (among other sediment sources), which have been deposited in downstream aquatic systems and bays, become a long-term, internal, source of P (via reductive dissolution)?

Furthermore, many milldams still exist in the mid-Atlantic with stores of legacy sediments upstream of these dams [17,65]. Backing up and pooling of stream waters in mill ponds encourage anoxic conditions in sediments. This raises the question of how do the anoxic conditions influence the fate and release of P associated with millpond and streambank sediments? Moreover, dams are increasingly being removed across the mid-Atlantic and elsewhere [66–69]. An important question that is on the mind of many watershed managers and environmental agencies is how does the removal of the dams influence particulate and dissolved fractions of P stored upstream of the dams in the stream channel and along the banks? Dam removals could reduce anoxic conditions and the associated reductive dissolution of P, but increase erosion of particulate P associated with in-stream and streambank sediments. Addressing these questions and determining the net balance of these fluxes and processes is important if we want to fully understand the complex interactions of human land use legacies on water quality and the health of our ecosystems.

Our observations on P source-sink mechanisms also have important implications for contemporary watershed management and restoration and watershed P budget and regulatory assessments. Many of the streams with legacy sediments, particularly in the mid-Atlantic, are incised and hydrologically disconnected from the streambanks or legacy sediment terraces [14]. Stream restoration efforts are currently underway that enhance the hydrologic connectivity (or exchange) of the stream with the streambanks or floodplains via bank grading, floodplain creation, and/or stream uplift [70–72]. Such restoration efforts should particularly try to leverage the untapped P sorption capacity of legacy sediments by increasing the surface area of floodplain sediments that are exposed to and interact with stream waters, maintaining the floodplains in oxic conditions (so that P is retained on iron oxides), and reducing the exports of fine legacy sediments to downstream flows. If streambank legacy sediments do not contain other contaminants (e.g., metals or organics), these sediments could serve as valuable P sinks and should be used on-site in floodplain creation and other restoration activities and not moved offsite to landfills for disposal (as happens in some restoration projects (personal observations of the first author)). If legacy sediments have to be removed from valley bottoms [73] and if they are not contaminated, one potential opportunity could be to spread them back on croplands. After all, legacy sediments are the silt and clay-rich topsoil of precolonial America. Their use would particularly be beneficial on cropland soils saturated with decades of fertilizer applications of P, also referred to as legacy P [62,74]. Legacy sediments may be able to mitigate the pollution potential of these P-rich soils via sorption and P sequestration.

Given the various source-sink P mechanisms highlighted in this study, watershed P budgets, models, and regulatory assessments will have to account for the inputs of sediment-bound P with bank erosion of legacy sediments, the sorption capacity of these sediments in the fluvial system under oxic conditions (thus removal of solution P via by sediments), stream water P concentrations and the P sorption-desorption potential, and the release of P from sediments into solution under anoxic conditions. The magnitude of each of these processes would likely vary spatially along the length of the drainage network, from the headwaters to the bays. Assessing the *net* effect of legacy sediments on aquatic P would thus require a spatially-distributed, integrated, quantitative assessment or model of each of these processes over the drainage network. Characterizing the role of legacy sediments for fluvial P budgets will become more important and urgent given that climate intensification of storm events and other processes could potentially affect the erosion and inputs of these sediments into aquatic ecosystems [75]. Understanding and quantifying this variability and targeting P hotspots with appropriate best management practices (BMPs) will be key to achieving the P reduction goals established by the Chesapeake Bay and other watershed programs.

4. Conclusions

Using measured P concentrations and a variety of P indices, we investigated the potential for legacy sediments to act as a source or a sink of P in aquatic ecosystems. This study shows that streambank legacy sediments have low P concentrations and large capacity for P sorption. Overall, under oxic conditions and in streams with moderate to elevated P (in excess of sediment EPC_0 values, e.g., agricultural and urban streams), legacy sediments will likely serve as a net sink for P. However, legacy sediments could release P and become a net source under low stream water P conditions and/or anoxic sediment conditions due to reductive dissolution of iron oxides. Thus, the net source-sink effect for P at the catchment or drainage network scale will have to be assessed by accounting for the spatial variability in source-sink behavior with stream water P and redox conditions. Understanding these processes and the balance of these fluxes is critical to understanding the impact of legacy sediments on aquatic ecosystems. Whether they serve as P sources or sink, legacy sediments and the processes described herein need to be considered in nutrient budgets and watershed models that are being implemented to assign BMPs and meet regulatory load reductions for water quality.

Author Contributions: Conceptualization, S.I. and N.S.; methodology, N.S., A.L., and G.J.; formal analysis, N.S.; writing—original draft preparation, S.I. and N.S.; writing—review and editing, S.I., N.S., A.L., G.J., and J.K.; funding acquisition, S.I. and J.K. All authors have read and agreed to the published version of the manuscript.

Funding: Funding for this project was provided by the US Department of Agriculture via grant # USDA-NIFA 2017-67019-26330.

Acknowledgments: We thank the Fair Hill Natural Resources Management Area staff and the University of Delaware Soils Testing Lab for their analytical support. We also recognize the guidance of Maria Pautler and Karen Gartley on the P test indices.

Conflicts of Interest: The authors declare no conflict of interest.

References

1. Correll, D.L. The Role of Phosphorus in the Eutrophication of Receiving Waters: A Review. *J. Environ. Qual.* **1998**, *27*, 261–266. [CrossRef]
2. Kleinman, P.J.A.; Fanelli, R.M.; Hirsch, R.M.; Buda, A.R.; Easton, Z.M.; Wainger, L.A.; Brosch, C.; Lowenfish, M.; Collick, A.S.; Shirmohammadi, A.; et al. Phosphorus and the Chesapeake Bay: Lingering Issues and Emerging Concerns for Agriculture. *J. Environ. Qual.* **2019**, *48*, 1191–1203. [CrossRef] [PubMed]
3. Smil, V. Phosphorus in the environment: Natural Flows and Human Interferences. *Annu. Rev. Energy Environ.* **2000**, *25*, 53–88. [CrossRef]
4. Henley, W.F.; Patterson, M.A.; Neves, R.J.; Lemly, A.D. Effects of Sedimentation and Turbidity on Lotic Food Webs: A Concise Review for Natural Resource Managers. *Rev. Fish. Sci.* **2000**, *8*, 125–139. [CrossRef]
5. Smith, V.H. Eutrophication of freshwater and coastal marine ecosystems a global problem. *Environ. Sci. Pollut. Res.* **2003**, *10*, 126–139. [CrossRef]
6. Goldes, S.A.; Ferguson, H.W.; Moccia, R.D.; Daoust, P. Histological effects of the inert suspended clay kaolin on the gills of juvenile rainbow trout, *Salmo gairdneri* Richardson. *J. Fish Dis.* **1988**, *11*, 23–33. [CrossRef]
7. Waters, T.F. *Sediment in Streams. Sources, Biological Effects, and Control*; American Fisheries Society: Bethesda, MD, USA, 1995; Monograph 7; p. 251.
8. Chesapeake Bay TMDL Document. Available online: https://www.epa.gov/chesapeake-bay-tmdl/chesapeake-bay-tmdl-document (accessed on 10 May 2020).
9. Kaufman, Z.; Abler, D.; Shortle, J.; Harper, J.; Hamlett, J.; Feather, P. Agricultural Costs of the Chesapeake Bay Total Maximum Daily Load. *Environ. Sci. Technol.* **2014**, *48*, 14131–14138. [CrossRef]
10. Fanelli, R.M.; Blomquist, J.D.; Hirsch, R.M. Point sources and agricultural practices control spatial-temporal patterns of orthophosphate in tributaries to Chesapeake Bay. *Sci. Total Environ.* **2019**, *652*, 422–433. [CrossRef]
11. Fox, G.A.; Purvis, R.A.; Penn, C.J. Streambanks: A net source of sediment and phosphorus to streams and rivers. *J. Environ. Manag.* **2016**, *181*, 602–614. [CrossRef]
12. Zaimes, G.N.; Schultz, R.C.; Isenhart, T.M. Streambank Soil and Phosphorus Losses under Different Riparian Land-Uses in Iowa1. *J. Am. Water Resour. Assoc.* **2008**, *44*, 935–947. [CrossRef]
13. James, L.A. Legacy sediment: Definitions and processes of episodically produced anthropogenic sediment. *Anthropocene* **2013**, *2*, 16–26. [CrossRef]
14. Walter, R.C.; Merritts, D.J. Natural Streams and the Legacy of Water-Powered Mills. *Science* **2008**, *319*, 299–304. [CrossRef]
15. Miller, A.; Baker, M.; Boomer, K.; Merritts, D.; Prestegaard, K.; Smith, S. *Legacy Sediment, Riparian Corridors, and Total Maximum Daily Loads*; STAC Publication Number 19-001: Edgewater, MD, USA, 2019; p. 64.
16. Johnson, K.M.; Snyder, N.P.; Castle, S.; Hopkins, A.J.; Waltner, M.; Merritts, D.J.; Walter, R.C. Legacy sediment storage in New England river valleys: Anthropogenic processes in a postglacial landscape. *Geomorphology* **2019**, *327*, 417–437. [CrossRef]
17. Merritts, D.; Walter, R.; Rahnis, M.; Hartranft, J.; Cox, S.; Gellis, A.; Poter, N.; Hilgartner, W.; Langland, M.; Manion, L.; et al. Anthropocene streams and base-level controls from historic dams in the unglaciated mid-Atlantic region, USA. *Philos. Trans. Math. Phys. Eng. Sci.* **2011**, *369*, 976–1009. [CrossRef] [PubMed]
18. Wegmann, K.; Lewis, R.; Hunt, M. Historic mill ponds and piedmont stream water quality: Making the connection near Raleigh, North Carolina. In *From the Blue Ridge to the Coastal Plain: Field Excursions in the Southeastern United States: Geological Society of America Field Guide*; Geological Society of America: Boulder, CO, USA, 2012.

19. Cashman, M.J.; Gellis, A.; Sanisaca, L.G.; Noe, G.B.; Cogliandro, V.; Baker, A. Bank-derived material dominates fluvial sediment in a suburban Chesapeake Bay watershed. *River Res. Appl.* **2018**, *34*, 1032–1044. [CrossRef]

20. Gellis, A.; Noe, G. Sediment source analysis in the Linganore Creek watershed, Maryland, USA, using the sediment fingerprinting approach: 2008 to 2010. *J. Soils Sediments* **2013**, *13*, 1735–1753. [CrossRef]

21. Jiang, G.; Lutgen, A.; Sienkiewicz, N.; Mattern, K.; Kan, J.; Inamdar, S. Streambank legacy sediment contributions to sediment-bound nutrient yields from a Mid-Atlantic, Piedmont Watershed. *J. Am. Water Resour. Assoc.* **2020**, in press.

22. Lutgen, A.; Jiang, G.; Siekiewicz, N.; Mattern, K.; Kan, J.; Inamdar, S. Nutrients and Heavy Metals in Legacy Sediments: Concentrations, Comparisons with Upland Soils, and Implications for Water Quality. *J. Am. Water Resour. Assoc.* **2020**, in press. [CrossRef]

23. Walter, R.; Merritts, D.; Rahnis, M. Estimating volume, nutrient content, and rates of stream bank erosion of legacy sediment in the Piedmont and Valley and Ridge Physiographic provinces of southeastern and central PA. In *Report to the Pennsylvania Department of Environmental Protection, Lancaster, Pennsylvania*; Franklin and Marshall: Lancaster, PA, USA, 2007.

24. Niemitz, J.; Haynes, C.; Lasher, G. Legacy sediments and historic land use: Chemostratigraphic evidence for excess nutrient and heavy metal sources and remobilization. *Geology* **2013**, *41*, 47–50. [CrossRef]

25. Lammers, R.W.; Bledsoe, B.P. What role does stream restoration play in nutrient management? *Crit. Rev. Environ. Sci. Technol.* **2017**, *47*, 335–371. [CrossRef]

26. Fleming, P.M.; Merritts, D.J.; Walter, R.C. Legacy sediment erosion hot spots: A cost-effective approach for targeting water quality improvements. *J. Soil Water Conserv.* **2019**, *74*, 67A–73A. [CrossRef]

27. Sims, J.T.; Maguire, R.O.; Leytem, A.B.; Gartley, K.L.; Pautler, M.C. Evaluation of Mehlich 3 as an Agri-Environmental Soil Phosphorus Test for the Mid-Atlantic United States of America. *Soil Sci. Soc. Am. J.* **2002**, *66*, 2016–2032. [CrossRef]

28. Sienkiewicz, N. Microbial Characteristics and Fate of Legacy Sediments in Mid-Atlantic Streams. Master's Thesis, University of Delaware, Newark, DE, USA, 2019.

29. Nair, V.D.; Reddy, K.R. Phosphorus Sorption and Desorption in Wetland Soils. In *Methods in Biogeochemistry of Wetlands*; American Society of Agronomy and Soil Science Society of America: Madison, WI, USA, 2013; pp. 667–681.

30. Haggard, B.; Sharpley, A. Phosphorus Transport in Streams: Processes and Modeling Considerations. In *Modeling Phosphorus in the Environment*; Radcliffe, D., Carbrera, M., Eds.; CRC Press: Boca Raton, FL, USA, 2007; pp. 105–130.

31. McDaniel, M.D.; David, M.B.; Royer, T.V. Relationships between Benthic Sediments and Water Column Phosphorus in Illinois Streams. *J. Environ. Qual.* **2009**, *38*, 607–617. [CrossRef]

32. Young, E.O.; Ross, D.S. Phosphate Release from Seasonally Flooded Soils: A Laboratory Microcosm Study. *J. Environ. Qual.* **2001**, *30*, 91–101. [CrossRef]

33. Sienkiewicz, N.; Bier, R.L.; Wang, J.; Zgleszewski, L.; Lutgen, A.; Jiang, G.; Mattern, K.; Inamdar, S.; Kan, J. Bacterial communities and nitrogen transformation genes in streambank legacy sediments and implications for biogeochemical processing. *Biogeochemistry* **2020**, in press. [CrossRef]

34. Kovar, J.; Pierzynski, G. *Methods of Phosphorus Analysis for Soils, Sediments, Residuals, and Waters*; Southern Cooperative Series Bulletin; Virginia Tech University: Blacksburg, VA, USA, 2009.

35. Bache, B.W.; Williams, E.G. A phosphate sorption index for soils. *J. Soil Sci.* **1971**, *22*, 289–301. [CrossRef]

36. Hongthanat, N.; Kovar, J.L.; Thompson, M.L.; Russell, J.R.; Isenhart, T.M. Phosphorus source-sink relationships of stream sediments in the Rathbun Lake watershed in southern Iowa, USA. *Environ. Monit. Assess.* **2016**, *188*, 453. [CrossRef]

37. House, W.A.; Denison, F.H. Factors influencing the measurement of equilibrium phosphate concentrations in river sediments. *Water Res.* **2000**, *34*, 1187–1200. [CrossRef]

38. Jiang, G. Stream Bank Legacy Sediment Contributions to Suspended Sediment and Nutrient Exports from a Mid-Atlantic Piedmont Watershed. Master's Thesis, University of Delaware, Newark, DE, USA, 2019.

39. McDowell, R.W.; Sharpley, A.N. A Comparison of Fluvial Sediment Phosphorus (P) Chemistry in Relation to Location and Potential to Influence Stream P Concentrations. *Aquat. Geochem.* **2001**, *7*, 255. [CrossRef]

40. Rahutomo, S.; Kovar, J.; Kovar, J.; Thompson, M.; Thompson, M. Phosphorus transformations in stream bank sediments in Iowa, USA, at varying redox potentials. *J. Soils Sediments* **2019**, *19*, 1029–1039. [CrossRef]

41. Perillo, V.L.; Ross, D.S.; Wemple, B.C.; Balling, C.; Lemieux, L.E. Stream Corridor Soil Phosphorus Availability in a Forested-Agricultural Mixed Land Use Watershed. *J. Environ. Qual.* **2019**, *48*, 185–192. [CrossRef]

42. Ishee, E.R.; Ross, D.S.; Garvey, K.M.; Bourgault, R.R.; Ford, C.R. Phosphorus Characterization and Contribution from Eroding Streambank Soils of Vermont's Lake Champlain Basin. *J. Environ. Qual.* **2015**, *44*, 1745–1753. [CrossRef]

43. McDowell, R.; Simpson, Z.; Simpson, Z.; Stenger, R.; Stenger, R.; Depree, C.; Depree, C. The influence of a flood event on the potential sediment control of baseflow phosphorus concentrations in an intensive agricultural catchment. *J. Soils Sediments* **2019**, *19*, 429–438. [CrossRef]

44. Reddy, K.R.; Kadlec, R.H.; Flaig, E.; Gale, P.M. Phosphorus Retention in Streams and Wetlands: A Review. *Crit. Rev. Environ. Sci. Technol.* **1999**, *29*, 83–146. [CrossRef]

45. Young, E.O.; Ross, D.S. Total and Labile Phosphorus Concentrations as Influenced by Riparian Buffer Soil Properties. *J. Environ. Qual.* **2016**, *45*, 294–304. [CrossRef]

46. Mozaffari, M.; Sims, J.T. Phosphorus availability and sorption in an Atlantic coastal plain watershed dominated by animal-based agriculture. *Soil Sci.* **1994**, *157*, 97–107. [CrossRef]

47. Sobotka, M. Legacy Sediments in Streams—Effects on Nutrient Partitioning during Simulated Re-Suspension Events. Master's Thesis, Virginia Commonwealth University, Richmond, VA, USA, 2011.

48. Grundtner, A.; Gupta, S.; Bloom, P. River bank materials as a source and as carriers of phosphorus to lake pepin. *J. Environ. Qual.* **2014**, *43*, 1991–2001. [CrossRef]

49. Roberts, E.J.; Cooper, R.J. Riverbed sediments buffer phosphorus concentrations downstream of sewage treatment works across the River Wensum catchment, UK. *J. Soils Sediments* **2018**, *18*, 2107–2116. [CrossRef]

50. Rahutomo, S.; Kovar, J.L.; Thompson, M.L. Varying redox potential affects P release from stream bank sediments. *PLoS ONE* **2018**, *13*, e0209208. [CrossRef]

51. Ekka, S.A.; Haggard, B.E.; Matlock, M.D.; Chaubey, I. Dissolved phosphorus concentrations and sediment interactions in effluent–dominated Ozark streams. *Ecol. Eng.* **2006**, *26*, 375–391. [CrossRef]

52. Jarvie, H.P.; Jürgens, M.D.; Williams, R.J.; Neal, C.; Davies, J.J.L.; Barrett, C.; White, J. Role of river bed sediments as sources and sinks of phosphorus across two major eutrophic UK river basins: The Hampshire Avon and Herefordshire Wye. *J. Hydrol.* **2005**, *304*, 51–74. [CrossRef]

53. Young, E.O.; Ross, D.S. Phosphorus Mobilization in Flooded Riparian Soils from the Lake Champlain Basin, VT, USA. *Front. Environ. Sci.* **2018**. [CrossRef]

54. Rodens, E.; Edmonds, J. Phosphate Mobilization in Iron-Rich Anaerobic Sediments: Microbial Fe(III) Oxide Reduction Versus Iron-Sulfide Formation. *Arch. Hydrobiol.* **1997**, *139*, 347–378.

55. Chambers, R.M.; Odum, W.E. Porewater oxidation, dissolved phosphate and the iron curtain. *Biogeochemistry* **1990**, *10*, 37–52. [CrossRef]

56. Kumaragamage, D.; Concepcion, A.; Gregory, C.; Goltz, D.; Indraratne, S.; Amarawansha, G. Temperature and freezing effects on phosphorus release from soils to overlying floodwater under flooded-anaerobic conditions. *J. Environ. Qual.* **2020**. [CrossRef]

57. Vidon, P.; Allan, C.; Burns, D.; Duval, T.; Gurwick, N.; Inamdar, S.; Lowrance, R.; Okay, J.; Scott, D.; Sebestyen, S. Hot Spots and Hot Moments in Riparian Zones: Potential for Improved Water Quality Management. *J. Am. Water Resour. Assoc.* **2010**, *46*, 278–298. [CrossRef]

58. Zhang, J.; Huang, X. Effect of Temperature and Salinity on Phosphate Sorption on Marine Sediments. *Environ. Sci. Technol.* **2011**, *45*, 6831–6837. [CrossRef]

59. Upreti, K.; Joshi, S.R.; McGrath, J.; Jaisi, D.P. Factors Controlling Phosphorus Mobilization in a Coastal Plain Tributary to the Chesapeake Bay. *Soil Sci. Soc. Am. J.* **2015**, *79*, 826–837. [CrossRef]

60. Baldwin, D.S.; Mitchell, A.M. The effects of drying and re-flooding on the sediment and soil nutrient dynamics of lowland river–floodplain systems: A synthesis. *Regul. Rivers Res. Manag.* **2000**, *16*, 457–467. [CrossRef]

61. Records, R.M.; Wohl, E.; Arabi, M. Phosphorus in the river corridor. *Earth Sci. Rev.* **2016**, *158*, 65–88. [CrossRef]

62. Sharpley, A.; Jarvie, H.P.; Buda, A.; May, L.; Spears, B.; Kleinman, P. Phosphorus Legacy: Overcoming the Effects of Past Management Practices to Mitigate Future Water Quality Impairment. *J. Environ. Qual.* **2013**, *42*, 1308–1326. [CrossRef]

63. Noe, G.B.; Boomer, K.; Gillespie, J.L.; Hupp, C.R.; Martin-Alciati, M.; Floro, K.; Schenk, E.R.; Jacobs, A.; Strano, S. The effects of restored hydrologic connectivity on floodplain trapping vs. release of phosphorus, nitrogen, and sediment along the Pocomoke River, Maryland USA. *Ecol. Eng.* **2019**, *138*, 334–352. [CrossRef]

64. Garnache, C. Solving the phosphorus pollution puzzle. *Am. J. Agric. Econ.* **2016**, *98*, 1334–1359. [CrossRef]
65. Merritts, D.; Walter, R.; Rahnis, M.; Cox, S.; Hartranft, J.; Scheid, C.; Potter, N.; Jenschke, M.; Reed, A.; Matuszewski, D.; et al. The rise and fall of Mid-Atlantic streams: Millpond sedimentation, milldam breaching, channel incision, and stream bank erosion. *Geol. Soc. Am. Rev. Eng. Geol.* **2013**, *XXI*, 183–203.
66. Foley, M.M.; Bellmore, J.R.; O'Connor, J.E.; Duda, J.J.; East, A.E.; Grant, G.E.; Anderson, C.W.; Bountry, J.A.; Collins, M.J.; Connolly, P.J.; et al. Dam removal: Listening in. *Water Resour. Res.* **2017**, *53*, 5229–5246. [CrossRef]
67. Ryan Bellmore, J.; Duda, J.J.; Craig, L.S.; Greene, S.L.; Torgersen, C.E.; Collins, M.J.; Vittum, K. Status and trends of dam removal research in the United States. *Wiley Interdiscip. Rev. Water* **2017**, *4*, e1164. [CrossRef]
68. Tonitto, C.; Riha, S. Planning and implementing small dam removals: Lessons learned from dam removals across the eastern United States. *Sustain. Water Resour. Manag.* **2016**, *2*, 489–507. [CrossRef]
69. Tullos, D.D.; Collins, M.J.; Bellmore, J.R.; Bountry, J.A.; Connolly, P.J.; Shafroth, P.B.; Wilcox, A.C. Synthesis of Common Management Concerns Associated with Dam Removal. *J. Am. Water Resour. Assoc.* **2016**, *52*, 1179–1206. [CrossRef]
70. Palmer, M.A.; Hondula, K.L.; Koch, B.J. Ecological Restoration of Streams and Rivers: Shifting Strategies and Shifting Goals. *Annu. Rev. Ecol. Evol. Syst.* **2014**, *45*, 247–269. [CrossRef]
71. Newcomer Johnson, T.; Kaushal, S.; Mayer, P.; Smith, R.; Sivirichi, G. Nutrient Retention in Restored Streams and Rivers: A Global Review and Synthesis. *Water* **2016**, *8*, 116. [CrossRef]
72. Filoso, S.; Palmer, M.A. Assessing stream restoration effectiveness at reducing nitrogen export to downstream waters. *Ecol. Appl.* **2011**, *21*, 1989–2006. [CrossRef] [PubMed]
73. Hartranft, J.; Merritts, D.; Walter, R.; Rahnis, M. The Big Spring Run Restoration Experiment: Policy, Geomorphology, and Aquatic Ecosystems in the Big Spring Run Watershed, Lancaster County, PA. *Sustain* **2011**, *24*, 24–30.
74. Jarvie, H. Murky waters: Phosphorus mitigation to control river eutrophication. *J. Environ. Qual.* **2013**, *42*, 295–304. [CrossRef] [PubMed]
75. Inamdar, S.; Johnson, E.; Rowland, R.; Warner, D.; Walter, R.; Merritts, D. Freeze–thaw processes and intense rainfall: The one-two punch for high sediment and nutrient loads from mid-Atlantic watersheds. *Biogeochemistry* **2018**, *141*, 333–349. [CrossRef]

 *soil systems*

Article

Speciation of Phosphorus from Suspended Sediment Studied by Bulk and Micro-XANES

Qingxin Zhang [1], Mackenzie Wieler [1], David O'Connell [2,*], Laurence Gill [2], Qunfeng Xiao [3] and Yongfeng Hu [1,3,*]

[1] Department of Chemical and Biological Engineering, University of Saskatchewan, Saskatoon, SK S7N 5A9, Canada; qiz393@mail.usask.ca (Q.Z.); mjw067@mail.usask.ca (M.W.)

[2] Department of Civil, Structural and Environmental Engineering, Trinity College Dublin, Dublin 2, Ireland; laurence.gill@tcd.ie

[3] Canadian Light Source University of Saskatchewan, Saskatoon, SK S7N 2V3, Canada; Qunfeng.xiao@lightsource.ca

* Correspondence: david.oconnell@tcd.ie (D.O.); Yongfeng.hu@lightsource.ca (Y.H.)

Received: 6 April 2020; Accepted: 16 August 2020; Published: 18 August 2020

Abstract: Mobilization, transformation, and bioavailability of fluvial suspended sediment-associated particulate phosphorus (PP) plays a key role in governing the surface water quality of agricultural catchment streams. Knowledge on the molecular P speciation of suspended sediment is valuable in understanding in-stream PP cycling processes. Such information enables the design of appropriate catchment management strategies in order to protect surface water quality and mitigate eutrophication. In this study, we investigated P speciation associated with fluvial suspended sediments from two geologically contrasting agricultural catchments. Sequential chemical P extractions revealed the operationally defined P fractions for the fluvial suspended sediments, with Tintern Abbey (TA) dominated by redox-sensitive P (P_{CBD}), Al, and Fe oxyhydroxides P (P_{NaOH}) and organic P (P_{Org}) while Ballyboughal (BB) primarily composed of acid soluble P (P_{Detr}), redox-sensitive P (P_{CBD}), and loosely sorbed P (P_{NH4Cl}). The dominant calcareous (Ca) elemental characteristic of BB suspended sediment with some concurrent iron (Fe) influences was confirmed by XRF which is consistent with the catchment soil types. Ca-P sedimentary compounds were not detected using bulk P K-edge XANES, and only P K-edge μ-XANES could confirm their presence in BB sediment. Bulk P K-edge XANES is only capable of probing the average speciation and unable to resolve Ca-P as BB spectra is dominated by organic P, which may suggest the underestimation of this P fraction by sequential chemical P extractions. Notably, μ-XANES of Ca K-edge showed consistent results with P K-edge and soil geochemical characteristics of both catchments where Ca-P bonds were detected, together with calcite in BB, while in TA, Ca-P bonds were detected but mostly as organic complexed Ca. For the TA site, Fe-P is detected using bulk P K-edge, which corresponds with its soil geochemical characteristics and sequential chemical P extraction data. Overall, P concentrations were generally lower in TA, which led to difficulties in Fe-P compound detection using μ-XANES of TA. Overall, our study showed that coupling sequential chemical P extractions with progressively more advanced spectroscopic techniques provided more detailed information on P speciation, which can play a role in mobilization, transformation, and bioavailability of fluvial sediment-associated P.

Keywords: phosphorus; sediment; chemical P extraction; microanalysis; X-ray absorption near-edge structure (XANES) spectroscopy

---

1. Introduction

There has been increasing awareness of the importance of fluvial suspended sediments in the transport of nutrients from agricultural catchments, which can degrade water quality and cause eutrophication [1]. A significant proportion of total phosphorus (TP) loads in agricultural streams is transported as particulate phosphorus (PP) or within the fluvial suspended sediment PP [2,3]. Suspended sediment-associated P is deposited on the riverbed channel or on floodplains and its subsequent remobilization must therefore have an important impact on the transport, delivery pathway and fate of P species within agricultural catchments. Deposition on the river channel bed or floodplains can result in short- or long-term P storage. Similarly, remobilization of riverbed sediments coupled with bank erosion can reintroduce P to the river channel. Hence, information on P fluxes, storage, mobilization, and bioavailability within agricultural streams is required for appropriate catchment scale management policies. While research has addressed the increasing recognition of the importance of sediment PP within freshwater systems, relatively little attention has been given to P in suspended and streambed sediments within agricultural catchments, with the majority of research focusing on P in soils and lentic sediments (lakes and reservoirs, etc.). Relatively few studies have specifically characterized and quantified P species within streambed sediments [4,5], and fewer still in fluvial suspended sediments [6]. Many studies have shown that in catchments with diffuse sources, episodic flooding events and surface water runoff control stream bed and suspended solid composition and fluxes over time [7–9]. In addition to P and organic matter, the concentration and form of metal complexes and their link to microbial mineralization processes are likely impacted by fluvial and stream bed sediments. For example, it has been shown that organic matter remineralization predominates P cycling in Chesapeake Bay sediments [10,11] and the formation of phosphate–Fe(III)–humic complexes significantly impact P cycling and sedimentary Fe(III) stabilization in organic matter Fe-rich lake sediments, which is rarely recognized [12,13]. Hence, the molecular forms of P and their association to metal species of fluvial and stream bed sediments are important for understanding P transformation, mobility, and the potential impact on surface water eutrophication.

Sequential chemical extractions (SCEs) have been widely used to assess and quantify P species with different binding mechanisms and bioavailability. The basis of such fractionation analyses is in the differential reactivity of solid substrates to various chemical extractants [14]. Hence, such chemical extraction methods provide only operationally defined P pools and do not directly determine P speciation [15]. Notwithstanding their widespread application, there are various limitations associated with many sequential chemical P extraction schemes, including: (1) the specificity of extracting agents for sedimentary chemical P forms is relative; (2) transformation processes during the extractions between fractions (i.e., P sorption on calcite from calcareous sediments may be extracted in the Fe-oxide CBD extractant step). For example, during extraction of organic matter-rich sediments using the original Psenner method, Al and Fe associated with humic acid complexes are often misappropriated [16–18]. In the original method, Fe-bound P is extracted using bicarbonate-buffered dithionite followed by extraction of Al-oxide-bound P using NaOH. Concurrently, within this NaOH treatment, significant organic matter-bound P is extracted. Hence an additional step involving the acidification of the NaOH treated sediments (pH ~1) precipitated humic acid (HA) associated Fe and Al resulting in a clear supernatant with precipitate containing up to 30% of the total sediment P [17]. Similarly, another SCE [19] for sedimentary P was modified for organic-rich sediments with insertion of an additional extraction step (Na_2CO_3) prior to the Fe-bound P focused on bicarbonate-buffered dithionite step in order to extract Fe and Al humic complexes [12,20]. Despite such limitations, SCEs are still useful to get initial estimates of sedimentary chemical P pools. However, more recently, synchrotron-based P K-edge X-ray absorption near-edge structure (XANES) spectroscopy [21,22] and solution ^{31}P nuclear magnetic resonance (NMR) [23,24] have been applied to directly probe and distinguish different P species in terms of the inorganic form (XANES) and organic P species (NMR). More importantly, the feasibility and advantages of the combined use of these techniques in soil and sediment speciation have been demonstrated [11,12,25–27]. While such advanced spectroscopic techniques are capable of providing bulk speciation information for soil and sediment P, microclusters of concentrated P may be overlooked. Such P speciation information on microclusters of concentrated P is important, particularly for heterogeneous samples where correlations between P and metal species are necessary to understand the transfer and transformation of P in dynamic systems [28,29]. In this study, our objectives were to: (i) couple chemical P fractionation with the bulk and P K-edge micro(μ)-XANES to show the tiered approach in studying P compositional dynamics in suspended fluvial sediments from geologically contrasting agricultural and; (ii) apply bulk and P K-edge μ-XANES to give progressively more accurate and detailed compositional information compared to chemical P fractionations; (iii) show how bulk and P K-edge μ-XANES can identify P species in detail and the correlation between P, Ca, and Fe elements from suspended solids from two geologically contrasting agricultural catchments in Ireland.

2. Materials and Methods

Suspended sediment samples were collected from two agricultural catchments located in the east and southeast of Ireland, namely, Ballyboughal (BB) and Tintern Abbey (TA), as shown in Figure 1. Both catchment streams are medium- to high-level eutrophic according to the 2015–2017 water quality map released by the Irish Environment Protection Agency [30], which makes them suitable locations to study the influences of agricultural catchment sediments to fluvial waters. The BB site has an area of 23 km^2, with soils primarily composed of river alluvium, fine loamy drift with limestones, and siliceous stones. TA has an area of 10 km^2, with soils composed of primary river alluvium and fine loamy drift with siliceous stones.

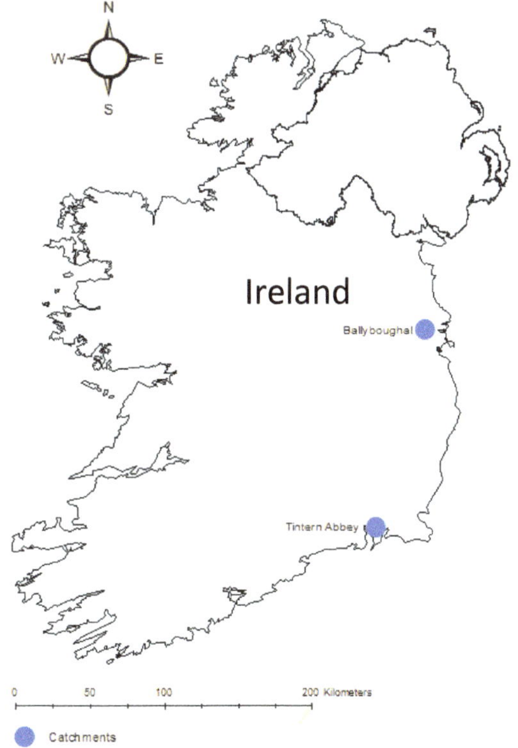

Figure 1. Location of the Ballyboughal (BB) and Tintern Abbey (TA) catchments in Ireland.

Fluvial sediment was gathered using time-integrated sediment traps, which were deployed to collect in-stream suspended solids over a 10-week period. One time period of samples was collected for this study, and taking into account the significant period of time (10 weeks) the traps were left in place, it is reasonable to consider the samples are representative of the catchment areas. Such sediment traps are known as Philips samplers [31], and their ability to collect suspended soils is based on slowing down the water velocity, resulting in sediment accumulation within the sediment trap tube. These samplers were installed horizontally in the middle of the channel and installed securely at approximately 60% of the average water depth using steel uprights and plastic cable ties to steel rebar embedded deep into the river channel [28]. The sediment traps were carefully removed from streams and emptied into clean 10 L containers which are kept as near as possible to the stream temperature while being transported back to the laboratory. The sediments were stored in a refrigerator for a few hours and processed by wet sieving the fluvial sediment through 63 μm steel stainless sieves, then centrifuged, and the supernatant decanted. The sediments were then freeze-dried for further analyses.

The sediment samples were sequentially extracted using a modified chemical extraction procedure [16,32], as shown in the flow diagram (Figure 2). Briefly, 1 M NH_4Cl was used to extract P adsorbed loosely to surfaces (P_{NH4Cl}); bicarbonate-dithionate (BD) extraction represents redox-sensitive P that is mainly bound to oxidized Fe and Mn compounds (P_{CBD}); 0.1 M NaOH was used to extract inorganic P compounds, such as Al and Fe oxyhydroxides (P_{NaOH}) and poly-P, P in detritus or complexes (P_{NRP}); humic P complexes (P_{hum}), acid-soluble P mainly bound to Ca (especially apatite), and Mg were differentiated using 0.5 M HCl (P_{Detr}); and, lastly, the residual P (P_{Res}) represents refractory organic P and non-extractable mineral P extracted using 1 M HCl at 120 °C. Phosphorus assumed to be associated with humic matter was precipitated by adding 2 M H_2SO_4 to a subsample of the NaOH extract [17]. Organic P (Figure 2) is calculated as the difference in TP between the first NaOH step and the second digested NaOH step [18].

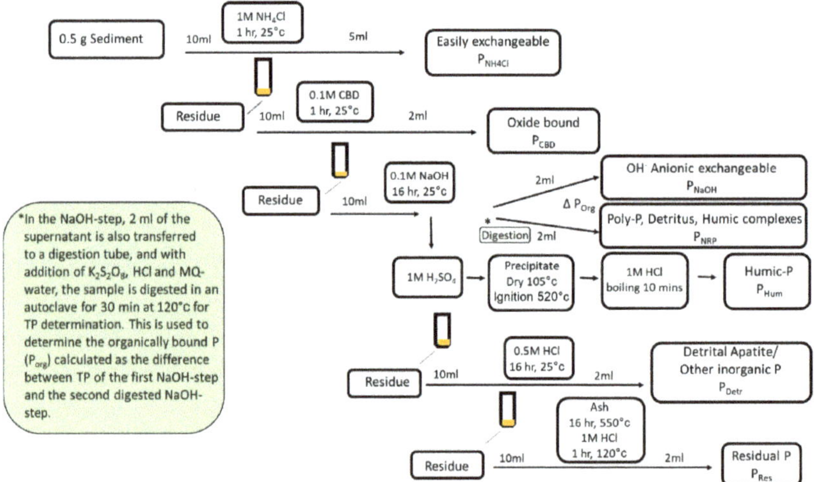

Figure 2. Modified Psenner sequential chemical P extraction procedure flow diagram [16,32].

The bulk and μ-XANES at P, Ca, and Fe K-edges were recorded using the Soft X-ray Microcharacterization Beamline (SXRMB) beamline at the Canadian Light Source. At SXRMB, a DCM monochromator using Si (111) crystals is used to cover an energy range of 2–10 keV. At the bulk station, the powder samples were mounted on the double-sided carbon tape and loaded into the vacuum chamber. The bulk spectra were recorded using a 7-element Si drift detector for sediment samples with low P concentration, and in TEY mode for reference samples. A pair of KB mirrors was used to focus the beam to a spot size of 10 μm × 10 μm with 10^9 photons/100mA/s flux [33]. A high-resolution and large area CCD camera is equipped to obtain the optical image of sample. A 4-element Si drift detector is used for μ-XANES analysis. A thin layer of sediment was spread on the carbon tape, and a large area of sample (3×3 mm^2) was first mapped with coarse resolution. Fine-resolution μ-XRF maps were acquired by selecting areas based on the elemental distribution and correlation. P, Ca, and Fe μ-XANES were acquired for selected hotspots. A photon energy of 7200 eV was used to record the XRF maps so that the distribution of P and other relevant elements can be tracked.

3. Results and Discussion

3.1. Total P and P Fractionation in Suspended Sediments

In this study, the overall median TP concentrations in the fluvial suspended sediments from the sum of all sedimentary P fractions at the Ballyboughal (BB) outflow was 3.4 mg·g^{-1}, and for the Tintern Abbey (TA) outflow, it was 0.9 mg·g^{-1} (Figure 3a). Such TP concentrations are comparable with previous studies on fluvial suspended and streambed sediments [5]. There is a relatively high TP concentration observed for the BB fluvial suspended sediment, likely due to additional sedimentary P contributions from domestic septic tanks and village wastewater treatment plant (WWTP) outflows [34] as the catchment is a relatively highly populated agricultural catchment in north county Dublin, Ireland.

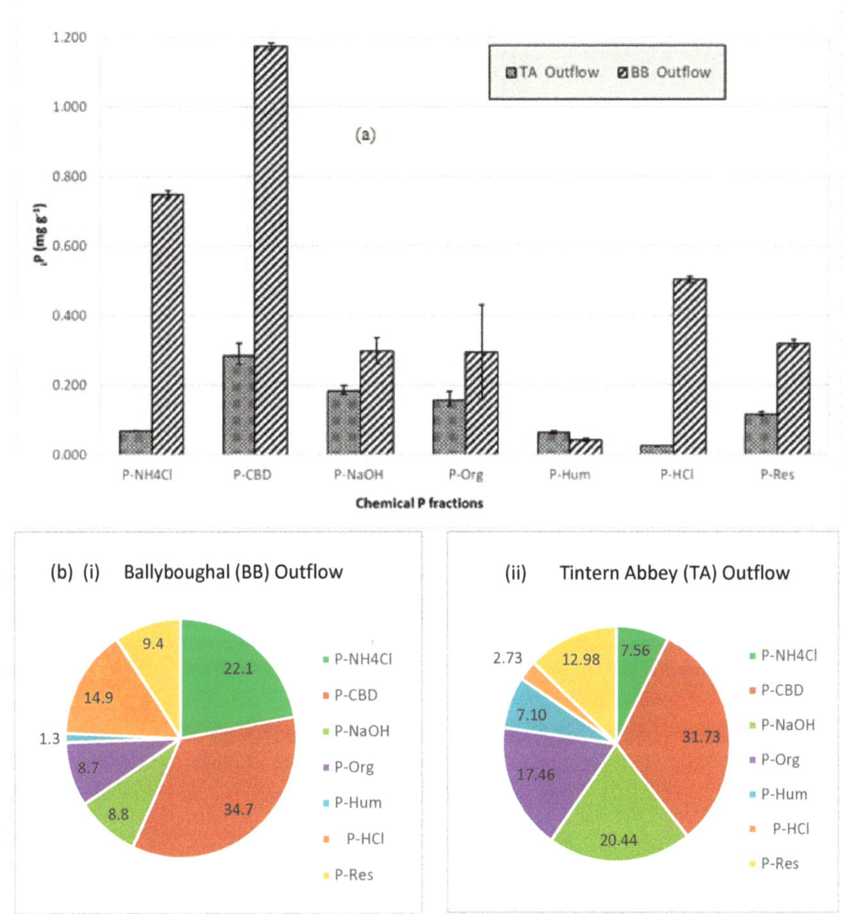

Figure 3. (**a**) Sequential chemical P fraction pools including total P (mg g^{-1} DW) for (**a**) Ballyboughal outflow (BB) and (**b**) Tintern Abbey outflow (TA); (**b**) sequential chemical P fractionations in relative percent (%) for (**i**) Ballyboughal (BB) outflow and (**ii**) Tintern Abbey (TA) outflow.

Within the BB suspended sediments, the most dominant P fractions included P_{CBD}, P_{Org}, and P_{NH4Cl} with concentrations of 1.18, 0.3, and 0.75 $mg \cdot g^{-1}$, respectively. By contrast, the most prevalent P fractions in the TA sediments were the P_{CBD}, P_{NaOH}, and P_{Org} with concentrations of 0.29, 0.18, and 0.16 $mg \cdot g^{-1}$ (Figure 3a). Figure 3b presents the relative proportional percentage chemical P fractionation results of representative fluvial suspended sediments from (i) BB and (ii) TA sites. The sequential chemical P fractionations separated the fluvial sediment TP into pools with diverse bioavailability. The distribution of P fractions differs in the suspended solids of the two geologically contrasting agricultural catchments. Loosely sorbed (P_{NH4Cl}) was elevated at 22% in BB while there was a lower percentage at 7.5% in TA. The BB agricultural catchment is typical of many Irish agricultural catchments in that additional P contributions, including P_{NH4Cl}, may come from domestic septic systems or from village WWTP outflows [34]. The relative P_{CBD} fraction which constitutes P bound to reducible species of Fe and Mn was almost the same in the suspended solids of both catchments being slightly elevated at the TA outflow at ~32% in comparison to ~35% at the BB outflow. Within the BB suspended solids, there is a relatively elevated percentage composition of P_{Org}, P_{Detr} (P bound to Ca and Mg), and P_{Res} at ~9% for all three. The relative percentage composition of organic P for TA is ~17%. A previous study on fluvial suspended solids from the mixed land-use Bras d'Henri River watershed in Quebec City, Canada, reported P_{Org} comprising up to 20% of TP using sequential chemical P fractionations, which was an order of magnitude greater than streambed sediments [5]. Previous studies have shown that mineralization of organic P (P_{Org}) to inorganic P ($_iP$) through enzymatic hydrolysis has a direct effect on P bioavailability in freshwater systems [35,36]. In addition, the TA outflow suspended sediment contains relatively elevated P_{Hum} and P_{NaOH} at ~7% and ~20.5%, which may be associated with the higher Fe and Al concentrations reflected in the catchment soils [34]. The P_{Hum} fraction may include Fe(III)-bearing colloids or organic matter–Fe(III)–P ternary complexes which can play an important role in P transport [8,12,17]. BB sediment samples clearly showed a higher relative fraction of P extracted by HCl (P_{Detr}) ~15%, indicating the presence of relatively elevated Ca-bound P, in agreement with the calcareous soils of the catchment [37]. The BB catchment does contain some areas rich in sandstone with elevated Fe, which may influence the relatively elevated percent composition redox-sensitive P (P_{CBD}) fraction. The relative percentage of redox-sensitive P (P_{CBD}) and NaOH-extracted P (P_{NaOH}) are both elevated in the TA sediment, consistent with the Fe-rich composition of the local mineralogy in the TA catchment [34]. The origin of such redox-sensitive P (P_{CBD}) and Al and Fe oxyhydroxide P (P_{NaOH}) fractions is associated with redox fluctuations impacting Fe(III)-oxides in hyporheic and riparian environments [38]. Organic matter degradation creates reducing anoxic conditions in such environments, particularly in summer, which can lead to reductive dissolution of Fe(III)-oxides to release Fe(II) and sorbed P [38,39].

3.2. Compositional Validation of Suspended Sediment Using X-Ray Fluorescence (XRF) and X-ray Absorption Near-Edge Structure (XANES)

The elemental compositions of these sediments as detected by X-ray fluorescence (XRF) (Figure 4) also show that the BB sediment has a much higher Ca concentration, while the TA sediment has a relatively higher Fe content. This is in general agreement with results from sequential chemical P fractionations which indicated more Ca-associated P within BB sediments and more Fe and/or redox associated P with TA sediments (Section 3.1; Figure 3). Previous studies have used synchrotron XRF coupled with operationally defined chemical P fractionation results to establish the dominant elements governing sediment composition [40–44].

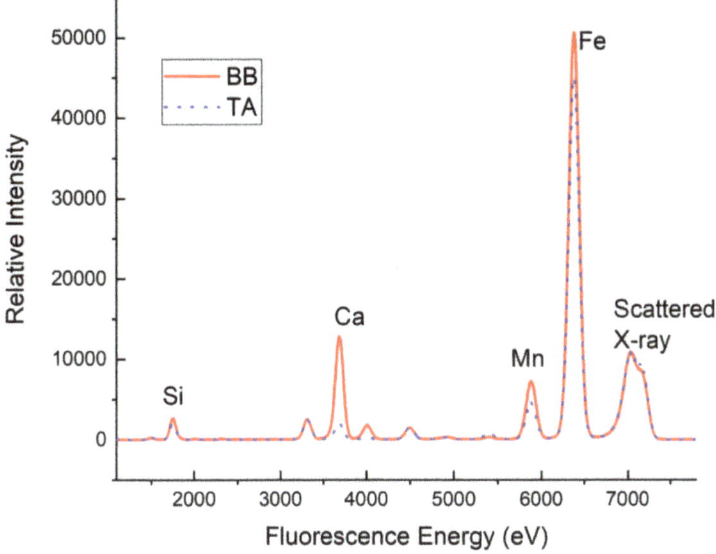

Figure 4. X-ray fluorescence spectra of bulk TA and BB sediments. The intensity was calibrated relative to the scattered X-ray (to an intensity of 10000 at 7000 eV), so that the comparison of elemental intensities is possible. The incident photon energy is 7200 eV.

Figure 5 shows the P K-edge spectra of (a) bulk TA and BB sediments and in comparison to (b) the spectra of selected P reference compounds. The Fe phosphate-related compounds have a unique pre-edge peak, as indicated by the arrow in the spectrum of $FePO_4.2H_2O$ shown in Figure 5b, while the Ca phosphate compounds, such as apatite, have several distinct shoulders in their P K-edge XANES [21,25]. P K-edge spectra of bulk sediment samples were dominated by featureless post-edge peak, similar to that of phytic acid [45]. This indicates the relatively significant presence of organic P (P_{Org}) in these samples, which may suggest the P_{Org} fraction of the chemical P fractionation results underestimate the significance of this P pool (Figure 3). The pre-edge peak, as indicated by the arrow, is clearly resolved in the spectrum of the TA sediment (Figure 5a), suggesting the presence of Fe-P in the bulk of TA sediment. This is in agreement with the chemical extraction and bulk XRF results (Figures 3 and 4). However, for Ca-rich BB sediment, no Ca-P related resonances are observed in the spectrum of BB sediment.

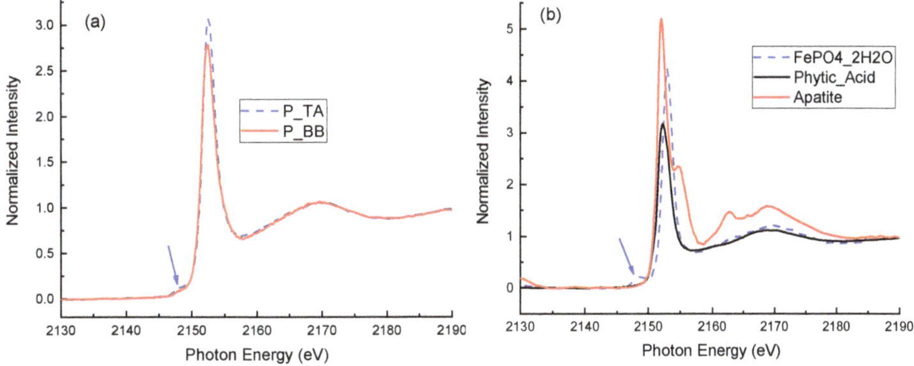

Figure 5. P K-edge XANES spectra of (**a**) bulk TA and BB sediments and (**b**) reference compounds.

This lack of Ca-P in the bulk spectrum of BB sediment implies the dominance of organic P and the inhomogeneous distribution of Ca-P in the BB sediment, which could be revealed by the microanalysis of these sediments. Chemical P fractionation results indicated a significant Ca-P pool normally associated with a recalcitrant mineral such as apatite. The tricolor μ-XAF maps of two sediments are shown in Figure 6. In this work, we chose to focus on the correlations between P, Fe, and Ca, as Fe oxides have been shown to complex with organic matter in sediments, thus impacting P mobility and surface water eutrophication [10–12]. The important role of Ca-P species in sediment weathering, P transformation in alkaline soil, and biogeochemical P cycling has been demonstrated by microscale XRF mapping, together with ^{31}P NMR and chemical extraction [27,42,43]. In Figure 6, elemental correlation maps or hotspot maps down to the micron scale are shown for P (blue), Ca (green), and Fe (red). These are elemental correlation maps with comparison of relative elemental concentration within a specified hotspot/area and not absolute concentrations. Where necessary, the hotspot can be identified down to 10 microns for μ-XANES. It is obvious that there are more Ca-rich hotspot correlations with P in the BB catchment. The BB sample also has very few and weak Fe spots, but there are identifiable Ca- and P-correlated spots (such as *A*). The TA sediment is generally Fe-rich, with only one Ca-rich hotspot. This is consistent with the XRF analysis and the mineralogy of these sites. A few hotspots (*A*, *B*, *C* as P-rich, Fe-rich, and Ca-rich for BB sediment and *D*, *E* as Fe-rich and Ca-rich for TA sediment, respectively) are selected for P, Ca, and Fe μ-XANES.

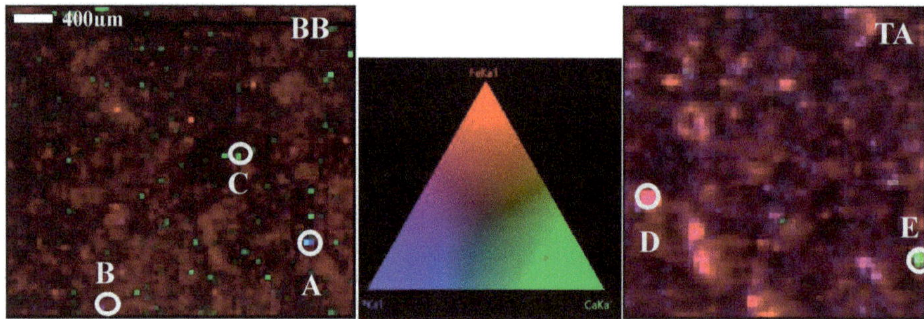

Figure 6. Elemental mapping for P (blue), Ca (green) and Fe (red) of BB and TA sediments with selected spots.

Figure 7a,b present the P K-edge μ-XANES of hotspots or region of interests (ROI) for BB and TA sediments. The P spectra for TA sediments (*D* and *E*) were quite noisy and without distinct features associated with inorganic P (Figure 7b), confirming its weak and significant organic P (Figure 3). The μ-XRF map of the TA sample revealed that ROI *D* had high Fe and P correlation (Figure 6); unfortunately, Fe-P could not be resolved in its μ-XANES (Figure 7b), likely due to its low P concentration and interference from the high Si content (Figure 4). For the Ca- and P-rich hotspots of BB sediment, the post-edge resonances were clearly resolved (Figure 7a, spots *A* and *C*), indicating the high Ca-P in the BB site. This is in agreement with the chemical extraction result, demonstrating the advantage of the μ-XANES, as no Ca-P was detected in the bulk P K-edge (Figure 5a). On the contrary, no Ca-P is detected in the μ-XANES of spot B, as it shows high Fe and P correlation (Figure 6). There might be a hint of the pre-edge peak in the P K-edge of spot B, implying the presence of Fe-P. Results of Ca K-edge μ-XANES (Figure 7c) are also in agreement with the P K-edge μ-XANES, as Ca in spot A matches well with that of apatite, spots *B* and *C* being mostly calcite, and spots *D* and *E* being mostly organic Ca [25]. The Fe K-edge μ-XANES of all spots (Figure 7d) are similar to each other and to those of bulk samples, as they are dominated by Fe hydroxyl oxide species [11,45]. No Fe-P can be resolved in the Fe K-edge μ-XANES due to its relatively low concentration.

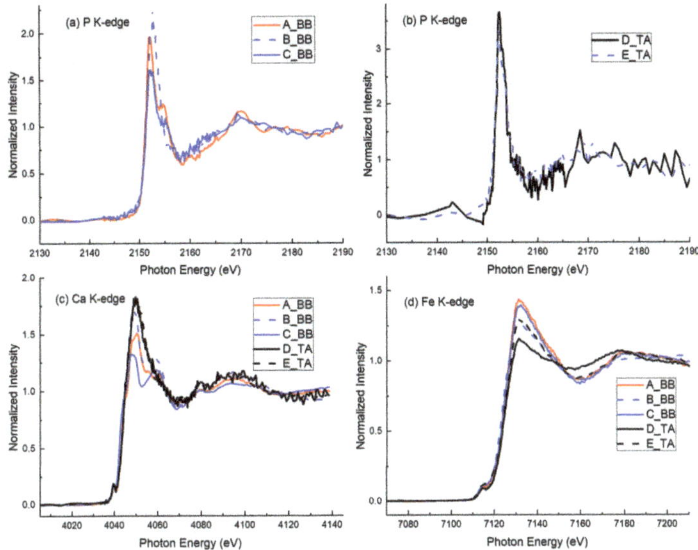

Figure 7. μ-XANES of selected spots identified in Figure 5: (**a,b**) P K-edge; (**c**) Ca K-edge and (**d**) Fe K-edge.

4. Conclusions

The chemical speciation of fluvial suspended sediment-associated PP from two geologically contrasting agricultural sites was studied by chemical extraction and bulk and μ-XANES. It was demonstrated that the microanalysis of P K-edge, combined with the μ-XANES of Ca and Fe K-edge, is critical to achieving a detailed characterization of the PP species. The limestone nature of BB was confirmed by XRF, which is consistent with the soil type at the BB site. The chemical confirmation of Ca-P-containing compounds in BB sediments was revealed only by P K-edge μ-XANES and not by its bulk spectrum, as the bulk XANES was only capable of probing the average speciation. Using bulk P K-edge XANES, it was difficult to resolve Ca-P peaks as the BB spectra is dominated by organic P. For the TA location, Fe-P containing compounds were detected by P K-edge XANES, which correlates with its soil geochemical characteristics and the sequential chemical P extraction data. Notably, the P concentration is generally lower in TA, which makes it difficult to detect Fe-P in the μ-XANES of TA. μ-XANES of Ca K-edge was consistent with the P K-edge and the geochemical characteristics of the study sites (i.e., both Ca-P associated with apatite and calcite detected in BB; TA has mostly organic Ca). A significant amount of organic P was detected in these sediments using P K-edge XANES which suggested the sequential chemical P extractions may have underestimated its importance. Overall, our study provided additional, detailed information on the P speciation associated with fluvial sediments by coupling sequential chemical P extractions with progressively more advanced spectroscopy, including XRF, bulk-XANES, and μ-XANES of multiple elements. This additional knowledge contributes to our understanding of the geochemical processes governing P mobilization, bioavailability, and transformation to potentially inform improved agricultural catchment management policies to protect the water quality of associated rivers and streams.

Author Contributions: D.O.; L.G. and Y.H. designed the study. Q.Z.; M.W.; D.O. collected the samples and performed most data measurement and analyses. Q.X. and Y.H. collected the XRF and XANES data. Y.H. and D.O. did most of the writing with inputs from all. All authors have read and agreed to the published version of the manuscript.

Funding: This work has been funded by the Natural Sciences and Engineering Research Council of Canada (NSERC) in collaboration with the Irish EPA (EPA), the Irish Department of Agriculture, Food and the Marine (DAFM) and WaterWorks Joint Call (WTWPJ 506072-2016).

Acknowledgments: We would like to acknowledge the support of numerous staff and scientists at the Canadian Light Source, which is supported by Canadian Foundation for Innovation, NSERC and the University of Saskatchewan. In addition, we acknowledge the support from the technical staff of the Environmental Engineering Research laboratory in Trinity College Dublin, Ireland.

Conflicts of Interest: The authors declare no conflict of interest. The funders had no role in the design of the study; in the collection, analyses, or interpretation of data; in the writing of the manuscript, or in the decision to publish the results.

References

1. Pulley, S.; Foster, I.D.L.; Antunes, A.P.M. The dynamics of sediment-associated contaminants over a transition from drought to multiple flood events in a lowland UK catchment. *Hydrol. Process.* **2016**, *30*, 704–719. [CrossRef]
2. Ballantine, D.J.; Waling, D.E.; Collins, A.L.; Leeks, G.J.L. The phosphorus content of fluvial suspended sediment in three lowland groundwater-dominated catchments. *J. Hydrol.* **2008**, *357*, 140–151. [CrossRef]
3. Sharpley, A.N.; Bergstrom, L.; Aronsson, H.; Bechmann, M.; Bolster, C.H.; Borling, K.; Djodjic, F.; Jarvie, H.P.; Schoumans, O.F.; Stamm, C.; et al. Future agriculture with minimized phosphorus losses to waters: Research needs and direction. *AMBIO* **2015**, *44*, S163–S179. [CrossRef] [PubMed]
4. Zhang, W.; Shan, B.; Zhang, H.; Tang, W. Assessment of preparation methods for organic phosphorus analysis in phosphorus-polluted Fe/Al-rich Haihe river sediments using solution ^{31}P-NMR. *PLoS ONE* **2013**, *8*, e76525. [CrossRef]
5. Su, J.; van Bochove, E.; Auclair, J.-C.; Theriault, G.; Denault, J.T.; Bosse, C.; Li, X.; Hu, C. Phosphorus algal availability and release potential in suspended and streambed sediments in relation to sediment and catchment characteristics. *Agricul. Ecosyst. Environ.* **2014**, *188*, 169–179. [CrossRef]
6. Stutter, M.I.; Langan, S.J.; Lumsdon, D.G.; Clark, L.M. Multi-element signatures of stream sediments and sources under moderate to low flow conditions. *Appl. Geochem.* **2009**, *24*, 800–809. [CrossRef]
7. Lamba, J.; Karthikeyan, K.G.; Thompson, A.M. Using radiometric fingerprinting and phosphorus to elucidate sediment transport dynamics in an agricultural watershed. *Hydrol. Process.* **2015**, *29*, 2681–2693. [CrossRef]
8. Bol, R.; Gruau, G.; Mellander, P.-E.; Dupas, R.; Bechmann, M.; Skarbovik, E.; Bieroza, M.; Djodjic, F.; Glendell, M.; Jordan, P.; et al. Challenges of reducing phosphorus based water eutrophication in the agricultural landscapes of Northwest Europe. *Front. Mar. Sci.* **2020**, *32*, 276. [CrossRef]
9. Gaspar, L.; Lizaga, I.; Blake, W.H.; Latorre, B.; Quijano, L.; Navas, A. Fingerprinting changes in source contribution for evaluating soil response during an exceptional rainfall in Spanish pre-pyrenees. *J. Environ. Manag.* **2019**, *240*, 136–148. [CrossRef]
10. Joshi, S.R.; Kukkadapu, R.K.; Burdige, D.J.; Bowden, M.E.; Sparks, D.L.; Jaisi, D.P. Organic matter remineralization predominates phosphorus cycling in the mid-bay sediments in the Chesapeake Bay. *Environ. Sci. Technol.* **2015**, *49*, 5887–5896. [CrossRef]
11. Li, W.; Joshi, S.R.; Hou, G.; Burdige, D.J.; Sparks, D.L.; Jaisi, D.P. Characterizing phosphorus speciation of Chesapeake Bay sediments using chemical extraction, ^{31}P NMR, and X-ray Absorption Fine Structure Spectroscopy. *Environ. Sci. Technol.* **2014**, *49*, 203–211. [CrossRef] [PubMed]
12. O'Connell, D.W.; Ansems, N.; Kukkadapu, R.K.; Jaisi, D.; Orihel, D.M.; Cade-Menun, B.J.; Hu, Y.; Wiklund, J.; Hall, R.I.; Chessell, H.; et al. Changes in sedimentary phosphorus burial following artificial eutrophication of Lake 227, Experimental Lakes Area, Ontario, Canada. *J. Geophys. Res. Biogeosci.* **2020**. [CrossRef]
13. Thandrup, B. Bacterial manganese and iron reduction in aquatic sediments. In *Advances in Microbiol Ecology*; Kluwer Academic/Plenum Publishing: New York, NY, USA, 2000; Volume 16, pp. 41–84.
14. Giguet-Covex, C.; Poulenard, J.; Chalmin, E.; Arnaud, F.; Rivard, C.; Jenny, J.-P.; Dorioz, J.-M. XANES spectroscopy as a tool to trace phosphorus transformation during soil genesis and mountain ecosystem development from lake sediments. *Geochim. Cosmochim. Acta* **2013**, *118*, 129–147. [CrossRef]
15. Gu, C.; Dam, T.; Hart, S.C.; Turner, B.L.; Chadwick, O.A.; Berhe, A.A.; Hu, Y.; Zhu, M. Quantifying uncertainties in sequential chemical extraction of soil phosphorus using XANES spectroscopy. *Environ. Sci. Technol.* **2020**, *54*, 2257–2267. [CrossRef]

16. Psenner, R.; Pucsko, R.; Sager, M. Fractionation of organic and inorganic phosphorus compounds in lake sediments: An attempt to characterize ecologically important fractions. *Archiv. Fuer. Hydrobiol.* **1984**, *70*, 111–155.

17. Paludan, C.; Jensen, H.S. Sequential extraction of phosphorus in freshwater wetland and lake sediment: Significance of humic acids. *Wetlands* **1995**, *15*, 365–373. [CrossRef]

18. Reitzel, K. Separation of aluminum bound phosphate from iron bound phosphate in freshwater sediments by a sequential extraction procedure. In *Phosphate in Sediments*; Serano, L., Golterman, H.L., Eds.; Backhuys Publishers: Leiden, The Netherlands, 2005; pp. 109–117.

19. Ruttenberg, K.C. Development of a sequential extraction method for different forms of phosphorus in 889 marine sediments. *Limnol. Oceanogr.* **1992**, *37*, 1460–1482. [CrossRef]

20. Baldwin, D.S. The phosphorus composition of a diverse series of Australian sediments. *Hydrobiologia* **1996**, *335*, 63–73. [CrossRef]

21. Hesterberg, D.; Zhou, W.Q.; Hutchison, K.J.; Beauchemin, S.; Sayers, D.E. XAFS study of adsorbed and mineral forms of phosphate. *J. Synchrotron Radiat.* **1999**, *6*, 636–638. [CrossRef]

22. Prietzel, J.; Dümig, A.; Wu, Y.; Zhou, J.; Klysubun, W. Synchrotron-based P K -edge XANES spectroscopy reveals rapid changes of phosphorus speciation in the topsoil of two glacier foreland chronosequences. *Geochim. Cosmochim. Acta* **2013**, *108*, 154–171. [CrossRef]

23. Cade-Menun, B.J. Characterizing phosphorus in environmental and agricultural samples by ^{31}P nuclear magnetic resonance spectroscopy. *Talanta* **2005**, *66*, 359–371. [CrossRef]

24. Cade-Menun, B.J. Improved peak identification in P-31-NMR spectra of environmental samples with a standardized method and peak library. *Geoderma* **2015**, *257*, 102–114. [CrossRef]

25. Ajiboye, B.; Akinremi, O.O.; Hu, Y.F.; Flaten, D.N. Phosphorus speciation of sequential extracts of organic amendments using NMR and XANES spectroscopies. *J. Environ. Qual.* **2007**, *36*, 1563–1576. [CrossRef] [PubMed]

26. Liu, J.; Hu, Y.; Yang, J.; Abdi, D.; Cade-Menun, B.J. Investigation of soil legacy phosphorus transformation in long- term agricultural fields using sequential fractionation, P K-edge XANES and solution P-NMR spectroscopy. *Environ. Sci. Technol.* **2015**, *49*, 168–176. [CrossRef] [PubMed]

27. Kar, G.; Hundal, L.S.; Schoenau, J.J.; Peak, D. Direct chemical speciation of P in sequential chemical extraction residues using P Kedge X-ray absorption near-edge structure spectroscopy. *Soil Sci.* **2011**, *176*, 589–595. [CrossRef]

28. Cooper, R.J.; Rawlins, B.G.; Kreuger, T.; Lézé, B.; Hiscock, K.M.; Pedentchouk, N. Contrasting controls on the phosphorus concentration of suspended particulate matter under base flow and storm event conditions in agricultural headwater streams. *Sci. Total Environ.* **2015**, *533*, 49–59. [CrossRef] [PubMed]

29. Rawlins, B.G. Controls on the phosphorus content of fine stream bed sediment in agricultural headwater catchments at the landscape scale. *Agric. Ecosyst. Environ.* **2011**, *144*, 352–363. [CrossRef]

30. Environment Protection Agency. *Water Quality in 2017: An Indicators Report*; Environment Protection Agency of Ireland: Dublin, Ireland, 2018.

31. Phillips, J.M.; Russell, M.A.; Walling, D.E. Time-integrated sampling of suspended sediment: A simple methodology for small catchments. *Hydrol. Process.* **2000**, *14*, 2589–2602. [CrossRef]

32. Hupfer, M.; Zak, D.; Rossberg, R.; Herzog, C.; Pothig, R. Evaluation of a well-established sequential phosphorus fractionation technique for use in calcite-rich lake sediments: Identification and prevention of artifacts due to apatite formation. *Limnol. Oceanogr. Methods* **2009**, *7*, 399–410. [CrossRef]

33. Xiao, Q.; Maclennan, A.; Hu, Y.; Hackett, M.; Leinweber, P.; Sham, T.-K. Medium-energy microprobe station at the SXRMB of the CLS. *J. Synchrotron Rad.* **2017**, *24*, 333–337. [CrossRef]

34. Fay, D.; Kramers, G.; Zhang, C.; McGrath, D.; Grennan, E. *Soils Geochemical Atlas of Ireland*; Teagasc and the Environmental Protection Agency: Dublin, Ireland, 2007; ISBN 1-84170-489-1.

35. Wang, J.Y.; Pant, H.K. Enzymatic hydrolysis of organic phosphorus in river bed sediments. *Ecol. Eng.* **2010**, *36*, 963–968. [CrossRef]

36. Zhang, W.; Tang, W.; Zhang, H.; Jianlin, B.; Jin, X.; Li, J.; Shan, B. Characterization of biogenic phosphorus in sediments from the multi-polluted Haihe River, China, using phosphorus fractionation and ^{31}P-NMR. *Ecol. Eng.* **2014**, *71*, 520–526. [CrossRef]

37. Fay, D.; Kramers, G.; Zhang, C. *Soil Geochemical Atlas of Ireland*; SAFER/EPA: Johnstown Castle, Ireland; Available online: http://erc.epa.ie/safer/iso19115/display?isoID=105 (accessed on 31 July 2020).

38. Van der Grift, B.; Behrends, T.; Osté, L.A.; Schot, P.P.; Wassen, M.J.; Griffioen, J. Fe hydroxyphosphate precipitation and Fe(II) oxidation kinetics upon aeration of Fe(II) and phosphate-containing synthetic and natural solutions. *Cosmochim. Acta Geochim.* **2016**, *186*, 71–90. [CrossRef]

39. Smolders, E.; Baetens, E.; Verbeeck, M.; Nawara, S.; Diels, J.; Verdievel, M.; Peeters, B.; De Cooman, W.; Baken, S. Internal loading and redox cycling of sediment iron explain reactive phosphorus concentrations in lowland rivers. *Environ. Sci. Technol.* **2017**, *51*, 2584–2592. [CrossRef] [PubMed]

40. Mockler, E.M.; Deakin, J.; Archbold, M.; Gill, L.; Daly, D.; Bruen, M. Sources of nitrogen and phosphorus emissions to Irish rivers and coastal waters: Estimates from a nutrient load apportionment framework. *Sci. Total Environ.* **2017**, *601–602*, 326–339. [CrossRef]

41. Joseph, M.M.; Kumar, C.S.R.; Renjith, K.R.; Kumar, T.R.G.; Chandramohanakumar, N. Phosphorus fractions in the surface sediments of three mangrove systems of southwest coast of India. *Environ. Earth Sci.* **2011**, *62*, 1209–1218. [CrossRef]

42. Liu, J.; Sui, P.; Cade-Menun, B.J.; Hu, Y.; Yang, J.; Huang, S.; Ma, Y. Molecular-level understanding of phosphorus transformation with long-term phosphorus addition and depletion in an alkaline soil. *Geoderma* **2019**, *353*, 116–124. [CrossRef]

43. Zimmer, D.; Kruse, J.; Siebers, N.; Panten, K.; Oelschlager, C.; Warkentin, M.; Hu, Y.; Zuin, L.; Leinweber, P. Bone char vs S-enriched bone char: Multi-method characterization of bone chars and their transformation in soil. *Sci. Total Environ.* **2018**, *643*, 145–156. [CrossRef]

44. Baumann, K.; Siebers, M.; Kruse, J.; Eckhardt, K.-U.; Hu, Y.; Michalik, D.; Siebers, N.; Kar, G.; Karsten, U.; Leinweber, P. Biological soil crusts as key player in biogeochemical P cycling during pedogenesis of sandy substrate. *Geoderma* **2019**, *338*, 145–158. [CrossRef]

45. Wang, X.; Hu, Y.; Tang, Y.; Yang, P.; Feng, X.; Xu, W.; Zhu, M. Phosphate and phytate adsorption and precipitation on ferrihydrite surfaces. *Environ. Sci. Nano* **2017**, *4*, 2193–2204. [CrossRef]

 *soil systems*

Article

Grazing Systems to Retain and Redistribute Soil Phosphorus and to Reduce Phosphorus Losses in Runoff

Anish Subedi, Dorcas Franklin *, Miguel Cabrera, Amanda McPherson and Subash Dahal

Department of Crop and Soil Sciences, University of Georgia, Athens, GA 30602, USA; anish.subedi@uga.edu (A.S.); mcabrera@uga.edu (M.C.); amanda.mcpherson@uga.edu (A.M.); subash.dahal@uga.edu (S.D.)
* Correspondence: dfrankln@uga.edu

Received: 11 September 2020; Accepted: 13 November 2020; Published: 16 November 2020

Abstract: A study of phosphorus accumulation and mobility was conducted in eight pastures in the Georgia piedmont, USA. We compared two potential grazing treatments: strategic-grazing (STR) and continuous-grazing-with-hay-distribution (CHD) from 2015 (Baseline) to 2018 (Post-Treatment) for (1) distribution of Mehlich-1 Phosphorus (M1P) in soil and (2) dissolved reactive phosphorus (DRP) and total Kjeldahl phosphorus (TKP) in runoff water. STR included rotational grazing, excluding erosion vulnerable areas, and cattle-lure management using movable equipment (hay-rings, shades, and waterers). After three years of treatment, M1P had significantly accrued 6- and 5-fold in the 0–5 cm soil layer and by 2- and 1.6-fold in the 5–10 cm layer for CHD and STR, respectively, compared to Baseline M1P. In STR exclusions, M1P also increased to 10 cm depth post-treatment compared to Baseline. During Post-Treatment, TKP runoff concentrations were 21% and 29% lower, for CHD and STR, respectively, in 2018 compared to 2015. Hot Spot Analysis, a spatial clustering tool that utilizes Getis-Ord Gi* statistic, revealed no change in Post-Treatment CHD pastures, while hotspots in STR pastures had moved from low-lying to high-lying areas. Exclusion vegetation retained P and reduced bulk density facilitating vertical transportation of P deeper into the soil, ergo, soil P was less vulnerable to export in runoff, retained in the soil for forage utilization and reduced export of P to aquatic systems

Keywords: soil P; vertical and horizontal P distribution; runoff water; exclusions; strategic grazing

---

1. Introduction

Improving availability of phosphorus (P) and at the same time reducing its loss to streams can contribute to sustainability of grazed pastures. Phosphorus is a vital nutrient for plants and animals, but it has been long identified as a major contributor to eutrophication. The underlying problem of the existing P-cycle is the failure to recover and reuse P in human waste, livestock manure, and food waste [1]. Grazing animals urinate and defecate around 81% of the P eaten [2]. Retention of P in animal waste, within the soils of grazing systems, will improve the efficiency of P use which could contribute to solving this problem. Management practices aimed toward vertical and horizontal distribution of P for better retention and utilization could reduce overall P losses from pastures. Distribution of P should focus on utilization of P for plant production while at the same time maintain soil surface properties that maximize infiltration, minimize over-land flow [3] and P improved water quality.

Internal inputs of P in pastures from weathered parent materials and external inputs from mineral fertilizers and dust deposited from the atmosphere are internally cycled through dung deposited by cattle, plant residues, and animal matter [4].

Surface application of phosphatic fertilizers exceeding the agronomic requirements of forages to kaolinitic soils has been shown to transfer readily soluble forms of inorganic phosphorus to lower depths in soil profile [5]. However, a study by McClaren et al. [6] reported total concentration of organic and inorganic forms of soil P in the 10–20 cm soil layer to be only 42% of the concentration measured in the 0–10 cm soil layer. Dung and plant residues are organic sources of P in soil that can reduce the need for external inputs of P, however, nutrient management strategies seldom consider both inorganic and organic forms of P [7]. Dung deposited by cattle during in-situ grazing can have cumulative benefits of improved pH and decreased P sorption, thus improving the efficiency of P cycle in the long term [8]. Adding P to the soil cannot be the sole strategy as research has shown increased losses of P with increased accumulation of P in soil. Phosphorus is relatively immobile in soil [9] until it reaches relatively high levels of soil saturation. Studies have shown that around 80% of manure P incorporated into soil by rain remains in the top 2 cm [10]. Surface soil P (2 cm) is a major contributor of P in runoff due to the desorption by runoff water [10], and this effect is exacerbated when vegetative cover is minimal [11]. Hence, P management should encompass strategies aimed to uniformly distribute available P to ensure better production of grass and forages and at the same time prevent addition of P in vulnerable areas with high transport and export potential to aquatic systems.

Managing P in pastures is particularly challenging given the diversity of landscapes under pastures and the complexity of P cycling, which is usually site specific. Dung produced by animals in pastures is deposited in patches that correlate with animal loci due to microtopography [12]. Dung returned to pasture soils is spatially heterogenous with greater dung P concentration near resting places (waterers, feeding stations, shades) [13,14].

Inherent vulnerability of sites based on proximity to streams and vulnerability to runoff and erosion should be considered [15]. The incidence of extreme events is expected to increase in the coming decades and research has reported increased rainfall intensity [16,17] and runoff [18,19] can increase P loss. Where P is deposited by cattle may greatly influence its retention within the pastures. In their study on grazed pastures, under contrasting grazing management and fertilizer applications, Bilotto et al. [20] reported greater mean annual changes in soil P on low slopes in comparison to high and medium slopes attributing the difference to movement of phosphorus in animal dung from higher slopes to the lower slopes. Nellesen et al. [21] indicated greater loss of P from pastures with unrestricted stream access compared to pastures with restricted stream access. Grazed grasslands are a significant source of P inputs to surface waters [16]. Additionally, Kurz et al. [22] suggested higher loss potential of organic P in pastures due to manure from grazing animals. In pastures under in-situ grazing, cattle dung stabilized P and increased soil pH [8]. Thus, effective P management strategies should involve use of techniques to reduce continuous treading of soil and excessive inputs of manure at vulnerable sites and maintain continuous vegetative soil cover to lower the losses as both particulate and dissolved P forms.

It is therefore important to generate strategic management systems that can uniformly distribute P inputs (dung and mineral fertilizers) in the pastures such that surface accumulation of P in erosion-prone regions of pastures is minimized. Areas with free access to cattle had 57–83% lower soil macroporosity and 8–17% greater bulk density when compared to areas where cattle were excluded [22]. Use of off-stream watering points (OSWPs) can have a potential benefit of reducing time spent by cattle in riparian areas, but, inclusions with shades, consideration of slope, size of paddock, and good grazing management practices could influence the effectiveness of OSWPs [23]. Rotational grazing in combination with a fenced riparian buffer can be effective in reducing runoff and erosion from pasture soils [24]. Management practices such as stocking rate and methods to manipulate distribution of shade structures, supplement feeding stations, waterers, fertilizers, and forage species diversity can affect efficiency of nutrient cycling in pastures [25].

Grasslands managed under different management practices have been extensively studied for nutrient cycling that encompasses nutrient composition of soils, input of nutrients, and management strategies to minimize losses and associated risks. Research on the relationships between soil test P and runoff P at the field- or pasture-scale is lacking. Most of the previous studies in grazing management have been small-scale plot studies and runoff simulation studies. Understanding the extent to which grazing management can vertically and horizontally retain and redistribute soil P concentrations and reduce P in runoff water at the pasture scale is needed. Strategies such as rotational grazing, exclusion of areas vulnerable to loss of P, and lure management of cattle have been identified as best management practices with the potential to improve P distribution, recycling, and retention in grazed pastures. This study compares two combinations of best management practices, strategic grazing (STR) or continuous-grazing-with-hay-distribution (CHD), to determine their impacts on distribution of soil P (Mehlich-1 P, M1P) and P loss as dissolved reactive P (DRP) and total Kjeldahl P (TKP) in runoff water, over a period of three years.

2. Materials and Methods

2.1. Study Sites

This study was conducted in eight pastures within the Georgia piedmont; four at the Animal and Dairy Science Eatonton Beef Research Unit (33.420759° N, 83.476555° W, Elevation 152–177 m, Eatonton) in Putnam County, GA, USA, and four at J Phil Campbell (JPC) Sr. Research and Education Center (33.887487° N, 83.420966° W; Elevation 213–259 m, Watkinsville) in Oconee County, GA (Figure 1). Pastures are characterized by moderate and wet winters and long and dry summers. Table 1 describes the study pastures at two locations, their respective area coverage, and number of sampling points in pastures (Matrix). Denser sampling indicates areas of interest (AOIs). The soil types [26] in the study locations, along with other hydrologically important characteristics, are summarized in Table 2.

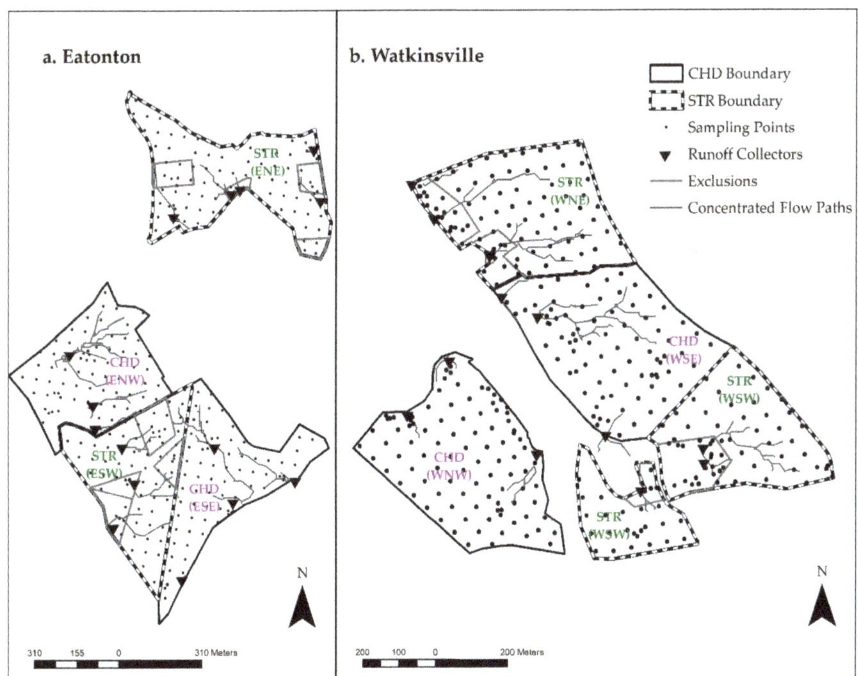

Figure 1. Maps showing; (**a**) Eatonton and (**b**) Watkinsville study sites with concentrated flow paths, pasture boundaries (CHD: continuous grazing with hay distribution, and STR: strategic grazing), runoff collectors, exclusions, and sampling locations (black dots).

Table 1. Area and number of soil sampling points in areas of interest (AOI) and matrix (50 m grid) for pastures at Eatonton and Watkinsville locations. Pasture abbreviations indicated locations relative to each other within each research location (ENE is east northeast, ENW is east northwest, etc.). Treatments are strategic grazing (STR) or continuous-grazing-with-hay-distribution (CHD).

Pastures	Treatment	Area (ha)	Number of Sampling Points	
			AOI	Matrix
Eatonton Beef Research Unit, Eatonton, Putnam county				
ENE	STR	21.53	12	74
ENW	CHD	18.42	14	64
ESE	CHD	17.96	11	64
ESW	STR	17.98	10	62
JPC, Watkinsville, Oconee county				
WNE	STR	15.06	11	70
WNW	CHD	16.97	18	72
WSE	CHD	18.4	19	55
WSW	STR	11.44	10	12

Table 2. Soil classifications of Eatonton and Watkinsville study pastures.

Location	Soil Series (% Area)	Class	Texture	Slope	Drainage
Eatonton	Davidson (60%)	Fine, kaolinitic, thermic, Rhodic Kandiudults	Loam to Clay Loam	2–15%	Well Drained
	Wilkes (17%)	Loamy, mixed, active, thermic, shallow Typic Hapludalfs	Loam to Sandy Loam	10–25%	Well Drained
	Iredell (12%)	Fine, mixed, active, thermic Oxyaquic Vertic Hapludalfs	Sandy Loam	<6%	Moderately Well Drained
	Enon (11%)	Fine, mixed, active, thermic Ultic Hapludalfs	Fine Sandy Loam	2–25%	Well Drained
Watkinsville	Cecil (60%)	Fine, kaolinitic, thermic Typic Kanhapludults	Sandy Loam	0–25%	Well Drained
	Pacolet (40%)	Fine, kaolinitic, thermic Typic Kanhapludults	Sandy Clay Loam	15–25%	Well Drained

2.2. Experimental Design

The study began in 2015 at both study sites with Baseline sampling carried out in spring/summer of 2015 (May–June). The 10-year legacy pasture management was continuous grazing management with free movement of cattle and fixed locations for hay feeding, waterers, and shade. All pastures were grazed continuously with 1.7–2.2 cattle head ha^{-1} prior to treatment application.

In May 2016, four replications of each of two grazing management treatments: (1) Continuous grazing with hay distribution (CHD) and (2) strategic rotational grazing (STR) were implemented. Cattle in CHD pastures had continued access to all locations inside pastures including areas with high P transport potential. Hay, however, was fed by distributing it to other locations in the pastures instead of conventional hay feeding at fixed locations in the pastures. This was done to prevent congregation of cattle at vulnerable locations during hay feeding. In this study, areas where cattle tended to congregate and that were also either close to the streams and/or areas in higher elevation, which had high P transport potential with steep slopes (6–27% slope), were designated as areas of interest (AOIs). In STR pastures, exclusions were setup by fencing the areas with high transport potential of P (Figure 1) and were over-seeded to maintain forage productivity and ground cover. Movable farm equipages: shades, waterers (with quick connects at different locations), and hay feeding rings were used to lure cattle to different locations strategically in STR pastures. Cattle density of all pastures during post-treatment was 1.1 cattle head ha^{-1}.

In STR pastures, exclusions were over-seeded with pearl millet (*Pennisetum glaucum*), crabgrass (*Digitaria* spp), and cow pea (*Vigna unguiculata*) in the spring, and with crimson clover (*Trifolium incarnatum*), canola (*Brassica napus*), ryegrass (*Lolium*) and cereal rye (*Secale cereale*) in the fall.

The exclusions were flash grazed based on the height of forage. Adaptive rotation strategy was used based on forage availability. Shade and waterers were moved to lure animals away from AOIs so that manure deposited by animals was less likely to be transported to the edge of fields or streams.

2.3. Sampling

Soils were sampled on a 50 m grid (matrix samples) in all pastures, with additional samples taken to ensure coverage with the AOIs. At each sampling point, two soil cores were taken, and each was divided into 0–5 cm, 5–10 cm, and 10–20 cm sections, bagged, placed in a dark cooler, and taken to the lab for processing. Soil sampling was carried out from mid-May to June in 2015 (Baseline) and 2018 (Post-Treatment). Within STR sites, sample points that were inside the excluded areas were termed "exclusions" while all other points outside the exclusions were termed "non-exclusions" to further understand the influence of the exclusions.

Pour-point runoff collectors were set up during Baseline (2015) at the drainage outlets of at least three watersheds within each pasture. Surface runoff samples generated by medium to high intensity rainfall (>20 mm) were collected from pour-point collectors, which had replaceable Nalgene bottles. Runoff samples were collected from June 2015 through April 2016 for Baseline (2015). Post-treatment runoff samples were collected from May 2016 to December 2018 as 2016, 2017, and 2018 sampling years. As 2016 was a drought year, there were no runoff samples for a 10-month period and this year was excluded from the analysis. Samples from each runoff event were brought to the lab and a 50 mL sub-sample was filtered (0.45 µm) within 48 h of the end of each event. The filtered runoff sub-samples and unfiltered runoff samples were stored in acid-washed Nalgene bottles at −10 °C for further analyses.

2.4. Analysis of Soil and Water Samples

Mehlich-1 phosphorus (M1P), which is a measure of plant available P in soil, was calculated using spectrophotometric analysis of soil extracts obtained from a two-component acid mixture [27]. Five grams of each soil sample was mixed with 20 mL of the double acid mixture (HCl and H_2SO_4) and shaken for 5 min, followed by filtering using a Whatman #42 filter. The extract was then analyzed for M1P employing the molybdenum blue dye procedure with spectrophotometric determination at 882 nm [28]. The analyses were expressed on a dry-weight basis by taking soil moisture into account. The concentrations of M1P were then converted to mass of M1P in kg P ha^{-1} using the bulk density of the soil samples calculated for another part of the study.

Dissolved reactive P (DRP) in runoff samples was measured in filtered runoff samples using spectrophotometric analysis on a Tecan Infinite Pro 200 (Tecan Group Ltd., Männedorf, Switzerland) with the Molybdenum-Blue dye [28]. Total Kjeldahl P (TKP) in runoff samples was determined from unfiltered runoff samples. A digestion solution containing $HgSO_4$, K_2SO_4 and concentrated H_2SO_4 was used to digest the unfiltered water samples at 114 °C for 14 h followed by digestion at 380 °C for another 2.5 h [29]. Phosphorus in the digests was determined using the Murphy–Riley procedure [28].

2.5. Conversion of Concentration to Loads of DRP and TKP in Runoff

Amount of runoff was estimated based on the amount of precipitation received, hydrologic condition of the soil and cover on the ground surface using the Curve Number (CN) method for estimating runoff [30]. It should be noted that ground cover conditions varied between post treatment calculations and the hydrologic conditions did not change for post treatment calculations. The CN is related to the potential maximum retention after runoff begins (S) as,

$$S = (1000/CN) - 10 \tag{1}$$

Initial abstraction (I_a), which represents all losses before runoff begins, is related to S as,

$$I_a = 0.2S \tag{2}$$

and was used to estimate the amount of runoff for each rainfall event using Equation (3),

$$Q = (P - 0.2S)^2/(P + 0.8S) \tag{3}$$

where Q is the amount of runoff, and P is the amount of precipitation. All values of DRP and TKP concentrations were converted to loads in runoff using the amount of runoff generated during each runoff event.

2.6. Spatial Analysis

A Trimble R10 GPS unit (Trimble, Sunnyvale, CA) was used to locate sampling points within the study pastures on a 50 m grid and to collect elevation every 2 m (4 cm resolution) for the development of digital elevation models (DEMs). ArcGIS 10.6.1 (ESRI, Redlands, CA, USA) was used to process the data by storing them in geodatabases. The DEMs were used to identify locations for pour-point collectors and delineation of watersheds. Values of the M1P concentrations in mg kg^{-1} of each soil depth (0–5, 5–10, and 10–20 cm) at point locations were converted into raster files using "Create TIN" and "TIN to Raster" tools to create a continuous surface (raster file) from the point file (ESRI). The raster files (M1P concentrations) were then used to visualize differences between 2015 and 2018 samples for M1P as 2018 minus 2015 raster for each depth. "Raster calculator" was used to carry out this analysis.

Spatial autocorrelation of M1P was estimated for each pasture under study using the Global Moran's I tool. Clusters of P in pastures were determined from the point values of M1P using the "Hot Spot Analysis" tool in ArcGIS (ESRI). Since sampling was conducted on a 50 m grid, 55 m was used to ensure more than one neighbor for each point, which would also ensure that not all points were neighbors of each other. The Hot Spot tool calculates the Getis-Ord GI* statistic [31,32], which is a z-score for each point under consideration, along with a *p*-value denoting significance. Higher z-values (0 to 1) would mean clustering of either high or low P concentrations.

2.7. Statistical Analysis

Comparison of the amount of M1P in soil between sampling periods, 2015 and 2018, was carried out for individual treatments (Baseline-CHD-2015 to Post-Treatment-CHD-2018, and Baseline-STR-2015 to Post-Treatment-STR-2018) using one-way analysis of variance (ANOVA) with differences analyzed by the Wilcoxon Test, which compared median M1P concentration. Comparison of the amount of M1P in soil between CHD and STR pastures at each sampling period (2015 or 2018) was also carried out using one-way ANOVA with differences analyzed by the Wilcoxon Test comparing median M1P concentrations. Comparisons between "exclusions" and "non-exclusions" in STR pastures for 2015 and 2018 sampling periods were carried out using one-way ANOVA. Non-exclusions are all other parts of STR pastures outside of fenced exclusions. The differences revealed were analyzed by the Wilcoxon Test. The Wilcoxon Test was used because of the non-normal M1P distribution.

Concentrations and loads of DRP and TKP in runoff were compared between sampling years (2015, 2017, and 2018) for each of the treatments (CHD or STR) and between CHD and STR treatments for each of the sampling years using one-way ANOVA. Differences were analyzed by the Wilcoxon Each Pair Test as DRP and TKP concentrations and loads did not meet the normal distribution assumption for parametric analysis. Simple linear regression model was used to determine the relationships between soil P (0–5 cm layer) and runoff P loads (DRP load, and TKP load) during Baseline and Post-Treatment. Event loads of DRP and TKP of the individual watersheds were fitted with the average soil-P values for each of the individual watersheds for the Baseline and Post-Treatment periods. Harmel et al. [33] did similar analysis on NO3 and PO4 losses from cropland and grazed pastures. The difference in slopes were analyzed by comparing fitted models with runoff P loads as response variables and soil P as the explanatory variable. Significance of interaction between sampling periods (Baseline vs. Post-Treatment) and soil P denotes difference in regression slopes between Baseline and Post-Treatment. Test of significance was conducted at 0.05 level of significance in all cases. All statistical analysis was carried out using the JMP software package (JMP®, Version 14. SAS Institute Inc., Cary, NC, USA, 1989–2019).

3. Results and Discussion

3.1. Changes in Vertical Distribution of Soil P

In this study, M1P increased significantly from 2015 to 2018 at 0–5 and 5–10 cm in both treatments (Table 3). These results indicate an accumulation of P in the top 10 cm layer of soil for both treatments, with 6.1 and 4.9 times increase in median M1P for the 0–5 cm layer and 2 and 1.6 times increase in median M1P for the 5–10 cm soil layer, CHD and STR pastures, respectively. Optimum M1P in Georgia Piedmont pasture soil is 30 mg P kg^{-1} [34]. In only a few years, both treatments were able to improve soil P concentrations of the 0–5 cm soil layer from low to moderate (half of the optimum M1P). The increase in M1P at 0–5 cm soil layer in both the treatments was due in part to the manure deposited by the grazing animals. P is relatively immobile in soil [9] and deposition of cattle dung accumulates inorganic P in the top 5 cm soil layer for a wide range of soils [35] while surface applied superphosphate accumulates inorganic P in the top 7.5 cm layer of a silt loam soil [5]. The unexpected increase in M1P at 5–10 cm depth could be explained by the decrease in bulk density [36]. Reduced bulk density results in greater porosity and movement (eluviation and illuviation) of clays and nutrients down the soil profile.

Table 3. Median Mehlich-1 phosphorus (M1P) in 2015 (Baseline) and 2018 (Post-Treatment) sampling dates in continuous grazing with hay distribution (CHD) and strategic grazing (STR) pastures.

Depth	Median CHD, 2015	Median CHD, 2018	Median STR, 2015	Median STR, 2018
cm		mg P kg^{-1}		
0–5	2.4Bb	14.7Aa	3.1Ba	15.1Aa
5–10	2.7Ba	5.4Aa	2.4Bb	3.9Ab
10–20	2.2Aa	2.3Aa	2.1Aa	1.6Aa

Medians separated by different upper-case letters denote significant difference between sampling dates within treatments (i.e., CHD, 2015, compared to CHD, 2018). Medians separated by different lower-case letters denote significant difference between treatments on the same sampling date (i.e., CHD, 2015, compared to STR, 2015). Difference is at 0.1 level of significance.

We speculate that soil biology and plant roots were responsible for transporting the available P from the top 5 cm to the 5–10 cm interval. However, this does not fully explain the increase in the 0–5 cm soil layer.

In CHD pastures, the increase at 0–5 cm may be partially explained by hay bales added during the drought in 2016. CHD pastures required 102 hay bales while STR pastures required only 34 hay bales due to sustained vegetation in exclusions late into a 2016 drought. Hay bales distributed at the Eatonton pastures had an average dry weight of 389 kg, 0.28% of which is P [37]. Assuming 1.1 kg P in each hay bale, the amount of P added would be 112 kg P in CHD and 37 kg P in STR pastures. As hay bales were distributed throughout the pasture during treatments and not in areas vulnerable to loss and because P increased in the top two layers, these results indicate that hay distribution can help retain P in pastures. Most P eaten by cattle returns to the soil in the form of manure adding the P eaten in hay and forage. In STR pastures, the increase in the 0–5 cm layer was likely due more to the retention of P deposited by grazing cattle than hay. STR pastures were effective in accumulating P due to redistribution, recycling, and retention of P.

3.2. Spatial Distribution of Phosphorus

With baseline hotspot analysis, all pastures and depths showed clusters of high M1P concentrations at low-lying areas that had high P transport potential (Figures 2a–c and 3a–c). Such hotspots must have been present due to continuous addition of manure and hay at congregation sites such as feeding areas and trees (natural shades). During Post-Treatment sampling, hotspots were still prevalent at similar locations in CHD pastures, but STR pastures showed such hotspots were more prevalent at higher elevations (Figures 2d–f and 3d–f). This has implications as to the reasons for reduced P losses in runoff from the STR pastures (see below).

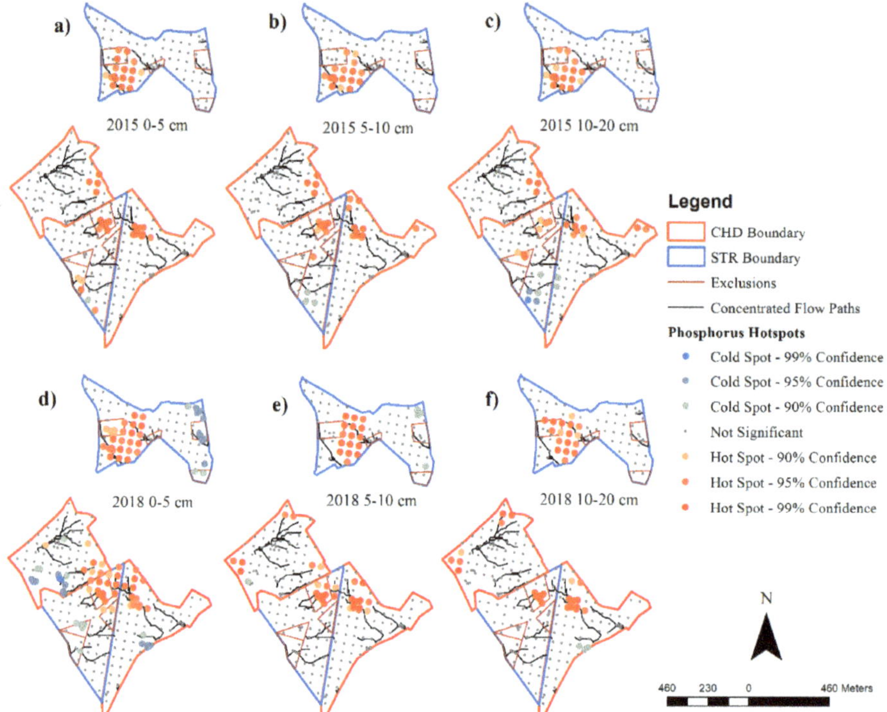

Figure 2. Hotspots of Mehlich-1 phosphorus (M1P) concentrations at 0–5, 5–10, and 10–20 cm soil depths during Baseline sampling (2015; (**a–c**)) and Post-Treatment (2018; (**d–f**)) at the Eatonton pastures.

Change in M1P distribution was mapped for the Eatonton (Figure S1) and the Watkinsville (Figure S2) study pastures calculated as 2018 raster–2015 raster. Difference maps of Eatonton and Watkinsville showed increase in M1P at locations of greater elevations in the pastures as compared to lower elevations. The high lying areas in pastures, denoted by yellowish to red color in the elevation model, showed greater increase in M1P at all three sampling depths in 2018 as compared to 2015. Increased availability of soil P at higher locations would mean greater time for runoff sediments to settle, possibly leading to lower losses of particulate P in runoff.

STR pastures showed lower availability of M1P at low lying exclusions close to streams. However, CHD pastures showed no change with higher M1P at low-lying, edge-of-stream areas making them more prone to lose P in soil as DRP and TKP. These areas were not excluded and over seeded to provide soil cover. It was interesting to see how M1P values were still high in areas at the previous hay-feeding locations in both CHD and STR pastures. STR pastures due to more uniform distribution of change (no hotspots) presented the potential of better redistribution of P compared to CHD.

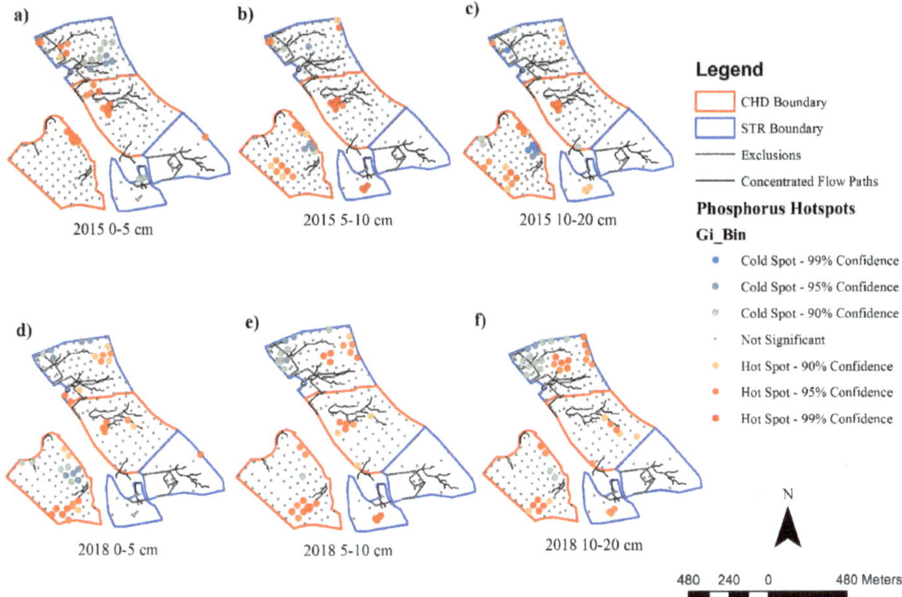

Figure 3. Hotspots of Mehlich-1 phosphorus (M1P) concentrations at 0–5, 5–10, and 10–20 cm soil depths during Baseline sampling (2015; (**a–c**)) and Post-Treatment (2018; (**d–f**)) at the Watkinsville pastures.

3.3. Effect of Exclusions

Both excluded and non-excluded areas, in STR pastures only, showed an almost five-fold increase from Baseline to Post-Treatment in the 0–5 cm layer. M1P in the 5–10 cm layer was almost twice as high in exclusions than in non-exclusions in 2018 (Table 4). Significant increases in M1P concentration in exclusions of the STR pastures at 0–5 and 5–10 cm layers suggest more P being captured by exclusions. The over-seeded exclusions could have helped slow runoff water and allowed for greater retention of particulate P even at 5–10 cm depth due to decreased bulk density and continued cover on the soil surface.

Table 4. Median Mehlich-1 phosphorus (M1P) in exclusions and non-exclusions in 2015 (Baseline) and 2018 (Post-Treatment) in strategic grazing (STR) pastures.

Depth	Non-Exclusions		Exclusions	
	2015	2018	2015	2018
cm	mg P kg^{-1}			
0–5	2.58B [1]a [2]	12.79Ab	3.48Ba	17.89Aa
5–10	2.28Aa	2.97Ab	2.6Ba	5.76Aa
10–20	2.14Aa	1.41Ba	2.03Aa	4.12Aa

[1] Medians separated by different upper-case letters denote significant difference between sampling dates in either non-exclusion (Non-Exclusions, 2015 vs. Non-Exclusion, 2018) or exclusion (Exclusions, 2015 vs. Exclusion, 2018). [2] Medians separated by lower-case letters denote significant difference between non-exclusion and exclusion at individual sampling dates (Non-Exclusions, 2015 vs. Exclusions, 2015). Difference is at 0.05 level of significance.

3.4. Changes in Runoff Water Phosphorus

Median concentrations and corresponding loads of DRP in runoff samples from CHD treatment was 0.14 mg P L^{-1} and 0.03 kg P ha^{-1} in 2015, 0.20 mg L^{-1} and 0.11 kg ha^{-1} in 2017, and 0.34 mg P L^{-1} and 0.12 kg P ha^{-1} in 2018, respectively. Similarly, in STR pastures, the median concentrations and corresponding loads of DRP were 0.38 mg P L^{-1} and 0.11 kg P ha^{-1} in 2015, 0.41 mg P L^{-1} and 0.17 kg

P ha^{-1} in 2017, and 0.32 mg P L^{-1} and 0.12 kg P ha^{-1} in 2018, respectively. The DRP concentrations and loads for the CHD treatments were significantly greater in 2018 when compared to 2015 and 2017 concentrations and loads (Figure 4). Stratification of P near the soil surface makes it more prone to loss in runoff water through desorption of P from soil surface and residues (litter and shoot) by water [38]. In the STR pastures no significant differences in either concentrations or loads of DRP were observed between sampling dates. For a given sampling period, the only significant difference between treatments for DRP concentrations and loads was noted in 2015 data, where STR 2015 was significantly greater than the CHD 2015. Distribution of DRP concentration in runoff samples from the two grazing treatments during the three sampling years revealed the absence of ≥ 2 mg P L^{-1} concentration in Post-Treatment runoff samples in the STR pastures. We make note of this as there were two tropical storms that occurred during the period 2017–2018 suggesting that, even during extreme events, the STR grazing system was able to reduce large pulses of DRP in runoff. For example, in 2015 the maximum rainfall event was 163.3 mm with a rainfall intensity of 2.0 mm hr^{-1} and in 2017 the maximum rainfall event (Hurricane Irma) was 103.6 mm with a rainfall intensity of 4.2 mm hr^{-1}. Yet the maximum concentration in runoff was <2 mg P L^{-1}.

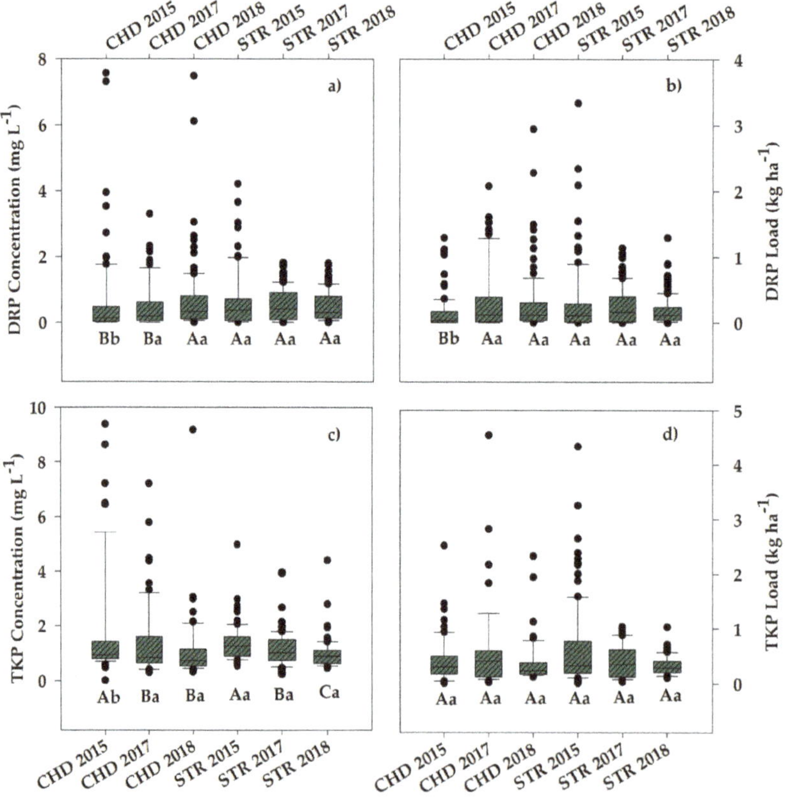

Figure 4. Comparison of dissolved reactive phosphorus (DRP). (**a**) DRP concentrations (mg L^{-1}), (**b**) DRP loads (kg ha^{-1}), (**c**) total Kjeldahl phosphorus (TKP) concentrations (mg L^{-1}), and (**d**) TKP loads (kg ha^{-1}) from Baseline (2015) to 2017 and 2018 in continuous grazing with hay distribution (CHD) and strategic grazing (STR) pastures. Upper-case letters denote comparison between sampling years for individual treatment (example: CHD, 2015 vs. CHD, 2017 vs. CHD, 2018) and lower-case letters denote comparison between CHD and STR treatments for individual sampling periods (example: CHD, 2015 vs. STR, 2015). Different letters denote significance at 0.05 level.

Median concentrations and corresponding loads of TKP in runoff samples from CHD treatment were 0.94 mg P L^{-1} and 0.30 kg P ha^{-1} in 2015, 0.85 mg L^{-1} and 0.40 kg ha^{-1} in 2017, and 0.74 mg P L^{-1} and 0.24 kg P ha^{-1} in 2018, respectively. Similarly, in STR pastures, the median concentrations and corresponding loads of TKP were 1.26 mg P L^{-1} and 0.33 kg P ha^{-1} in 2015, 1.03 mg P L^{-1} and 0.35 kg P ha^{-1} in 2017, and 0.90 mg P L^{-1} and 0.30 kg P ha^{-1} in 2018, respectively. Comparison of TKP concentrations in runoff water between sampling years revealed significantly lower TKP concentration in 2017 sampling compared to 2015 in both treatments. In 2018, however, only STR showed significant decrease in TKP concentration in comparison to both 2015 and 2017. Over-seeded exclusions in STR treatment could have helped reduce the concentration of TKP in runoff water through retention.

3.5. Relationship between Soil Phosphorus and Phosphorus in Runoff

The relationship between M1P, the plant available fraction of soil P and DRP and TKP loads in surface runoff was studied for CHD and STR grazing treatments over a three-year study period. We found that M1P at 0–5 cm depth was significantly correlated with DRP load in runoff water for both grazing treatments during both sampling dates (Figure 5). As for DRP loads, regression slopes derived from the relationship, DRP versus M1P (0–5 cm soil depth), during Baseline and Post-Treatment were compared using a simple regression model. No differences in the slopes were indicated for CHD pastures between Baseline and Post-Treatment, while the slope in STR pasture was significantly lower for Post-Treatment as compared to Baseline.

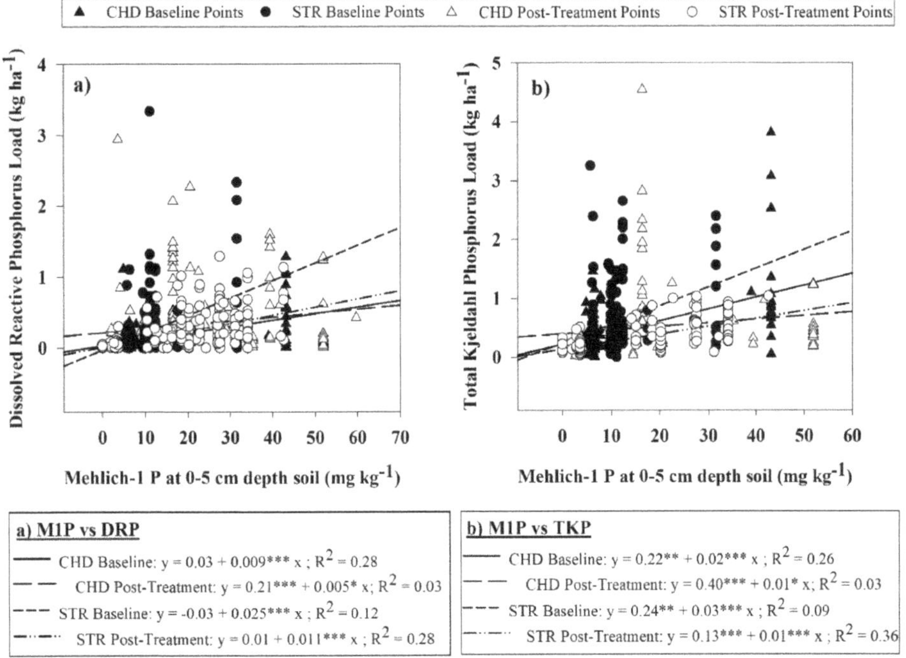

Figure 5. Relationship of (**a**) dissolved reactive phosphorus (DRP) load and (**b**) total Kjeldahl phosphorus (TKP) in runoff water and Mehlich-1 Phosphorus (M1P) in 0–5 cm depth soil compared between Baseline (2015) and Post-Treatment (2016–2018) sampling dates in continuous grazing with hay distribution (CHD) and strategic (STR) grazing treatments. *, **, and *** represent statistical significance at ≤0.05, ≤0.01, and ≤0.001, respectively.

Similar relationships between the M1P in soil and TKP in runoff water were studied for the two grazing systems during the two sampling dates. Results revealed significant positive correlation between the soil P and TKP in runoff water. Comparison of regression slopes for the two sampling periods indicated that the slopes were three-fold and two-fold lower during Post-Treatment as compared to Baseline in STR and CHD, respectively. Load of DRP in runoff water per the amount of M1P found in the soil was near half during Post-Treatment sampling compared to Baseline in STR. The intercept of lines relating soil P with TKP loads was significant for both treatments during the Baseline and Post-Treatment samplings. Positive intercept values suggest that the source of total P, other than M1P, could be organic matter at the soil surface [39], in our case, very likely dung and hay residues.

Transport of different P fractions in runoff and interactions with soil P, in exclusion and non-exclusion areas, provide additional insights into loss and transport mechanisms, especially in areas with high P transport potential. P application in areas with high P transport potential should be avoided [40]. More explicitly, areas of high P transport potential need to be managed to reduce P losses in runoff [15]. These would include: (1) steep areas close to streams and prone to erosion, (2) low-lying areas close to the streams, (3) concentrated flow-paths, and (4) high elevation areas with greater slopes. Exclusions placed in areas with high P transport potential can help reduce the P loss from these areas. In our study, exclusions at low-lying areas aided in the retention of particulate P and its vertical movement deeper into the rhizosphere. Vegetation in the exclusions served as buffers that interrupted the direct interaction of runoff water with the soil and slowed runoff water down, therefore facilitating deposition of particulate P. Furthermore, allowance of flash grazing of these areas resulted in continued use of these areas as grazing lands, while still reducing the amount of time cattle were present in the vulnerable areas causing chronic, direct deposition of manure. Vegetated exclusions acted as buffers and provided opportunity for the P in runoff to settle and infiltrate 10 cm into the soil while also reducing the amount of particulate P lost in runoff.

4. Conclusions

After three years of application of STR grazing system, soil P in the 0–5 and 5–10 cm depths increased significantly compared to historically continuously grazed systems that had hay-feeding and watering at the same locations yearly. The CHD grazing system with hay distribution at different locations also increased soil P. In STR pastures, exclusions increased soil P at 0–5 and 5–10 cm depths in Post-Treatment showing retention of P and reduction in total P losses in runoff water. Exclusions also provided added benefits of forage availability during drought, and reduced interaction of animals with vulnerable low-lying locations in pastures. Combining rotational grazing (every 5 to 10 days) and lure management of cattle with movable equipages aided in recycling of soil P to less vulnerable high-lying portions of the pastures, which was demonstrated by hotspot analysis. Hotspots of soil P prevalent at vulnerable areas during the baseline period moved to higher locations in STR pastures allowing greater retention of P. With the increase in soil P, DRP losses increased in both treatments, however, less occurrence of larger P losses during extreme events in the STR treatment suggests that STR grazing systems could be more resistant to extreme weather events. TKP losses were reduced significantly with CHD and STR grazing management systems as they ensured distribution of P in areas less vulnerable to loss in runoff. Hence, CHD and STR grazing managements present considerable potential to retain P for forage use rather than being exported to aquatic systems. Further research is needed to determine the effectiveness of these management practices on pastures fertilized with broiler litter and in other landscapes beyond the Georgia Piedmont.

Supplementary Materials: The following are available online at http://www.mdpi.com/2571-8789/4/4/66/s1, Figure S1: Spatial distribution of change in Mehlich-1 phosphorus (change mg kg^{-1} = 2018 raster−2015 raster) at Eatonton location. Figure S2: Difference and DEM of Watkinsville pastures: Spatial distribution of change in Mehlich-1 phosphorus (change mg kg^{-1} = 2018 raster−2015 raster) at Watkinsville location.

Author Contributions: Conceptualization, D.F.; methodology, D.F. and A.S.; formal analysis, A.S.; investigation, A.S., D.F., S.D., and A.M.; writing—original draft preparation, A.S.; writing—review and editing, A.S., D.F., and M.C.; funding acquisition, D.F. All authors have read and agreed to the published version of the manuscript.

Funding: This research was funded by NRCS-USDA, Conservation Innovation Grant. Grant number 69-3A75-14-251.

Acknowledgments: The authors are grateful to NRCS-USDA for their assistance with the first order soil survey, and to the Sustainable Agriculture Laboratory team, and Charles Trumbo at the University of Georgia for their support in lab and in field.

Conflicts of Interest: The authors declare no conflict of interest.

References

1. Sharpley, A. Managing agricultural phosphorus to minimize water quality impacts. *Sci. Agric.* **2016**, *73*, 1–8. [CrossRef]
2. Rothwell, S.A.; Doody, D.G.; Johnston, C.; Forber, K.J.; Cencic, O.; Rechberger, H.; Withers, P.J.A. Phosphorus stocks and flows in an intensive livestock dominated food system. *Resour. Conserv. Recycl.* **2020**, *163*, 105065. [CrossRef]
3. McDowell, R.W.; Nash, D.M.; Robertson, F. Sources of phosphorus lost from a grazed pasture receiving simulated rainfall. *J. Environ. Qual.* **2007**, *36*, 1281–1288. [CrossRef] [PubMed]
4. Chapin, F.S.; Matson, P.A.; Vitousek, P.M. Nutrient cycling. In *Principles of Terrestrial Ecosystem Ecology*; Springer: New York, NY, USA, 2011; pp. 259–296. [CrossRef]
5. Tian, J.; Boitt, G.; Black, A.; Wakelin, S.; Condron, L.M.; Chen, L. Accumulation and distribution of phosphorus in the soil profile under fertilized grazed pasture. *Agric. Ecosyst. Environ.* **2017**, *239*, 228–235. [CrossRef]
6. McLaren, T.I.; Smernik, R.J.; Simpson, R.J.; McLaughlin, M.J.; McBeath, T.M.; Guppy, C.N.; Richardson, A.E. The chemical nature of organic phosphorus that accumulates in fertilized soils of a temperate pasture as determined by solution 31P NMR spectroscopy. *J. Plant Nutr. Soil Sci.* **2017**, *180*, 27–38. [CrossRef]
7. Dodd, R.J.; Sharpley, A.N. Recognizing the role of soil organic phosphorus in soil fertility and water quality. *Resour. Conserv. Recycl.* **2015**, *105*, 282–293. [CrossRef]
8. During, C.; Weeda, W.C. Some effects of cattle dung on soil properties, pasture production, and nutrient uptake: I. Dung as a source of phosphorus. *N. Z. J. Agric. Res.* **1973**, *16*, 423–430. [CrossRef]
9. Coleman, T.L.; Eke, A.U.; Bishnoi, U.R.; Sabota, C. Nutrient losses in eroded sediment from a limited-resource farm. *Field Crop. Res.* **1990**, *24*, 105–117. [CrossRef]
10. Vadas, P.A.; Gburek, W.J.; Sharpley, A.N.; Kleinman, P.J.; Moore, P.A., Jr.; Cabrera, M.L.; Harmel, R.D. A model for phosphorus transformation and runoff loss for surface-applied manures. *J. Environ. Qual.* **2007**, *36*, 324–332. [CrossRef]
11. Butler, D.M.; Franklin, D.H.; Ranells, N.N.; Poore, M.H.; Green, J.T., Jr. Ground cover impacts on sediment and phosphorus export from manured riparian pasture. *J. Environ. Qual.* **2006**, *35*, 2178–2185. [CrossRef]
12. Yoshitake, S.; Soutome, H.; Koizumi, H. Deposition and decomposition of cattle dung and its impact on soil properties and plant growth in a cool-temperate pasture. *Ecol. Res.* **2014**, *29*, 673–684. [CrossRef]
13. Hirata, M.; Sugimoto, Y.; Ueno, M. Return of dung to bahiagrass (Paspalum notatum Flugge) pasture by dairy cattle. *Jpn. J. Grassl. Sci.* **1990**, *35*, 350–357. [CrossRef]
14. Franzluebbers, A.J.; Poore, M.H.; Freeman, S.R.; Rogers, J.R. Soil-surface nutrient distributions in grazed pastures of North Carolina. *J. Soil Water Conserv.* **2019**, *74*, 571–583. [CrossRef]
15. Sharpley, A.; Helmers, M.J.; Kleinman, P.; King, K.; Leytem, A.; Nelson, N. Managing crop nutrients to achieve water quality goals. *J. Soil Water Conserv.* **2019**, *74*, 91A–101A. [CrossRef]
16. Haygarth, P.M.; Jarvis, S.C. Soil derived phosphorus in surface runoff from grazed grassland lysimeters. *Water Res.* **1997**, *31*, 140–148. [CrossRef]
17. Zhang, R.; Li, M.; Yuan, X.; Pan, Z. Influence of rainfall intensity and slope on suspended solids and phosphorus losses in runoff. *Environ. Sci. Pollut. Res.* **2019**, *26*, 33963–33975. [CrossRef] [PubMed]
18. Miller, J.J.; Curtis, T.; Chanasyk, D.S.; Willms, W.D. Influence of cattle trails on runoff quantity and quality. *J. Environ. Qual.* **2017**, *46*, 348–355. [CrossRef]
19. Shigaki, F.; Sharpley, A.; Prochnow, L.I. Rainfall intensity and phosphorus source effects on phosphorus transport in surface runoff from soil trays. *Sci. Total Environ.* **2007**, *373*, 334–343. [CrossRef]
20. Bilotto, F.; Vibart, R.; Mackay, A.; Costall, D. Modelling long-term changes in soil phosphorus and carbon under contrasting fertiliser and grazing management in New Zealand hill country. *J. N. Z. Grassl.* **2019**, *81*, 171–178. [CrossRef]
21. Nellesen, S.; Kovar, J.; Haan, M.; Russell, J. Grazing management effects on stream bank erosion and phosphorus delivery to a pasture stream. *Can. J. Soil Sci.* **2011**, *91*, 385–395. [CrossRef]

22. Kurz, I.; O'Reilly, C.D.; Tunney, H. Impact of cattle on soil physical properties and nutrient concentrations in overland flow from pasture in Ireland. *Agric. Ecosyst. Environ.* **2006**, *113*, 378–390. [CrossRef]

23. Malan, J.A.C.; Flint, N.; Jackson, E.L.; Irving, A.D.; Swain, D.L. Offstream watering points for cattle: Protecting riparian ecosystems and improving water quality? *Agric. Ecosyst. Environ.* **2018**, *256*, 144–152. [CrossRef]

24. Pilon, C.; Moore, P.A.; Pote, D.H.; Pennington, J.H.; Martin, J.W.; Brauer, D.K.; Raper, R.L.; Dabney, S.M.; Lee, J. Long-term effects of grazing management and buffer strips on soil erosion from pastures. *J. Environ. Qual.* **2017**, *46*, 364–372. [CrossRef] [PubMed]

25. Dubeux, J.C.; Sollenberger, L.E. Nutrient cycling in grazed pastures. In *Management Strategies for Sustainable Cattle Production in Southern Pastures*; Academic Press: Cambridge, MA, USA, 2020; pp. 59–75. [CrossRef]

26. Hendricks, T.; Franklin, D.; Dahal, S.; Hancock, D.; Stewart, L.; Cabrera, M.; Hawkins, G. Soil carbon and bulk density distribution within 10 Southern Piedmont grazing systems. *J. Soil Water Conserv.* **2019**, *74*, 323–333. [CrossRef]

27. Mehlich, A. Determination of P, Ca, Mg, K, Na, and NH_4. *N. Carol. Soil Test Div.* **1953**, 23–89.

28. Murphy, J.; Riley, J.P. A modified single solution method for the determination of phosphate in natural waters. *Anal. Chim. Acta* **1962**, *27*, 31–36. [CrossRef]

29. Jirka, A.M.; Carter, M.J.; May, D.; Fuller, F.D. Ultramicro semiautomated method for simultaneous determination of total phosphorus and total Kjeldahl nitrogen in waste waters. *Environ. Sci. Technol.* **1976**, *10*, 1038–1044. [CrossRef]

30. Cronshey, R.G.; Roberts, R.T.; Miller, N. Urban hydrology for small watersheds (TR-55 Rev.). In *Hydraulics and Hydrology in the Small Computer Age*; ASCE: Reston, VA, USA, 1985; pp. 1268–1273.

31. Getis, A.; Ord, J.K. The analysis of spatial association by use of distance statistics. *Geogr. Anal.* **1992**, *24*, 189–206. [CrossRef]

32. Ord, J.K.; Getis, A. Local spatial autocorrelation statistics: Distributional issues and an application. *Geogr. Anal.* **1995**, *27*, 286–306. [CrossRef]

33. Harmel, R.D.; Smith, D.R.; Haney, R.L.; Dozier, M. Nitrogen and phosphorus runoff from cropland and pasture fields fertilized with poultry litter. *J. Soil Water Conserv.* **2009**, *64*, 400–412. [CrossRef]

34. Kissel, D.; Sonon, L. *Soil Test Handbook for Georgia. Georgia Cooperative Extension Service*; Special Bulletin 62; College of Agricultural and Environmental Sciences, University of Georgia: Athens, GA, USA, 2015; p. II-13A.

35. Aarons, S.R.; O'Connor, C.R.; Gourley, C.J.P. Dung decomposition in temperate dairy pastures. I. Changes in soil chemical properties. *Soil Res.* **2004**, *42*, 107–114. [CrossRef]

36. Subedi, A.; Franklin, D.; Dahal, S.; Cabrera, M. Strategic grazing management effects on loss-on-ignition carbon and bulk density in soil and total suspended sediments, carbon in sediment, and dissolved organic carbon in runoff water. *J. Environ. Qual.* **2020**, under review.

37. Cabrera, M.L.; University of Georgia, Athens, GA, USA. Personal communication, 2020.

38. Ryan, M.H.; Tibbett, M.; Lambers, H.; Bicknell, D.; Brookes, P.; Barrett-Lennard, E.G.; Ocampo, C.; Nicol, D. Pronounced surface stratification of soil phosphorus, potassium and sulfur under pastures upstream of a eutrophic wetland and estuarine system. *Soil Res.* **2017**, *55*, 657–669. [CrossRef]

39. Schroeder, P.D.; Radcliffe, D.E.; Cabrera, M.L.; Belew, C.D. Relationship between soil test phosphorus and phosphorus in runoff: Effects of soil series variability. *J. Environ. Qual.* **2004**, *33*, 1452–1463. [CrossRef]

40. Sharpley, A.N.; Daniel, T.; Gibson, G.; Bundy, L.; Cabrera, M.; Sims, T.; Stevens, R.; Lemunyon, J.; Kleinman, P.; Parry, R. *Best Management Practices to Minimize Agricultural Phosphorus Impacts on Water Quality. ARS-163 2006*; USDA-ARS: Washington, DC, USA, 2006.

Publisher's Note: MDPI stays neutral with regard to jurisdictional claims in published maps and institutional affiliations.

 *soil systems*

Article

Stratified Soil Sampling Improves Predictions of P Concentration in Surface Runoff and Tile Discharge

William Osterholz [1,*], Kevin King [1], Mark Williams [2], Brittany Hanrahan [1] and Emily Duncan [3]

[1] United States Department of Agriculture-Agricultural Research Service (USDA-ARS) Soil Drainage Research Unit, 590 Woody Hayes Dr., Columbus, OH 43210, USA; kevin.king@usda.gov (K.K.); brittany.hanrahan@usda.gov (B.H.)

[2] USDA-ARS National Soil Erosion Research Laboratory, 275 South Russell Street, West Lafayette, IN 47907, USA; mark.williams2@usda.gov

[3] Los Angeles Regional Water Quality Control Board, Los Angeles, CA 90013, USA

* Correspondence: will.osterholz@usda.gov; Tel.: +1-614-292-0954

Received: 21 August 2020; Accepted: 13 November 2020; Published: 19 November 2020

Abstract: Phosphorus (P) stratification in agricultural soils has been proposed to increase the risk of P loss to surface waters. Stratified soil sampling that assesses soil test P (STP) in a shallow soil horizon may improve predictions of P concentrations in surface and subsurface discharge compared to single depth agronomic soil sampling. However, the utility of stratified sampling efforts for enhancing understanding of environmental P losses remains uncertain. In this study, we examined the potential benefit of integrating stratified sampling into existing agronomic soil testing efforts for predicting P concentrations in discharge from 39 crop fields in NW Ohio, USA. Edge-of-field (EoF) dissolved reactive P (DRP) and total P (TP) flow-weighted mean concentrations in surface runoff and tile drainage were positively related to soil test P (STP) measured in both the agronomic sampling depth (0–20 cm) and shallow sampling depth (0–5 cm). Tile and surface DRP and TP were more closely related to shallow depth STP than agronomic STP, as indicated by regression models with greater coefficients of determination (R^2) and lesser root-mean square errors (RMSE). A multiple regression model including the agronomic STP and P stratification ratio (P_{strat}) provided the best model fit for DRP in surface runoff and tile drainage and TP in tile drainage. Additionally, STP often varied significantly between soil sampling events at individual sites and these differences were only partially explained by management practices, highlighting the challenge of assessing STP at the field scale. Overall, the linkages between shallow STP and P transport persisted over time across agricultural fields and incorporating stratified soil sampling approaches showed potential for improving predictions of P concentrations in surface runoff and tile drainage.

Keywords: dissolved reactive phosphorus; total phosphorus; soil test phosphorus; soil stratification; water quality

---

1. Introduction

Phosphorus (P) losses from croplands are a major driver of hypoxia and harmful and nuisance algal blooms in waterbodies worldwide [1,2]. Dissolved reactive P (DRP) is the P fraction that primarily drives these algal blooms as it is readily bioavailable, though other fractions of total P (TP) also contribute bioavailable P and so play a secondary role [3]. Phosphorus lost from croplands can be directly derived from recent fertilizers (i.e., incidental P) [4] or from P stored in soil pools (i.e., legacy soil P) [5], but recent research suggests legacy soil P is the dominant P source in the Western Lake Erie Basin (WLEB) [6]. Previous studies have shown that DRP and TP are readily transported via both the artificial subsurface tile drainage network (i.e., tile drainage) as well as overland surface runoff in the

WLEB [7–9]. Thus, to improve predictions of environmental P losses it is necessary to further our understanding of how soil P pools relate to surface and subsurface edge-of-field P losses.

Soil P is routinely measured in agricultural systems as a component of agronomic soil testing to estimate crop fertility requirements [10]. Agronomic soil testing was designed to estimate potential crop nutrient demand and yield response to fertilizer applications, so sampling depths typically target the crop's primary rooting zone. For example, in Ohio, Michigan, and Indiana, USA, the 0–20 cm depth is the typical agronomic soil testing depth and is the basis of university extension fertilizer recommendations [11]. Results of these tests are also useful for predicting the potential risk of P loss; P risk assessment tools typically use STP as a risk prediction factor [12–14], and previous studies have reported positive relationships between agronomic soil test P (STP) and P concentrations in surface runoff and tile drainage [15–20]. However, agronomic testing programs were not designed to monitor potential water quality impacts of soil P, so modifications to these agronomic sampling approaches hold the potential to improve soil testing for environmental purposes.

Depth of sampling is one component of agronomic soil testing that could be adapted to better understand environmental P losses. Soil P can be highly stratified with greater P concentrations in the surface layer, particularly in no-till or reduced tillage systems [21–23]. In the WLEB, stratification of soil P in croplands was shown to be prevalent across the Sandusky river watershed, and was hypothesized to be a contributing factor to increasing DRP concentration trends in surface waters [18]. The shallow surface soil horizon is the dominant zone of interaction between soil P and surface runoff water and thus has a disproportionate influence on surface runoff P concentrations [24–27]. In contrast, water discharged via subsurface tile drains passes through the soil matrix thereby expanding the zone of interaction and increasing opportunities for P sorption or assimilation. However, the interaction between water and the soil matrix decreases where preferential macropore flow dominates, particularly in medium- and fine-textured soils, resulting in elevated P concentrations in tile discharge [28–31]. In these situations, tile drainage water chemistry may largely reflect the surface soil characteristics. Stratified soil sampling that quantifies soil P in the shallow zone of interaction (0–5 cm) in addition to the agronomic depth (0–20 cm) has the potential to better predict environmental P loss in surface runoff and tile drainage as compared to traditional soil sampling approaches that only quantify soil P in the agronomic depth.

The objective of this study was to determine whether a stratified soil sampling regime could explain more variability in environmental P losses than traditional agronomic depth samples. Stratified soil sampling was conducted on 39 fields distributed throughout NW Ohio, USA instrumented with edge-of-field (EoF) water quality monitoring, which enabled examination of relationships between STP and EoF P losses across a broad range of soil characteristics and management regimes. The hypothesis tested was that STP from shallow (0–5 cm) soil samples will better predict DRP and TP concentrations in both surface runoff and tile drainage compared to STP from agronomic (0–20 cm) soil samples. In addition, repeated soil sampling in individual fields enabled assessments of changes in STP and P stratification ratio (0–5 cm STP/0–20 cm STP; P_{strat}) over the course of a 3-year study period.

2. Materials and Methods

2.1. Experimental Sites

Edge-of-field water quality monitoring was used to assess relationships between STP and surface runoff and tile drainage P concentrations across a network of 39 crop fields in the NW quadrant of Ohio underlain with artificial subsurface tile drainage systems. Field locations and management were previously described in detail [32,33]. Soils were medium to fine textured, with drainage classification of somewhat poorly drained to very poorly drained. Fields were nearly level to gently sloping (average slope range 0.4–5.1%; mean 1.6%), and were generally representative of regional topography, soils, and management practices (i.e., nutrient management, tillage, and subsurface drainage). General soil characteristics (textural class, slope, pH, soil organic matter) of the fields in this study are presented in Table S1. Field management was performed by farmers and followed common practices in the region. Soybean (*Glycine max*, (L.) Merr)-corn (*Zea mays*, L) rotations were the primary cropping system, but winter wheat (*Triticum aestivum*, L.), alfalfa (*Medicago sativa*, L.), and winter cover crops (various species) were also included in a subset of fields. Tillage practices ranged from multiple tillage passes each year to long-term no tillage. Soil fertility management typically consisted of springtime N fertilizer applications to corn and wheat and fall broadcast applications of P and K fertilizers once per crop rotation. Starter fertilizers containing N, P, and K were commonly applied at crop planting. Additionally, several fields received manure applications within 2 years of this study.

2.2. Runoff Phosphorus Concentrations

Surface and subsurface water quality was monitored from outlets at the edge of fields; sampling approach and instrumentation were described in depth by Williams et al. 2016 [32]. Contributing drainage areas for surface and subsurface flow were between 1.1–18.5 ha (mean = 8.1 ± 4.6 ha). Water quality data were collected via automated measurement of flow rate at 10-min intervals and automated collection of water samples using both time- and flow-weighted basis. Tile drain flow was monitored with compound weirs (Thel-Mar, Brevard NC, USA) and bubbler modules (Teledyne ISCO, Lincoln NE, USA), while surface runoff was measured with 0.6 m H-flumes (Tracom, Alpharetta GA, USA) and bubbler modules. Water samples were collected with Teledyne Isco (Lincoln, NE, USA) automated samplers. Event-based water samples were automatically collected on a flow-weighted basis from surface flumes for the entire period of record. For tile drainage and prior to 2015, samples were automatically collected on a time interval approach (aliquots collected every 6 h and composited for a 24 h period). Starting in 2015, the time interval samples were supplemented with additional event-based samples collected over the rise and fall of the hydrograph to better represent the discharge events. Event sampling was triggered by increased discharge, with a 200 mL aliquot taken for each 1 mm of volumetric depth. Ten aliquots were combined into a single 2 L sample. Event sampling ceased when flow ceased (surface runoff) or when flow declined to baseline levels (tile drainage). Water samples were retrieved from the field at least weekly and were stored at 4 °C until laboratory analysis.

Water samples were analyzed for both DRP and TP concentrations. Briefly, samples were split into two aliquots, with one filtered at 0.45 um for DRP analysis, and an unfiltered sample used for TP analysis. The unfiltered sample underwent alkaline-persulfate digestion prior to TP analysis [34]. The filtered (DRP) and digested (TP) samples were analyzed for orthophosphate concentration using a flow injection analyzer (Lachat Instruments, Loveland, CO, USA) via the ascorbic acid reduction method. The resulting discrete P concentration data were converted into 10 min P concentrations by linear interpolation. The 10 min constructed P concentration values were then multiplied by the corresponding 10 min flow values to calculate 10 min P loads [35]. Resulting load data were summed into daily cumulative P loads. Daily estimates of flow and P load were summed into total flow and P load for the relevant period of each soil sampling event (described below). The total P load over the period of a given soil sampling event was then divided by total flow from that period to calculate the flow-weighted mean DRP (FWM DRP) and TP concentrations (FWM TP) associated with the soil sampling event.

2.3. Soil Test Phosphorus

Stratified soil samples were collected from contributing field areas of each monitored outlet between December 2014 and December 2017. The frequency of soil sampling events within a given field depended on crop rotation, establishment of EoF water quality monitoring instrumentation, and resource availability; 14 EoF sites were sampled on three occasions, 24 sites were sampled twice, and two were sampled once. The surface runoff dataset included a total of 52 individual soil sampling events, while 86 soil sampling events were included in the tile drainage dataset. Each soil sampling event consisted of taking three to nine samples at discrete locations within a given field. Sampling locations were selected based on USDA-NRCS soil maps and local topography to ensure the sampling captured the variability in soils across the areas contributing to discharge at the field outlets. On average, one sample was collected for each 1.5 ha of contributing field area. Individual soil sampling locations were somewhat consistent from year to year, but limited precision of GPS coordinates meant subsequent samples were likely >10 m apart. Samples collected from 2014–2016 were taken with hand-held push probes (2 cm diameter) and the 2017 samples were collected using a hydraulic soil probe (5 cm diameter). At each location, five individual cores, distributed within ~2 m of a central point, were collected, split into 0–5 cm and 5–20 cm depths, and combined into one sample. Soil samples were air dried, ground, and analyzed for STP with Mehlich-3 extractant by the Ohio State University Service Testing and Research laboratory (Wooster, OH, USA). A simple average of the discrete STP data was used to estimate the field average STP concentration values, and within field variation in STP was characterized with the coefficient of variation (CV). The field average STP data were used to calculate the P_{strat} for each field according to Equation (1):

$$P_{strat} = \frac{[\text{STP}]0-5\ cm}{[\text{STP}]5-20\ cm} \tag{1}$$

Management information was used to assign a range of dates for each soil sampling event for which the STP measurements were considered most relevant for predicting EoF runoff P concentrations. The period of relevance for a given soil sampling event could extend up to one year before and after the sampling date, which would correspond to soil sampling occurring once in a 2-year crop rotation. This period was truncated to less than 2 years if the field was subjected to either a tillage operation or a P fertilizer or manure application as these operations were expected to alter the STP concentrations and P_{strat}. Additionally, if a P fertilizer or manure application occurred prior to the soil sampling event, the first 2 weeks following the application were also excluded from date range to restrict the influence of short-lived direct P fertilizer losses [4]. Thus, no P applications or tillage operations occurred during the period of relevance for the soil sampling events. Resulting lengths of the date ranges were from 159 to 673 days, with an average length of 384 days.

2.4. Statistical Analysis

Relationships between STP and FWM P concentrations were examined with ordinary least squares regressions. Prior to regression analysis, the FWM DRP and TP concentrations were natural log transformed to comply with normality assumptions. The goodness of fit of the resulting regressions was assessed using the coefficient of determination (R^2) and root mean square error (RMSE; the standard deviation of the model residuals). The field average STP and the maximum STP value in each field were both tested as predictors of FWM DRP and TP, and regression model fits were compared. Residuals from the field average agronomic STP-FWM P concentration regressions were extracted and subsequently correlated with P_{strat}, with Pearson's *r* used to assess correlation strength. Multiple linear regressions were constructed in a stepwise manner from two predictor variables: agronomic STP and P_{strat}. These predictor variables were checked for multicollinearity. The improvement in model fit provided by the addition of P_{strat} to the agronomic STP model was further examined by extracting residuals from the simple regressions of agronomic STP vs. FWM-P concentration, as well as the multiple linear regression. For a given observation, the magnitude of the two model residuals were compared and the difference between residuals was analyzed for simple linear relationships to soil textural class, field average slope, agronomic STP, and average daily discharge.

In fields with more than one soil sampling event, the influence of management practices on changes in STP and P_{strat} within the fields was examined. The effects of management factors on changes in STP and P_{strat} (e.g., STP in soil sampling period 1—STP in soil sampling period 2) were tested with t-tests for three class variables (+/− P fertilizer, +/− manure, +/− tillage) and with ordinary least squares regression for the rate of P applied (only for fields that received manure or P fertilizer). Analyses were conducted in SAS v. 9.4 (SAS Institute, Cary NC, USA) and Sigmaplot (Systat Software, San Jose, CA, USA).

3. Results

3.1. Soil Test P

Shallow depth (0–5 cm) STP averaged 61 mg P kg^{-1} across the fields and agronomic depth (0–20 cm) STP averaged 40 mg P kg^{-1} for the tile drainage dataset (Table 1). Soil test P of individual fields ranged from 19–202 and 12–150 mg P kg^{-1} for the shallow and agronomic depths, respectively. Variability of STP within fields was high, with an average CV of 32–39% across all the sampling events and both depths, while 14% of the sampling events surpassed a CV of 50% (data not shown). The field average P_{strat} was 1.88, with large variation among the studied fields (range 1.18–3.35).

Table 1. Soil test P (STP) concentrations and P stratification ratios (P_{strat}) across soil sampling events at 39 fields, and edge-of-field discharge and dissolved reactive P (DRP) and total P (TP) flow weighted mean (FWM) concentrations during the relevant sampling windows.

		Surface Runoff *				Tile Drainage *			
		Mean	Median	Min.	Max.	Mean	Median	Min.	Max.
STP (mg P	0–5 cm	61	54	19	145	66	57	19	202
kg^{-1})	0–20 cm	40	35	12	121	44	37	12	150
STP coefficient	0–5 cm	35	31	6	83	32	29	2	83
of variation (%)	0–20 cm	39	34	7	108	36	33	1	108
P_{strat} [†]		1.98	1.90	1.18	3.35	1.88	1.80	1.25	3.35
FWM DRP Conc (mg L^{-1})		0.19	0.15	0.02	0.66	0.07	0.05	0.01	0.27
FWM TP Conc (mg L^{-1})		0.65	0.53	0.25	1.91	0.28	0.24	0.08	0.64
Total discharge (mm)		283	242	20	735	87	64	9	279

* Surface runoff soil sampling windows n = 52; tile drainage n = 86 [†] P stratification ratio.

Soil test P and P_{strat} within individual fields showed a high degree of variability between subsequent soil sampling events (Table 2). The agronomic STP in a given field increased or decreased between subsequent soil sampling events by >15 mg P kg^{-1} in 12% of cases (average STP change: 9.4 ± 1.1 mg P kg^{-1}). Shallow STP changed by >15 mg P kg^{-1} in 23% of cases (average STP change: 13.6 ± 2.1 mg P kg^{-1}). Averaged across all fields and sampling events, the absolute change in STP was an increase of 1.8 ± 1.7 and 4.3 ± 3.1 mg P kg^{-1} for agronomic and shallow depths, respectively. Furthermore, in many fields the within field variability in STP demonstrated large changes between subsequent soil sampling events; for example, the CV for agronomic STP differed by >20% in 15 of the 48 soil sampling event comparisons (average CV change: 15.1 ± 2.1% for agronomic depth, 13.6 ± 1.8% for shallow depth). Similarly, the P_{strat} often demonstrated significant within field changes, with 21% of cases changing by >0.50 between soil sampling events (average change: 0.43 ± 0.07).

Table 2. Summary of changes in soil test P (STP), STP coefficient of variation, and P stratification ratio (P_{strat}) between subsequent soil sampling events in individual fields.

		Observed Increases (#)	Observed Decreases (#)	Mean of Absolute Values *	Largest Decrease *	Largest Increase *
STP	0–5 cm	30	18	14	−52	102
	0–20 cm	31	17	9	−32	25
STP coefficient of	0–5 cm	30	18	14	−53	37
variation	0–20 cm	25	23	15	−78	53
P_{strat} [†]		25	23	0.43	−1.55	1.87

* STP = mg P kg^{-1}, coefficient of variation = % [†] P stratification ratio.

Management practices occurring between soil sampling events generally did not consistently explain the observed changes in STP or P_{strat} (Table S2). Manure application between soil sampling events was the only management factor that significantly influenced STP changes, where shallow and agronomic STP were 13.3 and 10.6 mg kg^{-1} greater, respectively, in fields with manure compared to those without (*t* test; *t*-statistics -2.3 and -3.4, P $= 0.029$ and P $= 0.0014$ for shallow and agronomic depths, respectively). However, neither chemical P fertilizer application (*t* test; *t*-statistics 0.41 and 1.1, P $= 0.68$ and P $= 0.26$ for shallow and agronomic depths, respectively) nor the amount of P applied between soil sampling events (regression; t-statistics 1.04 and 1.04, P $= 0.3$ and P $= 0.3$ for shallow and agronomic depths, respectively) influenced changes in STP. Similarly, changes in STP were not different in fields that were tilled between sampling events compared to fields that did not undergo a tillage operation (*t* test, *t*-statistic -0.35 and 0.01, P $= 0.7$ and P $= 0.9$ for shallow and agronomic depths, respectively). Additionally, changes in P$_{strat}$ were not related to form or amount of P applied or tillage (t-tests and regression, *p*-values > 0.52 for all).

3.2. Surface Runoff and Tile Drainage Phosphorus Concentrations

The average FWM DRP concentration in surface runoff across all soil sampling windows was 0.19 ± 0.02 mg DRP L^{-1}, and the FWM TP concentration averaged 0.65 ± 0.05 mg TP L^{-1} (Table 1). Phosphorus concentrations were lower in tile drainage and averaged 0.066 ± 0.008 mg DRP L^{-1} and 0.28 ± 0.02 mg TP L^{-1}. There was a high degree of variability in FWM P concentrations between the sampling windows, for example DRP concentrations ranged from 0.02–0.66 mg DRP L^{-1} for surface runoff and 0.01–0.27 mg DRP L^{-1} for tile drainage.

3.3. Relationships between STP and FWM P Concentrations

Positive relationships were observed between STP in both sampling depths and the FWM DRP concentrations in both surface runoff and tile drainage (Figure 1). For surface runoff DRP, similar regression slopes (0.015 vs. 0.016) were observed from relationships with shallow vs. agronomic STP, indicating that increases in agronomic or shallow STP resulted in similar increases in FWM DRP concentration (Table 3). However, shallow STP explained more variation in surface DRP concentration (R^2 of 0.31 vs. 0.19) and improved the predictions of surface DRP concentration (RMSE of 0.66 vs. 0.72) compared to agronomic STP. For tile drainage DRP, regression slopes were also similar for relationships using both soil sampling depths (0.014 vs. 0.016). Greater variation in tile drainage DRP was explained with shallow STP (R^2 of 0.44 vs. 0.32) and predictions likewise improved (0.56 vs. 0.62) (Table 2). Comparing R^2 and RMSE values between surface and tile DRP models demonstrated that STP, measured from shallow and agronomic depths, was a stronger predictor of tile drainage DRP than surface DRP.

Flow-weighted mean TP concentrations in surface runoff and tile drainage was also positively related to STP in both sampling depths (Figure 2). However, the R^2 of the regression models indicated that less variation in TP concentrations was explained by shallow and agronomic STP compared to DRP (Table 3). As observed with DRP, the slopes of both the surface runoff and tile drainage TP regressions were similar for agronomic and shallow depth samples. More variation in surface runoff TP was explained with shallow STP compared to agronomic STP (R^2 0.26 and 0.21, respectively), but prediction accuracy similar (RMSE 0.44 and 0.45, respectively). The tile drainage TP showed similar patterns as surface runoff, with greater variation explained by shallow STP compared to agronomic STP (R^2 0.19 and 0.11) and similar prediction accuracy (RMSE of 0.50 and 0.53). In general, differences between the predictive power of agronomic and shallow depth samples were lesser for TP than those observed for DRP.

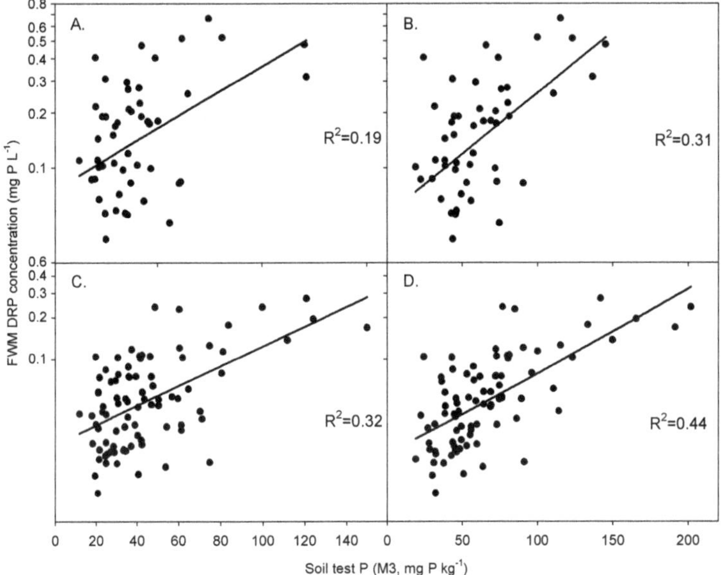

Figure 1. Regression relationships between soil test P (STP) and flow-weighted mean dissolved reactive P (FWM DRP) concentrations. Surface runoff FWM DRP vs. STP in the (**A**) agronomic soil horizon (0–20 cm) and (**B**) the shallow soil horizon (0–5 cm). Tile drainage FWM DRP vs. STP in the (**C**) agronomic soil horizon and the (**D**) shallow soil horizon. Regressions were performed on log transformed runoff P concentrations; data are plotted in original units with log-scale x-axis.

Table 3. Regression results for field average soil test P vs. flow weighted mean dissolved reactive P (DRP) and total P (TP) concentrations. Regressions were performed on natural log transformed FWM P concentrations.

	DRP				TP			
	Regression Slope *	Regression Intercept *	R^2	RMSE	Regression Slope *	Regression Intercept *	R^2	RMSE
Surface runoff								
0–5 cm samples	0.015	−2.90	0.31	0.66	0.009	−1.11	0.26	0.44
0–20 cm samples	0.016	−2.59	0.19	0.72	0.01	−0.98	0.21	0.45
Tile drainage								
0–5 cm samples	0.014	−3.90	0.44	0.56	0.007	−1.87	0.19	0.50
0–20 cm samples	0.016	−3.73	0.32	0.62	0.007	−1.75	0.11	0.53

* P concentration data were log transformed prior to regression; all regression slopes and intercepts were significant at the $p < 0.01$ level.

Additionally, the maximum agronomic and shallow STP values in each field were also tested as a predictors of EoF P concentrations (Table S3). Maximum STP at both sampling depths was positively and significantly related to DRP and TP concentrations in surface runoff and tile drainage (regressions; R^2 range 0.07 to 0.38, $p < 0.05$ for all). However, the field average shallow and agronomic STP model explained more variation and improved predictions of EoF P concentration data in all cases.

Residuals of the regression of DRP concentration against agronomic STP were positively correlated with P_{strat} for both surface runoff (r = 0.34, $p = 0.01$) and tile drainage (r = 0.36, $p < 0.001$; Figure 3). A significant positive correlation between residual TP and P_{strat} was also observed in tile drainage (r = 0.28, $p < 0.01$) but not surface runoff (Figure 4). The positive correlations indicate that fields with greater P_{strat} tended to also have more positive residuals, i.e., runoff P concentrations were prone to underprediction by the agronomic STP. Likewise, runoff P concentrations in fields with lesser P_{strat} were prone to over-prediction by agronomic STP.

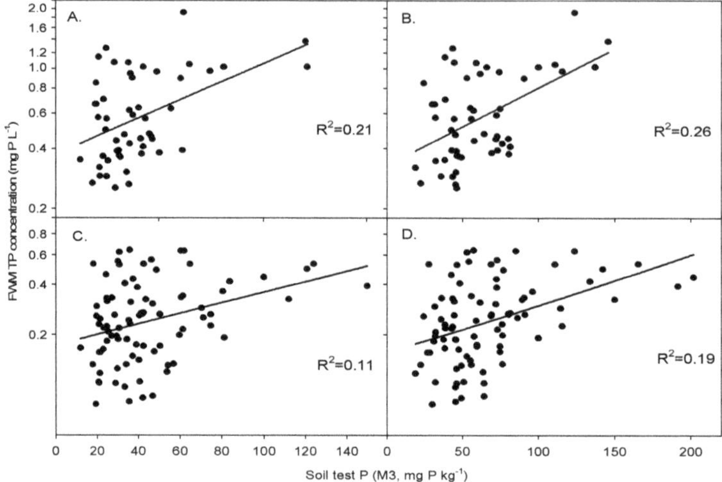

Figure 2. Regression relationships between soil test P (STP) and flow weighted mean total P (FWM TP) concentrations (natural log transformed). Surface runoff FWM TP vs. STP for the (**A**) agronomic soil horizon (0–20 cm) and the (**B**) surface cm soil horizon (0–5 cm). Tile drainage FWM TP vs. STP for the (**C**) agronomic soil horizon and the (**D**) surface soil horizon. Regressions were performed on log transformed runoff P concentrations; data are plotted in original units with log-scale x-axis.

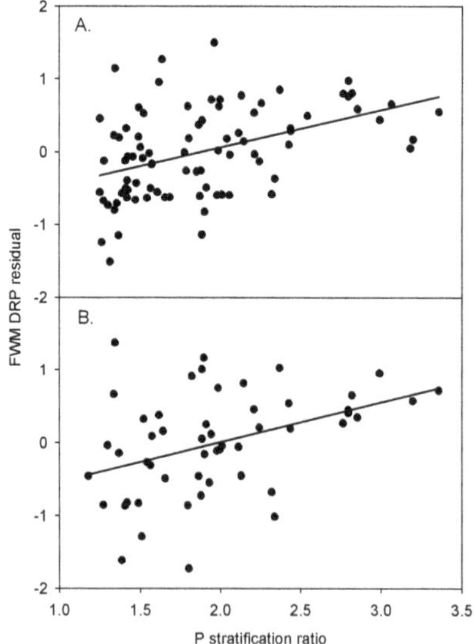

Figure 3. Correlation between residual FWM DRP (from regression of agronomic STP (0–20 cm depth) vs. natural log transformed FWM DRP concentrations) and the P stratification ratio for tile drainage (**A**) and surface runoff (**B**).

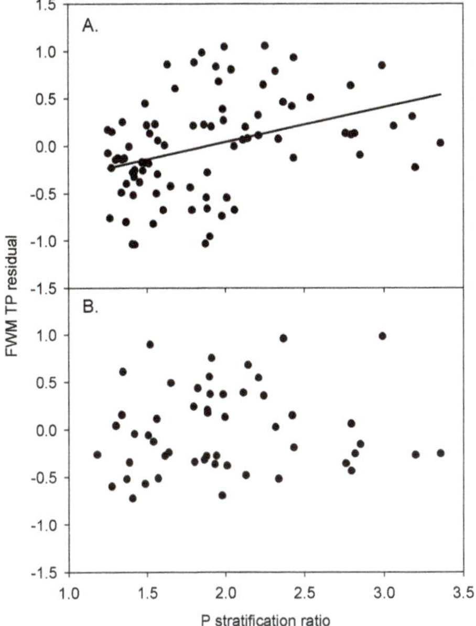

Figure 4. Correlation between residual FWM TP (from regression of agronomic STP (0–20 cm depth) vs. natural log transformed FWM TP concentrations) and the P stratification ratio for tile drainage (**A**) and surface runoff (**B**).

Phosphorus stratification ratio was a significant factor when added to the regressions of agronomic STP vs. EoF P concentrations for surface DRP, tile DRP, and tile TP concentrations (Table 4). Interactions between STP and P_{strat} were not significant and so were not included in the final model. However, P_{strat} was not significant when added to the surface TP model. Model fit and explanatory power of the two factor models was similar or slightly better to that of the single factor shallow sample models (Table 3). In addition, comparison of residuals from the agronomic STP models to residuals from the agronomic STP + P_{strat} models showed that the improvement in fit provided by addition of P_{strat} was not related to soil texture class, field slope, STP, or average daily discharge for either DRP or TP (ANOVA and regressions; $p > 0.05$ for all; data not shown).

Table 4. Multiple linear regression results predicting edge-of-field FWM P concentrations with agronomic soil test P (STP) and P stratification ratio (P_{strat}). Regressions were performed on natural log transformed FWM P concentrations.

	Model *	R^2	RMSE
Surface runoff			
FWM DRP	$0.020 \times STP + 0.59 \times P_{strat} - 3.92$	0.34	0.66
FWM TP	$0.010 \times STP - 0.98$	0.21	0.45
Tile drainage			
FWM DRP	$0.019 \times STP + 0.53 \times P_{strat} - 4.81$	0.46	0.56
FWM TP	$0.009 \times STP + 0.38 \times P_{strat} - 2.53$	0.24	0.49

* Model factors were significant at the $p < 0.01$ level.

4. Discussion

4.1. Soil Test P

Soil test P values and P_{strat} of individual fields were highly dynamic over the 3 years of this study, highlighting the importance of frequent soil sampling. A large portion of the observed STP changes between sampling events was presumably due to inherent within-field spatial variability in STP. This study used field average STPs derived from multiple samples that were taken at a sampling intensity (average of 1 sample per 1.5 ha) similar to that of commercial agronomic soil sampling in the WLEB. Other studies have shown agronomic STP frequently varies dramatically at the scale of 10 s of meters [36–38], so more intensive sampling may be required to improve the accuracy of calculated field average STP. This study was not designed to quantify the relationship between sampling intensity and STP variability, so further research will be required to quantify the soil sampling intensity needed to achieve acceptable precision of field average STP values.

Changes in STP and P_{strat} can also be caused by management activities such as tillage and P fertilizer management (e.g., P rate, fertilizer placement) [21,39], and soil sampling schemes should take into account these management activities. However, in this study we did not see a large influence of management practices on changes in STP or P_{strat}. For example, the amount of P applied between two soil sampling events, which ranged from 0–119 kg P ha^{-1} (Table S2), was not associated with statistically significant changes in STP, indicating that recent (i.e., within the past 3 years) P application rate was not a primary driver of STP over that period. Relatively large additions of P fertilizer are required to substantially increase STP. For instance, annual P applications of 44 kg ha^{-1} increased STP by only 2.5 mg kg^{-1} yr^{-1} at a location in Iowa [40] and, in Ohio, P fertilizer rates double the crop removal rate did not substantially increase STP after 9 years at three locations [41]. It is likely that within field variability in STP was much greater than changes induced by P fertilizer applications and consequently overwhelmed observation of these changes. In contrast, manure application was associated with increases in STP in subsequent soil sampling events, likely due in part to the relatively large amounts of P (average 44 kg P ha^{-1}) added to the manured fields. Additionally, compared to chemical P fertilizer, manure may maintain a greater proportion of P in labile forms that are extracted by the Mehlich-3 extractant over a period of at least several months after application [42]. Manure application has been previously identified as a major factor driving P losses in runoff and tile drainage [43,44]. Interestingly, an earlier analysis of EoF water quality from the same fields used this study found that P losses were significantly greater in fields that received manure applications [45]. These results highlight the importance of manure management for addressing agricultural P losses, and affirm that the connections between manure application, STP, and P losses should be a priority for future research.

Tillage operations prior to a soil sampling event was not related to changes in STP or P_{strat}. However, it is important to note that the tillage operations used were either vertical, non-inversion tillage or shallow tillage. Such conservation tillage operations limit the mixing of surface and deeper soil horizons, and thus typically maintain significant soil stratification [46,47]. A previous study suggested that one-time inversion tillage that fully eliminated P stratification could greatly reduce EoF P losses [18]. Our results provide evidence that non-inversion tillage practices will not substantially mitigate P stratification, but more intensive inversion tillage will be required to reduce the level of P stratification.

The limited influence of management on STP changes in this study is likely due in part to the soil sampling intensity and frequency. More intensive or frequent sampling may have strengthened our ability to identify management influences, but this study used a soil sampling intensity similar to commercial farms in the region so these results may be a reasonable approximation for the strength of individual fertilization and non-inversion tillage effects on STP that could be expected from commercially collected STP data across individual crop fields.

4.2. Relationships between STP and FWM P Concentrations

Stratified soil sampling provided improved prediction of environmental P losses compared to typical agronomic soil sampling. Shallow sample STP accounted for more variability in both surface runoff and tile drainage FWM P concentrations (DRP and TP) compared to agronomic sample STP. Additionally, the best model fits for surface runoff DRP and tile drainage DRP and TP were obtained with regressions using agronomic sample STP combined with P_{strat}. Thus, widespread testing of stratified soil samples could be used to improve identification of fields in the WLEB with increased risk of environmental P losses due to high P stratification. Agronomic soil testing is currently widely employed by producers in the WLEB [48], so stratified soil sample collection and analysis could be readily integrated into existing soil testing efforts. However, producer incentives may be required to encourage widespread implementation since stratified soil sampling increases soil testing costs and provides little direct benefit to producers.

This study provided new evidence that stratified sampling was useful for improving predictions of P loss risk for soil types and geography common in the WLEB, but previous research from small experimental plots has provided mixed evidence that stratified soil sampling can improve relationships between environmental P losses and STP. For example, a rainfall simulation experiment on 4 soils in Texas showed that 0–5 cm soil samples produced stronger relationships between surface runoff DRP and STP compared to 0–15 cm samples for two soil types, but on two other soils the 0–15 cm samples produced the stronger relationships [26]. In Manitoba, surface runoff DRP from snowmelt was better predicted by 0–5 cm STP than 0–15 cm STP ([49] Wilson et al., 2019). Conversely, a rainfall simulation study in Wisconsin showed that shallow samples (0–2 cm) did not provide consistent improvements in relationships between surface runoff DRP and STP, relative to 0–15 cm samples [15]. Similarly in pasture soils shallow (0–2 cm) samples were not better predictors of surface runoff DRP than deeper (0–10 cm) samples [50]. Finally, sandy pasture soils under simulated rainfall showed no change in the relationship between surface runoff P concentrations and STP between several sample depths (0–2, 0–5, or 0–10 cm), but the soils in that study had relatively little P stratification [51]. The depth of agronomic sampling in this study (20 cm) likely caused greater distinction between sampling depths than studies with shallower sampling depths, enabling differentiation between these soil depths effects on EoF P concentrations. Additionally, removal of EoF observations immediately following P applications may have reduced variability in EoF P concentrations and enhanced our ability to identify differences in the relationships between STP and EoF P concentrations. Furthermore, the finely textured soils that dominate NW Ohio may favor the enhanced importance of surface soil layers to EoF water quality, as the zone of interaction with water is likely more limited than in coarsely textured soils with greater infiltration rates. Finally, this study encompassed a relatively large number of fields with a wide range of STP and P_{strat}, which provided a sufficient range of conditions from which we were able to observe significant relationships between STP and EoF P concentrations. The predictive benefit of adding P_{strat} to the agronomic STP models was not related to differences in soil texture, slope, or hydrologic conditions within this study, suggesting that soil P stratification was important across the full range of conditions in the studied fields. However, it should be noted that the patterns observed here may not extend to regions with differing climates, management regimes, and soil characteristics.

Concentrations of DRP were more closely related to STP and P stratification than TP concentrations, indicating that STP is relatively more important for understanding DRP losses compared to TP losses. Agronomic soil tests use extractants, including the Mehlich-3 extractant used in this study, that aim to indicate the plant availability of orthophosphate over the period of a growing season, and thus can be expected to relate relatively closely to DRP concentrations in runoff [51,52]. In contrast, TP in runoff typically includes a significant portion of sediment-bound particulate P that is not measured by soil tests. Surface runoff TP loss has been shown to be closely related to sediment loss which is controlled by multiple factors in addition to STP, such as soil erodibility, ground cover, and conservation practices [15,53]. Incorporating these factors into runoff P concentration predictions (in addition to STP) could further improve model predictive power, particularly for TP loss.

The relationships between STP and P concentrations in runoff reported in this work were based on EoF monitoring, and were unsurprisingly weaker than STP-runoff P concentration relationships previously reported from more controlled rainfall simulator studies (e.g., [15,26]). In this study, the variability in runoff P concentrations that was unexplained by STP is likely in large part a result of the EoF data collection effort occurring over the broad range of environmental conditions and management practices included in the USDA EoF network. Management and environmental factors are known to influence DRP and TP concentrations in runoff, and recent research has identified crop rotation [54], soil texture [55,56], tillage [57,58], and precipitation characteristics [59–61] as important factors that can influence surface runoff or tile drainage P concentrations. Regardless, this study demonstrated that the influence of STP on EoF P concentrations was readily observed despite the variability in management and environmental characteristics over time and space. Similarly, the benefits of stratified soil sampling were robust enough to be observed across a wide range of crop production scenarios. Thus, predictions of P losses, whether made using empirical relationships or process-based models, could be improved by enhanced efforts to gather and use information on soil P stratification.

An additional challenge of developing relationships at the field scale is the limited understanding of the variability in contributions to EoF water quality across a large field area with variable soil properties and topography. Improving the accuracy of predictions of environmental P losses from heterogeneous fields can be achieved through soil testing regimes that account for disproportionate contributions of field areas P losses, particularly through surface runoff [62,63]. Furthermore, in soils where macropores play a role in drainage, subsurface tile drains have been shown to have direct connection with surface soils above the drains [64,65]. Additionally, small "hotspots" of high STP within fields could potentially play a disproportionate role in determining EoF P concentrations [38,66]. In this study, the maximum observed STP did not provide better predictions of EoF P concentrations than the field average STP for either surface runoff or tile discharge, but a more intensive soil sampling regime may be necessary to effectively characterize P hotspots. While this study did not take into account any spatial variation in contributions to EoF water quality across the field areas, future research should investigate the feasibility of designing targeted environmental soil testing to account for spatial differences in contributions to EoF water quality in tile drained landscapes.

Soil P and runoff P concentrations were measured in a relatively large number of crop fields (39) in this study, yet how closely these fields represent the broader landscape of the WLEB remains an important question. A recent study of STP stratification in croplands in the WLEB presented results of over 140,000 soil samples [18]. The region-wide average STP presented in that study was similar to the average of the fields studied in this research; the agronomic (0–20 cm) STP averaged 48.1 mg kg^{-1} compared to 44 mg kg^{-1} in this study. Furthermore, in Baker et al., (2017) [18] 71% of agronomic STPs were <47 mg P kg^{-1}, i.e., the state extension recommended range for "build-up" and "maintenance" of STP for corn, whereas in this study 69% of field average STPs were also below that threshold. Finally, on average surface (0–5 cm) STP was 68% higher compared to the 5–20 cm depth, resulting in an average P_{strat} of 1.68 in Baker et al. (2017) [18], which was somewhat less than in this study (1.88). The greater P stratification in this study may be due to the high prevalence of no-till and non-inversion tillage in the fields included in the USDA-ARS EoF network [9]. However, the relatively close agreement between our study and the findings of Baker et al. (2017) [18] indicates that the relationships observed in this study should be expected to hold true across the broader WLEB.

5. Conclusions

Robust relationships between agronomic STP and P concentrations were observed across 39 production crop fields in Ohio. Phosphorus stratification varied widely across the fields, and P concentrations in both tile discharge and surface runoff were found to be related more closely to STP of shallow samples (0–5 cm) compared to the agronomic samples (0–20 cm). In fields with greater P_{strat}, predicted EoF P concentrations using the agronomic sample STP resulted in systematic underestimation of tile discharge DRP and TP concentrations and surface runoff DRP concentration. The improvement in model predictive power from using shallow sample STP rather than agronomic sample STP was greater for DRP compared to TP. Additionally, both STP and P_{strat} varied significantly within fields and were dynamic over time, highlighting the need for frequent and intensive soil sampling to accurately estimate the P status and risk of environmental P loss of fields. Overall, our results suggest stratified soil sampling can be a readily implemented method to improve understanding of the risk of environmental P losses in the WLEB.

Supplementary Materials: The following are available online at http://www.mdpi.com/2571-8789/4/4/67/s1, Table S1: Soil characteristics of fields monitored for edge-of-field P losses; Table S2: Initial values and changes (delta) in soil test phosphorus (STP), CV of STP, and P stratification ratio (P_{strat}); Table S3: Linear regression results for field maximum STP vs. FWM P concentrations.

Author Contributions: K.K. conceived and designed the experiments; K.K., M.W., and E.D. performed the experiments; W.O., and E.D. analyzed the data; W.O. and B.H. wrote the paper. All authors have read and agreed to the published version of the manuscript.

Funding: The Ohio State University helped support this research through hosting the USDA ARS Soil Drainage Research Unit and providing academic access to faculty resources, libraries, and students. Funding for this project was provided by several sources including the 4R Research Fund (IPNI-2014- USA-4RN09), USEPA (DW-12-92342501-0), Conservation Innovation Grants (The Ohio State University: 69-3A75-12-231; Heidelberg University: 69-3A75-13-216), the Ohio Corn and Wheat Growers Association, the Ohio Soybean Association, the Ohio Farm Bureau, the NRCS Mississippi River Basin Initiative, the Nature Conservancy, the NRCS Cooperative Conservation Partnership Initiative, and the USDA Conservation Effects Assessment Project.

Acknowledgments: The authors would like to thank the landowners of the study sites who provided access to the fields and management data; Jedediah Stinner, Katie Rumora, Marie Pollock, Phil Levison and Sara Henderson for their help in data collection and site maintenance; and Eric Fischer for analytical expertise. This research was a contribution from the Long-term Agroecosystem Research (LTAR) network. The LTAR network is supported by the USDA. Mention of trade names or commercial products in this publication is solely for the purpose of providing specific information and does not imply recommendation or endorsement by the U.S. Department of Agriculture.

Conflicts of Interest: The authors declare no conflict of interest.

References

1. Maccoux, M.J.; Dove, A.; Backus, S.M.; Dolan, D.M. Total and soluble reactive phosphorus loadings to Lake Erie. *J. Great Lakes Res.* **2016**, *42*, 1151–1165. [CrossRef]

2. Scavia, D.; David Allan, J.; Arend, K.K.; Bartell, S.; Beletsky, D.; Bosch, N.S.; Brandt, S.B.; Briland, R.D.; Daloğlu, I.; DePinto, J.V.; et al. Assessing and addressing the re-eutrophication of lake erie: Central basin hypoxia. *J. Great Lakes Res.* **2014**, *40*, 226–246. [CrossRef]

3. Stumpf, R.P.; Johnson, L.T.; Wynne, T.T.; Baker, D.B. Forecasting annual cyanobacterial bloom biomass to inform management decisions in Lake Erie. *J. Great Lakes Res.* **2016**, *42*, 1174–1183. [CrossRef]

4. Withers, P.J.A.; Ulén, B.; Stamm, C.; Bechmann, M. Incidental phosphorus losses–are they significant and can they be predicted? *J. Plant Nutr. Soil Sci.* **2003**, *166*, 459–468. [CrossRef]

5. Rowe, H.; Withers, P.J.A.; Baas, P.; Chan, N.I.; Doody, D.; Holiman, J.; Jacobs, B.; Li, H.; MacDonald, G.K.; McDowell, R.; et al. Integrating legacy soil phosphorus into sustainable nutrient management strategies for future food, bioenergy and water security. *Nutr. Cycl. Agroecosyst.* **2016**, *104*, 393–412. [CrossRef]

6. King, K.W.; Williams, M.R.; Johnson, L.T.; Smith, D.R.; LaBarge, G.A.; Fausey, N.R. Phosphorus availability in western lake erie basin drainage waters: Legacy evidence across spatial scales. *J. Environ. Qual.* **2017**, *46*, 466–469. [CrossRef]

7. Smith, D.R.; King, K.W.; Johnson, L.; Francesconi, W.; Richards, P.; Baker, D.; Sharpley, A.N. Surface runoff and tile drainage transport of phosphorus in the midwestern USA. *J. Environ. Qual.* **2015**, *44*, 495–502. [CrossRef]

8. Van Esbroeck, C.J.; Macrae, M.L.; Brunke, R.I.; McKague, K. Annual and seasonal phosphorus export in surface runoff and tile drainage from agricultural fields with cold temperate climates. *J. Great Lakes Res.* **2016**, *42*, 1271–1280. [CrossRef]

9. Pease, L.A.; King, K.W.; Williams, M.R.; LaBarge, G.A.; Duncan, E.W.; Fausey, N.R. Phosphorus export from artificially drained fields across the eastern corn belt. *J. Great Lakes Res.* **2018**, *44*, 43–53. [CrossRef]

10. Beetstra, M.A.; Tellez, C.; Wilson, R. *4R Nutrient Stewardship in the Western Lake Erie Basin Part II: A Panel Study*; The Ohio State University: Columbus, OH, USA, 2017.

11. Vitosh, M.L.; Johnson, J.W.; Mengel, D.B. *Tri-State Fertilizer Recommendations for Corn, Soybeans, Wheat, and Alfalfa*; Ohio State University Agricultural Extension Bulletin E-2567: Columbus, OH, USA, 1995.

12. Williams, M.R.; King, K.W.; Dayton, E.; LaBarge, G.A. Sensitivity analysis of the ohio phosphorus risk index. *Trans. ASABE* **2015**, *58*, 93–102. [CrossRef]

13. Kleinman, P.J.A.; Sharpley, A.N.; Buda, A.R.; Easton, Z.M.; Lory, J.A.; Osmond, D.L.; Radcliffe, D.E.; Nelson, N.O.; Veith, T.L.; Doody, D.G. The promise, practice, and state of planning tools to assess site vulnerability to runoff phosphorus loss. *J. Environ. Qual.* **2017**, *46*, 1243–1249. [CrossRef] [PubMed]

14. Reid, K.; Schneider, K.; McConkey, B. Components of phosphorus loss from agricultural landscapes, and how to incorporate them into risk assessment tools. *Front. Earth Sci.* **2018**, *6*, 135. [CrossRef]

15. Andraski, T.W.; Bundy, L.G. Relationships between phosphorus levels in soil and in runoff from corn production systems. *J. Environ. Qual.* **2003**, *32*, 310–316. [CrossRef] [PubMed]

16. Djodjic, F. Phosphorus leaching in relation to soil type and soil phosphorus content. *J. Environ. Qual.* **2004**, *33*, 7. [CrossRef]

17. Vadas, P.A.; Kleinman, P.J.A.; Sharpley, A.N.; Turner, B.L. Relating soil phosphorus to dissolved phosphorus in runoff: A single extraction coefficient for water quality modeling. *J. Environ. Qual.* **2005**, *34*, 572–580. [CrossRef]

18. Baker, D.B.; Johnson, L.T.; Confesor, R.B.; Crumrine, J.P. Vertical stratification of soil phosphorus as a concern for dissolved phosphorus runoff in the Lake Erie basin. *J. Environ. Qual.* **2017**, *46*, 1287–1295. [CrossRef]

19. Duncan, E.W.; King, K.W.; Williams, M.R.; LaBarge, G.; Pease, L.A.; Smith, D.R.; Fausey, N.R. Linking soil phosphorus to dissolved phosphorus losses in the midwest. *Agric. Environ. Lett.* **2017**, *2*, 170004. [CrossRef]

20. Osterholz, W.R.; Hanrahan, B.R.; King, K.W. Legacy Phosphorus concentration–discharge relationships in surface runoff and tile drainage from Ohio crop fields. *J. Environ. Qual.* **2020**, *49*, 675–687. [CrossRef]

21. Sharpley, A.N. Soil mixing to decrease surface stratification of phosphorus in manured soils. *J. Environ. Qual.* **2003**, *32*, 1375. [CrossRef]

22. Messiga, A.J.; Ziadi, N.; Morel, C.; Grant, C.; Tremblay, G.; Lamarre, G.; Parent, L.-E. Long term impact of tillage practices and biennial p and n fertilization on maize and soybean yields and soil P status. *Field Crops Res.* **2012**, *133*, 10–22. [CrossRef]

23. Smith, D.R.; Huang, C.; Haney, R.L. Phosphorus fertilization, soil stratification, and potential water quality impacts. *J. Soil Water Conserv.* **2017**, *72*, 417–424. [CrossRef]

24. Ahuja, L.R.; Sharpley, A.N.; Yamamoto, M.; Menzel, R.G. The depth of rainfall-runoff-soil interaction as determined by [32] P. *Water Resour. Res.* **1981**, *17*, 969–974. [CrossRef]

25. Daniel, T.C.; Edwards, D.R.; Sharpley, A.N. Effect of extractable soil surface phosphorus on runoff water quality. *Trans. ASAE* **1993**, *36*, 1079–1085. [CrossRef]

26. Torbert, H.A.; Daniel, T.C.; Lemunyon, J.L.; Jones, R.M. Relationship of soil test phosphorus and sampling depth to runoff phosphorus in calcareous and noncalcareous soils. *J. Environ. Qual.* **2002**, *31*, 1380–1387. [CrossRef]

27. Kleinman, P.; Sharpley, A.; Buda, A.; McDowell, R.; Allen, A. Soil Controls of phosphorus in runoff: Management barriers and opportunities. *Can. J. Soil Sci.* **2011**, *91*, 329–338. [CrossRef]

28. Gächter, R.; Ngatiah, J.M.; Stamm, C. Transport of phosphate from soil to surface waters by preferential flow. *Environ. Sci. Technol.* **1998**, *32*, 1865–1869. [CrossRef]

29. Williams, M.R.; King, K.W.; Ford, W.; Buda, A.R.; Kennedy, C.D. Effect of tillage on macropore flow and phosphorus transport to tile drains: Tillage, macropore flow, and phosphorus. *Water Resour. Res.* **2016**, *52*, 2868–2882. [CrossRef]

30. Plach, J.M.; Macrae, M.L.; Williams, M.R.; Lee, B.D.; King, K.W. Dominant glacial landforms of the lower great lakes region exhibit different soil phosphorus chemistry and potential risk for phosphorus loss. *J. Great Lakes Res.* **2018**, *44*, 1057–1067. [CrossRef]

31. Williams, M.R.; King, K.W.; Duncan, E.W.; Pease, L.A.; Penn, C.J. Fertilizer placement and tillage effects on phosphorus concentration in leachate from fine-textured soils. *Soil Tillage Res.* **2018**, *178*, 130–138. [CrossRef]

32. Williams, M.R.; King, K.W.; Ford, W.; Fausey, N.R. Edge-of-Field research to quantify the impacts of agricultural practices on water quality in Ohio. *J. Soil Water Conserv.* **2016**, *71*, 9A–12A. [CrossRef]

33. King, K.W.; Williams, M.R.; LaBarge, G.A.; Smith, D.R.; Reutter, J.M.; Duncan, E.W.; Pease, L.A. Addressing agricultural phosphorus loss in artificially drained landscapes with 4R nutrient management practices. *J. Soil Water Conserv.* **2018**, *73*, 35–47. [CrossRef]

34. Patton, C.J.; Kryskalla, J.R. *Methods of Analysis by the U.S. Geological Survey National Water Quality Laboratory: Evaluation of Alkaline Persulfate Digestion as an Alternative to Kjeldahl Digestion for the Determination of Total and Dissolved Nitrogen and Phosphorus in Water*; Water Resources Investigations Report 03-4174; United States Geological Survey: Denver, CO, USA, 2003.

35. Williams, M.R.; King, K.W.; Macrae, M.L.; Ford, W.; Van Esbroeck, C.; Brunke, R.I.; English, M.C.; Schiff, S.L. Uncertainty in nutrient loads from tile-drained landscapes: Effect of sampling frequency, calculation algorithm, and compositing strategy. *J. Hydrol.* **2015**, *530*, 306–316. [CrossRef]

36. Mallarino, A.P. Spatial variability patterns of phosphorus and potassium in no-tilled soils for two sampling scales. *Soil Sci. Soc. Am. J.* **1996**, *60*, 1473–1481. [CrossRef]

37. Lauzon, J.D.; O'Halloran, I.P.; Fallow, D.J.; von Bertoldi, A.P.; Aspinall, D. Spatial variability of soil test phosphorus, potassium, and ph of ontario soils. *Agron. J.* **2005**, *97*, 524–532. [CrossRef]

38. Grandt, S.; Ketterings, Q.M.; Lembo, A.J.; Vermeylen, F. In-field variability of soil test phosphorus and implications for agronomic and environmental phosphorus management. *Soil Sci. Soc. Am. J.* **2010**, *74*, 1800–1807. [CrossRef]

39. Mallarino, A.P.; Borges, R. Phosphorus and potassium distribution in soil following long-term deep-band fertilization in different tillage systems. *Soil Sci. Soc. Am. J.* **2006**, *70*, 702–707. [CrossRef]

40. Dodd, J.R.; Mallarino, A.P. Soil-Test phosphorus and crop grain yield responses to long-term phosphorus fertilization for corn-soybean rotations. *Soil Sci. Soc. Am. J.* **2005**, *69*, 1118–1128. [CrossRef]

41. Fulford, A.M.; Culman, S.W. Over-fertilization does not build soil test phosphorus and potassium in ohio. *Agron. J.* **2018**, *110*, 56–65. [CrossRef]

42. Laboski, C.A.M.; Lamb, J.A. Changes in soil test phosphorus concentration after application of manure or fertilizer. *Soil Sci. Soc. Am. J.* **2003**, *67*, 544–554. [CrossRef]

43. Kleinman, P.J.A.; Sharpley, A.N.; Moyer, B.G.; Elwinger, G.F. Effect of mineral and manure phosphorus sources on runoff phosphorus. *J. Environ. Qual.* **2002**, *31*, 2026–2033. [CrossRef]

44. Macrae, M.L.; English, M.C.; Schiff, S.L.; Stone, M. Intra-annual variability in the contribution of tile drains to basin discharge and phosphorus export in a first-order agricultural catchment. *Agric. Water Manag.* **2007**, *92*, 171–182. [CrossRef]

45. Hanrahan, B.R.; King, K.W.; Williams, M.R.; Duncan, E.W.; Pease, L.A.; LaBarge, G.A. Nutrient balances influence hydrologic losses of nitrogen and phosphorus across agricultural fields in northwestern Ohio. *Nutr. Cycl. Agroecosyst.* **2019**, *113*, 231–245. [CrossRef]

46. Deubel, A.; Hofmann, B.; Orzessek, D. Long-term effects of tillage on stratification and plant availability of phosphate and potassium in a loess chernozem. *Soil Tillage Res.* **2011**, *117*, 85–92. [CrossRef]

47. Büchi, L.; Wendling, M.; Amossé, C.; Jeangros, B.; Sinaj, S.; Charles, R. Long and short term changes in crop yield and soil properties induced by the reduction of soil tillage in a long term experiment in Switzerland. *Soil Tillage Res.* **2017**, *174*, 120–129. [CrossRef]

48. Wilson, R.S.; Beetstra, M.A.; Reutter, J.M.; Hesse, G.; Fussell, K.M.D.; Johnson, L.T.; King, K.W.; LaBarge, G.A.; Martin, J.F.; Winslow, C. Commentary: Achieving phosphorus reduction targets for Lake Erie. *J. Great Lakes Res.* **2019**, *45*, 4–11. [CrossRef]

49. Wilson, H.; Elliott, J.; Macrae, M.; Glenn, A. Near-surface soils as a source of phosphorus in snowmelt runoff from cropland. *J. Environ. Qual.* **2019**, *48*, 921–930. [CrossRef]

50. Hart, M.R.; Cornish, P.S. Soil sample depth in pasture soils for environmental soil phosphorus testing. *Commun. Soil Sci. Plant Anal.* **2010**, *42*, 100–110. [CrossRef]

51. Schroeder, P.D.; Radcliffe, D.E.; Cabrera, M.L. Rainfall timing and poultry litter application rate effects on phosphorus loss in surface runoff. *J. Environ. Qual.* **2004**, *33*, 2201–2209. [CrossRef]

52. Pote, D.H.; Daniel, T.C.; Sharpley, A.N.; Moore, P.A.; Edwards, D.R.; Nichols, D.J. Relating extractable soil phosphorus to phosphorus losses in runoff. *SOIL Sci. Soc. Am. J.* **1996**, *60*, 855–859. [CrossRef]

53. Renard, K.; Foster, G.; Weesies, G.; McCool, D.; Yoder, D. *Predicting Soil Erosion by Water: A Guide to Conservation Planning with the Revised Universal Soil Loss Equation (RUSLE)*; Agriculture Handbook; U.S. Department of Agriculture: Washington DC, USA, 1997.

54. Smith, D.R.; Francesconi, W.; Livingston, S.J.; Huang, C. Phosphorus losses from monitored fields with conservation practices in the Lake Erie basin, USA. *AMBIO* **2015**, *44*, 319–331. [CrossRef]

55. Cox, F.R.; Hendricks, S.E. Soil test phosphorus and clay content effects on runoff water quality. *J. Environ. Qual.* **2000**, *29*, 1582–1586. [CrossRef]

56. Penn, C.J.; Mullins, G.L.; Zelazny, L.W. Mineralogy in relation to phosphorus sorption and dissolved phosphorus losses in runoff. *Soil Sci. Soc. Am. J.* **2005**, *69*, 1532–1540. [CrossRef]

57. Gaynor, J.D.; Findlay, W.I. Soil and phosphorus loss from conservation and conventional tillage in corn production. *J. Environ. Qual.* **1995**, *24*, 734–741. [CrossRef]

58. Jarvie, H.P.; Johnson, L.T.; Sharpley, A.N.; Smith, D.R.; Baker, D.B.; Bruulsema, T.W.; Confesor, R. Increased soluble phosphorus loads to Lake Erie: Unintended consequences of conservation practices? *J. Environ. Qual.* **2017**, *46*, 123–132. [CrossRef] [PubMed]

59. Vidon, P.; Cuadra, P.E. Phosphorus dynamics in tile-drain flow during storms in The US Midwest. *Agric. Water Manag.* **2011**, *98*, 532–540. [CrossRef]

60. Williams, M.R.; King, K.W.; Baker, D.B.; Johnson, L.T.; Smith, D.R.; Fausey, N.R. Hydrologic and biogeochemical controls on phosphorus export from western Lake Erie tributaries. *J. Great Lakes Res.* **2016**, *42*, 1403–1411. [CrossRef]

61. Pease, L.A.; Fausey, N.R.; Martin, J.F.; Brown, L.C. Weather, landscape, and management effects on nitrate and soluble phosphorus concentrations in subsurface drainage in the western Lake Erie basin. *Trans. ASABE* **2018**, *61*, 223–232. [CrossRef]

62. Buda, A.R.; Kleinman, P.J.A.; Srinivasan, M.S.; Bryant, R.B.; Feyereisen, G.W. Effects of hydrology and field management on phosphorus transport in surface runoff. *J. Environ. Qual.* **2009**, *38*, 2273–2284. [CrossRef]

63. Sharpley, A.; Jarvie, H.P.; Buda, A.; May, L.; Spears, B.; Kleinman, P. Phosphorus legacy: Overcoming the effects of past management practices to mitigate future water quality impairment. *J. Environ. Qual.* **2013**, *42*, 1308–1326. [CrossRef]

64. Shipitalo, M.J.; Gibbs, F. Potential of earthworm burrows to transmit injected animal wastes to tile drains. *Soil Sci. Soc. Am. J.* **2000**, *64*, 2103–2109. [CrossRef]

65. Frey, S.K.; Rudolph, D.L.; Conant, B. Bromide and chloride tracer movement in macroporous tile-drained agricultural soil during an annual climatic cycle. *J. Hydrol.* **2012**, *460–461*, 77–89. [CrossRef]

66. Penn, C.J.; Bryant, R.B.; Needelman, B.; Kleinman, P. Spatial distribution of soil phosphorus across selected New York dairy farm pastures and hay fields. *Soil Sci.* **2007**, *172*, 797–810. [CrossRef]

Publisher's Note: MDPI stays neutral with regard to jurisdictional claims in published maps and institutional affiliations.

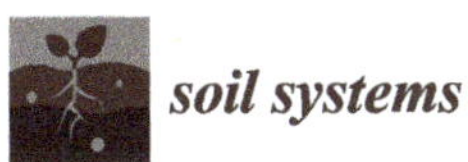 *soil systems*

Article

Weathering Intensity and Presence of Vegetation Are Key Controls on Soil Phosphorus Concentrations: Implications for Past and Future Terrestrial Ecosystems

Rebecca M. Dzombak and Nathan D. Sheldon *

Department of Earth & Environmental Sciences, University of Michigan, Ann Arbor, MI 48109, USA; rdzombak@umich.edu
* Correspondence: nsheldon@umich.edu

Received: 2 September 2020; Accepted: 13 December 2020; Published: 15 December 2020

Abstract: Phosphorus (P) is an essential limiting nutrient in marine and terrestrial ecosystems. Understanding the natural and anthropogenic influence on P concentration in soils is critical for predicting how its distribution in soils may shift as climate changes. While it is known that P is sourced from bedrock weathering, relationships between weathering, P, and other soil-forming factors have not been quantified at continental scales, limiting our ability to predict large-scale changes in P concentrations. Additionally, while we know that Fe oxide-associated P is an important P phase in terrestrial environments, the range in and controls on soil Fe concentrations and species (e.g., Fe in oxides, labile Fe) are poorly constrained. Here, we explore the relationships between soil P and Fe concentrations, soil order, climate, and vegetation in over 5000 soils, and Fe speciation in ca. 400 soils. Weathering intensity has a nuanced control on P concentrations in soils, with P concentrations peaking at intermediate weathering intensities (Chemical Index of Alteration, CIA~60). The presence of vegetation (but not plant functional types) affected soils' ability to accumulate P. Contrary to expectations, P was not more strongly associated with Fe in oxides than other Fe phases. These results are useful both for predicting changes in potential P fluxes from soils to rivers under climate change and for reconstructing changes in terrestrial nutrient limitations in Earth's past. In particular, soils' tendency to accumulate more P with the presence of vegetation suggests that biogeochemical models invoking the evolution and spread of land plants as a driver for increased P fluxes in the geological record may need to be revisited.

Keywords: soil fertility; phosphorus cycling; weathering; iron speciation; biogeochemistry

---

1. Introduction

Phosphorus (P) is an essential, often limiting nutrient in ecosystems across the globe [1,2]. P has been studied in plants, soils, rivers, lakes, and oceans—both modern and ancient—and is at the center of critical questions about Earth's past and future. Today, concerns around P in terrestrial settings focus on soil fertility and crop yields, as well as on agricultural runoff and its effect on eutrophication [3–5]. Climate, vegetation, and weathering and erosion rates are inextricably linked, so in order to improve quantitative constraints on potential P fluxes from soils, each of these factors must be considered. The redox-sensitive metal iron (Fe) is often associated with both P and organic matter/carbon (C) in soils (e.g., [6–9]). The P-Fe oxide pathway is critical for terrestrial P transport; however, it is currently less well-understood than other aspects of terrestrial P cycling, such as P-C associations. Additionally, while much research linking P, weathering, and soil age has been done, such work has typically done on relatively small scales (e.g., chronosequences in Hawaii) relative to the global scale of the questions that we ask here. Therefore, constraining climate, vegetation, and weathering controls on soil P and Fe is critical for predicting changes to fluxes of P from soils to rivers and beyond, and to improving our understanding of how regional soil fertility may evolve in response to a changing climate.

Prior to the evolution and expansion of land plants, changes in the flux of P from continents to oceans are thought to have been a primary control on marine productivity, and consequently, on the concentration of oxygen in the atmosphere [10,11]. It remains debated whether land plants' colonization of continents would have increased [12] or decreased [13] that P flux. The better we understand the modern distribution of and controls on P in soils, the better we can model global biogeochemical changes in Earth's past.

1.1. P, Fe, and Weathering and Erosion

P is released from bedrock via the weathering of P-bearing minerals (primarily apatites, which are Ca minerals relatively resistant to weathering [4]). Weathering depends on all soil-forming factors (climate, biology, topography, time, parent material), but is primarily controlled by climate (e.g., precipitation, temperature, seasonality) and time (surface age, length of exposure [14]). Organic matter also serves as an important pool of P in soils ([4,5]), and vegetation, fungi, and microbial life play an important role in both P and Fe liberation and mobilization through symbiotic (mycorrhizal) relationships [15,16]. The presence of plants, then, changes how both P and Fe are distributed in soils—but the links between landscape-scale soil chemistry, plant functional type, climate, and weathering intensity have not been quantified.

P can be depleted from a soil profile within thousands to tens of thousands of years [4,17–20]. In terrestrial vegetation (biomass), P's residence time is ~13 years, and residence in soil is ~600 years without P replenishment by dust [4]. Soil P can also be replenished through dust deposition (e.g., [17,21,22]). The natural progression of a soil, with a slow accumulation of Fe and Al oxides and a loss of mobile cations and nutrients, means that P is typically very low in old and/or intensely weathered soils [19]. Other soil properties, such as clay content, are also linked to weathering and vegetation, and could affect the concentrations of both P and Fe. Clay minerals and clay grain-size fractions can sorb considerable amounts of P and Fe, and clay content typically increases with weathering time and/or intensity [23,24]. Weathering intensity is expected to decrease with latitude due to colder temperatures, drier precipitation regimes, and less vegetation and shallower rooting depths [23,24]. Additionally, because the potential volume of erodible material generally increases as weathering intensity increases (though varies with factors such as slope and uplift rate, e.g., [25–29], weathering is linked to P fluxes not only through direct apatite dissolution, but also through erosion rates, e.g., [30,31]). Understanding how weathering intensity is related to soil P (and Fe) concentrations is important for predicting how their erosional fluxes and transport may change in response to climate change (e.g., [30,31]).

After weathering, dust deposition is an important secondary source of P, particularly for old soils whose P has been depleted through weathering, erosion, and biological use. For example, chronosequence studies in Hawaii have shown that for older soils, dust replenishment of P is critical to maintaining fertility once bedrock sources have been depleted [17,32,33]. Dust is also a critical source of P in the Amazon basin, which receives significant dust fluxes from Africa [18,34–36], compensating for P loss in deeply weathered tropical soils. Dust-sourced P is less significant for areas that receive little dust deposition, for younger soils that still have a regular input of P from bedrock weathering, or for shorter timescales than either for significant dust accumulation or timescales of soils losing their bedrock-sourced P (i.e., anthropogenic timescales). Arvin et al. [21] found that dust deposition is an important P source in montane soils, despite their relatively "fresh" bedrock. The balance between weathering rate and dust accumulation is also relevant for determining how important the latter is in soil P [22]. Dust is also commonly a source of Fe oxides, increasing soil fertility [35,37,38].

1.2. P and Fe Kinetics in Soils

P and Fe are chemically associated in soils in part because of P's affinity to sorb to Fe (oxyhydr)oxides (e.g., [4,9,39–42]). Orthophosphate, the common ion form of free P in soils, is reactive towards those oxides and becomes immobile (not readily bioavailable) when they associate into neoformed minerals or are sorbed onto existing mineral surfaces. Fe-bound P may be a critical mode of P transport among terrestrial ecosystems and from terrestrial/aquatic to marine environments, as oxyhydroxide-bound P is one of the major bioavailable/exchangeable forms of P in fluvial systems [5].

While we know that Fe oxide-adsorption of P is kinetically favorable in soils and other environments where P is being exchanged (e.g., [43–45]), our understanding of first-order controls on Fe concentration and speciation (i.e., Fe^{3+} in oxides versus labile Fe^{2+}) in soils could be improved. Examining a full suite of extractable Fe (rather than only dithionite-citrate and/or ammonium oxalate) allows a more nuanced picture of Fe phases in soils, which can ultimately be linked to variable P mobility in different soil redox conditions [43,46,47]. It is possible that regional climate factors (precipitation, temperature) exert control on Fe speciation (i.e., reduced vs. oxidized) in soils via soil moisture. Controls on soil moisture are debated and likely vary from profile to profile [48–53], and pH (related to soil pore space saturation; often microbially-mediated) is a primary control on reduction/oxidation of Fe phases in soils (e.g., [54]). Additionally, soil porewaters can serve as an intermediate step between recalcitrance and fluvial suspension for dissolved constituents (e.g., [54,55]). Exploring these relationships is key for understanding the likelihood of P mobilization under different moisture situations.

1.3. P and Fe in the Geologic Record

Geologists are interested in constraining many of the processes described above in ancient soils and ecosystems. Weathered P, transported to the oceans via fluvial networks and continental drainage, is typically invoked as a first-order control on marine primary productivity, which is associated with C sedimentation, atmospheric CO_2 drawdown, and oxygen production (e.g., [10,11,56]). Similar to P, Fe can stimulate marine productivity on short timescales (e.g., [57–60]), but it is of additional interest because in the past, it served as a P sorption site in marine systems, effectively sequestering P so that it is not bioavailable (the so-called "Fe-P trap") and limiting marine productivity [10].

With these processes in mind, geologists are interested in continent-to-ocean fluxes and mobility of Fe and P through time. However, the continent-to-ocean transport of P, although central to many biogeochemical models, is usually prescribed and not based on actual changes in fossil soil (paleosol) P values or weathering intensity data. Rather, it has been generalized based on crustal P values and fluxes controlled by large-magnitude changes, such as global glaciations or limited modern observations (e.g., [61]). The value in providing robust constraints on the range and distribution of P in soils, as well as climatic and environmental controls on its mobility and bioavailability, will vastly improve the parameters for these models.

In addition to biogeochemical models, geologists are interested in how the evolution of terrestrial vegetation and mycorrhizae (fungal root symbionts) affected P bioavailability, terrestrial mobility, and fluxes to the oceans. Researchers have argued for both an increase [12,56,62,63] and decrease [13,15] in P fluxes to the oceans following the spread of land plants. Constraining how modern vegetation relates to soil P mobility and distributions will help to answer this question.

Finally, the geochemical composition of fossil soils is used to reconstruct a range of climatic and environmental changes; however, these tools are primarily based on small datasets. This work provides ranges of reasonable values for soil chemistries under known environmental and atmospheric conditions, providing critical background information to improve our paleoclimate and paleoenvironment reconstructions.

1.4. This Work

In the context of these concerns, several big-picture questions remain. Despite much research on P in modern soils, ecosystem (vegetation, anthropogenic impact) and climate (precipitation, weathering) controls on its concentration have not been quantified on landscape to continental scales for generalizable conclusions. Additionally, although P is known to have a strong affinity towards Fe (oxyhydr)oxides in soils, Fe species and soil redox conditions have not been considered as potential first-order controls on landscape-scale P bioavailability. Questions about how climate (precipitation, temperature), soil factors (drainage, soil moisture, clay content), and overall weathering intensity affect soil Fe-P associations needs to be interrogated, to see if Fe species ultimately serves as a mediator for P bioavailability. Moreover, once the distribution of and controls on soil P have been quantified, how do those new constraints affect or constrain our understanding of biogeochemical P models in Earth's past?

We address these questions, along with hypotheses about P in the geologic record, using a large geochemical database of modern soils (n > 5000). The results can inform models of both past and future terrestrial biogeochemical cycles.

2. Materials and Methods

2.1. Sampling

Soil samples were collected in the field by the authors, and received from the USDA KSSL soil repository, and from contributors (total n = 404). A majority (n = 247) of these samples are B horizons, and therefore, reflect longer-term, stable pools of P (e.g., P sorbed to clays or Fe (oxyhydr)oxides, which we test for here, or recalcitrant organic matter) and Fe. These soils include Entisols, Inceptisols, Alfisols, Ultisols, Histosols, Mollisols, and Aridisols and provide coverage from low (Hawaii) to high (Iceland) northern latitudes (Supplemental Figure S1). Soils collected in the field or from repositories were screened for anthropogenic influence (i.e., on developed or disturbed land), with anthropogenically-altered soils noted as such and binned separately in our treatments of the data here. To complement these soils, we used a United States Geological Survey (USGS) soil geochemistry interactive report [64] and collated geochemical data for the Top 5 cm, A horizons, and C horizons of all the soils included in that work (n = 4857). See Supplemental Table S1 for summary descriptions of these two datasets, and Supplemental Table S2 for complete sample information for the new dataset presented here. Geochemical data for the USGS dataset [64] are available at https://doi.org/10.3133/sir20175118.

Combining these soil databases yields excellent spatial, climatological, profile depth, drainage, vegetation (as plant functional type/generalized biome), and soil order coverage (Supplemental Figure S1). While the samples are mostly from the United States, variation in soil forming factors across the continent reflect most of the potential global variation. The tropics are somewhat under-sampled, but data from Hawaii are broadly reflective of those latitudes. While there are species-specific relationships between plants and soil geochemistry, generalized biome is acceptable for the continental scale of questions examined in this work, as well as the comparative nature of our inquiries (e.g., [65–68]). Differences between plant functional types typically swamp differences within one plant functional type. Additionally, because biomes on sub-continental scales are typical for global biogeochemical models, and because more specific data are not available for the latitudinal scale of this work, generalizing by biome scale/plant functional type is appropriate.

2.2. Geochemical Analyses

All analyses were performed on dry soils within the Sheldon lab, as were samples during the USGS analyses. Soils were dried per USDA APHIS regulations upon import and ground to <70 μm for homogeneity during analyses; samples were stored dry in oxic (ambient room) conditions (during the experimental design, tests of pre- and post-drying, with drying temperatures varying between 25, 50, and 100 °C, showed no significant geochemical differences). Bulk elemental geochemistry was performed at ALS Laboratories in Vancouver, BC, Canada, using a four-acid digestion and Inductively Coupled Plasma Mass Spectrometer (ICP-MS) analysis. External precision was 0.01% for all major elements and 10 ppm for P_2O_5, analyzed at ALS. For samples from the USGS dataset, 1σ sample errors were 488 ppm and 1.39 wt% for P_2O_5 and total Fe, respectively) [64]. The 1σ sample errors for total C and organic C were 4.6% (n.d. for inorganic C [64]).

To analyze Fe speciation in soils, we used a four-step sequential Fe extraction following Poulton and Canfield [69] and Raiswell et al. [70]. This extraction protocol separates operationally-defined pools as ascorbic acid (labile Fe; Fe_{asc}), dithionite- (crystalline Fe oxyhydroxides; Fe_{dith}), ammonium oxalate- (highly crystalline Fe oxyhydroxides; Fe_{ox}), and sodium acetate-extractable Fe (Fe in carbonates, analyzed separately from the sequential extraction; Fe_{acet}). Initially, we included a chromium-reducible sulfur (CRS [71]) extraction separately to analyze for Fe sulfides (e.g., pyrite), but those concentrations were below detection limit for all non-wetland soils. CRS was limited to wetland soils after the initial exploratory extractions, and raw values were normalized to average soil order densities to account for the large difference in density between Histosols and denser, more mineral-rich soil orders [72]. Sample supernatants were centrifuged and diluted 50×, and Fe was measured via ICP-Optical Emission Spectroscopy and ICP-MS at the University of Michigan.

An internal standard of a local soil, dried at three different temperatures, was included in all runs for inter-run consistency checks and to determine reproducibility. The absolute standard deviation for a sample's reproducibility depends on the concentration of each Fe species, so the relative percent error $((\sigma/\mu) \times 100)$ is used instead. Repeated analyses show that Fe_{asc} is reliable within 3.7% of the mean, Fe_{dith}, within 1.7% of the mean, and Fe_{ox}, within 7.0% of the mean. Reproducibility on the Fe_{acet} samples was larger, at 15% of the mean, because the means in many of the samples we analyzed were near machine detection limits (~1000 ppm), where a difference of several hundred ppm yields far larger standard deviations than species with higher concentrations. The Fe_{asc} pool is considered to be a low constraint because a larger proportion of labile Fe may be in a soil's porewater (rather than solids). Because here we are primarily interested in relative differences between oxide and labile Fe species, and because species' concentrations tend to differ by orders of magnitude, we have deemed this approach acceptable. For further reading on nuances in sample preparation and extraction protocols, see two recent reviews by Raiswell et al. [73] and Algeo and Liu [74].

Fe speciation data were filtered for outliers by removing any samples where Fe_{spec} weight percent > total Fe weight percent (i.e., the concentration of a measured Fe pool was higher than the measured total Fe). These discrepancies (n = 107, 95, 139, 106, for Fe_{asc}, Fe_{acet}, Fe_{dith}, and Fe_{ox}, respectively) likely arose from the 50×-dilution being insufficient, with high Fe concentrations overwhelming the ICP-OES, leading to erroneous measurements. Future work using Fe speciation on soils should consider using stronger dilution factors for high-Fe soils to circumvent instrument limitations.

2.3. Other Data Collated

Geochemical data for the Top 5 cm, A horizons, and C horizons from 4857 soils around the U.S. are available on an interactive portal [64]. Weathering-relevant cations (Al, Ca, Na, K) and the nutrients of interest (P, Fe) along with latitude and total clay content (clay: A and C horizons only) were compiled. For the USGS dataset, clay content was determined by X-Ray diffraction (for 10 and 14Å clays only). Organic, inorganic, and total carbon data were available for a majority of soils in A and C horizons in the USGS dataset (not all had inorganic carbon). It should be noted that a dry soil sample will have a higher relative weight percent organic carbon due to water loss during drying. Soil orders and drainage capability are not given for soils in this database, but because they span the conterminous U.S., it is a reasonable assumption that they include a variety of soil orders. Clay content as grain-size fraction, parent material, and soil moisture regime were also gathered for the newly-collected dataset from USDA Soil Series reports when available (Supplemental Table S2).

Climate data were gathered from Soil Series pages; when these were not available, the PRISM 30-year averages (1981–2010) were used. If neither Soil Series nor PRISM climate data were available (i.e., for non-US samples), country-specific governmental meteorological data or journal articles were used (see Supplemental Table S2 for details).

The Chemical Index of Alteration (CIA) [75] was calculated to assess the degree of weathering for each soil, which reflects all of the soil-forming factors to some degree, but predominantly soil age and climate [14]. The CIA is thought to reflect the weathering of feldspars to form clay minerals where high values represent more intense weathering. Molar oxide ratios are used (Equation (1)). The CIA of an individual soil may not correspond directly with the CIA values of sediments in rivers (e.g., [76]). However, because our soils' average CIA values (59 ± 15, 42 ± 32, and 57 ± 21 for A, B, and C horizons, respectively) are within 1σ of the North American rivers' average from that study (n = 7; mean CIA = 66 ± 8.5, range 53–73), though with larger ranges (0.5 to 99 for all three horizons), we proceed with its use here.

$$CIA = (Al_2O_3/(Al_2O_3 + CaO + NaO + K_2O)) \times 100 \tag{1}$$

2.4. Statistical Analyses

To examine how the total Fe and P correlate with CIA (weathering intensity), a running average with a 10-CIA-unit window was used. Linear least-squares regressions were used to test for significant linear correlations where appropriate, based on the hypotheses that were being tested. Due to the large size of the USGS dataset, p-values are extremely low ($<10^{-16}$; essentially 0), meaning it is highly likely that the two variables are not randomly correlated. Strength of correlation should be primarily considered before taking any relationships to be predictive, but for simplicity's sake, coefficients of determination (R^2 values) are reported. Principle Components Analysis (PCA) was performed for USGS database soils by horizon (Top 5 cm, A, and C). These analyses are used to test the expected relationships (e.g., that Fe and P will be positively and linearly related), rather than to build predictive models. All of the statistical analyses were performed in Matlab.

3. Results

Complete sample information and all geochemical results are found in Supplemental Tables S2–S5 and are available in the online version of this paper. For the larger USGS dataset [64], we have geochemical, drainage, and vegetation (generalized plant functional type) data, but no soil order or Fe speciation information. Results for soil order, drainage, parent material, and Fe speciation analyses are based only on the newly collected dataset (n = 404).

3.1. Fe and P: Concentrations, Soil Order, and Vegetation

3.1.1. P

The average total P concentration (referred to simply as "P" here onwards) in the continental crust is ca. 870 ppm P, and there is relatively little variation between bedrock lithologies [77]. The maximum soil P concentration in this compilation was >20,000 ppm (>2 wt%; C horizon of an anthropogenically-modified soil), an order of magnitude greater than the crustal average. Mean P concentrations for each horizon studied were depleted relative to the crustal average: Top 5 cm, A, and C horizons means were 660, 632, and 508 ppm P, respectively (USGS data), and B horizons from the new dataset had a mean of 937 ppm P. P concentrations vary spatially throughout continental U.S. (Supplemental Figure S2), with hotspots in a variety of different geologic, climatic, and biologic provinces. P concentrations varied among soil orders (Figure 1), with the highest concentrations in Inceptisols (Figure 1A), a relatively weakly-developed soil type.

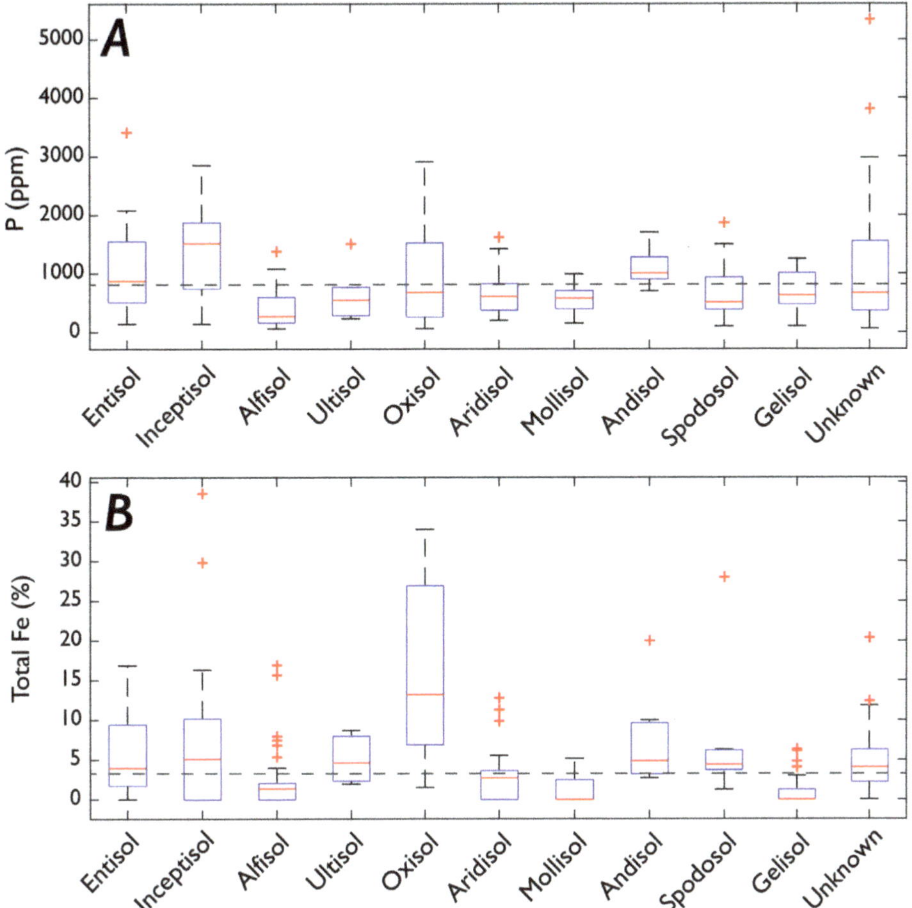

Figure 1. P and Fe concentrations in modern soils' B horizons, binned by soil order. Red lines are median values, blue bins are 25th and 75th quartile, and red crosses are beyond the 75th quartile; the horizontal dashed gray line is the crustal average (870 ppm P and 3.5% Fe) [77]. (**A**) P concentrations, showing that younger soils (Entisols, Inceptisols) tend to have higher P than older soils (Alfisols, Ultisols). Oxisols have high ranges of P, but a much smaller bin size than other orders (n = 8). Andisols have high inherited P from volcanic parent materials. (**B**) Fe concentrations, showing consistent Fe concentrations between ca. 1 and 10 wt% with the exception of Oxisols, which are defined by their high rates of Fe oxide accumulation. Bin sizes: Entisols (44), Inceptisols (75), Alfisols (42), Ultisols (6), Oxisols (8), Aridisols (76), Mollisols (60), Andisols (15), Spodosols (9), Gelisols (30), Unknown (39).

The vegetation groups are Barren (lacking vegetation), Altered (i.e., affected by human activity), Forest (no differentiation between deciduous and coniferous), Herbaceous/Grassland, Developed/Cultivated (i.e., occupied or manicured), Shrubland, and Other (mostly including early-succession plants or microbial earths). For USGS samples, protocols dictated that roads, buildings, and industrial sites be avoided [64]. Among vegetation types, there is little variability between P concentrations and vegetation types (Figure 2); while Barren landscapes' B horizons show higher mean P, this is likely due to small bin size (n = 5; Figure 2C). Overall, variability in concentrations in different vegetation types and soil orders was minimal, with most soils being depleted in P relative to the crustal average. Vegetated soils (both natural and anthropogenically-altered) had higher ranges of P than unvegetated (Barren). There was a weak positive correlation with mean annual precipitation (Supplemental Figure S3A), but the range of P values at a given precipitation amount is typically too large (~1500 ppm) to be of predictive use. This trend could also be explained by related factors, such as vegetation coverage (cover versus bare ground) and weathering intensity, which are associated with precipitation.

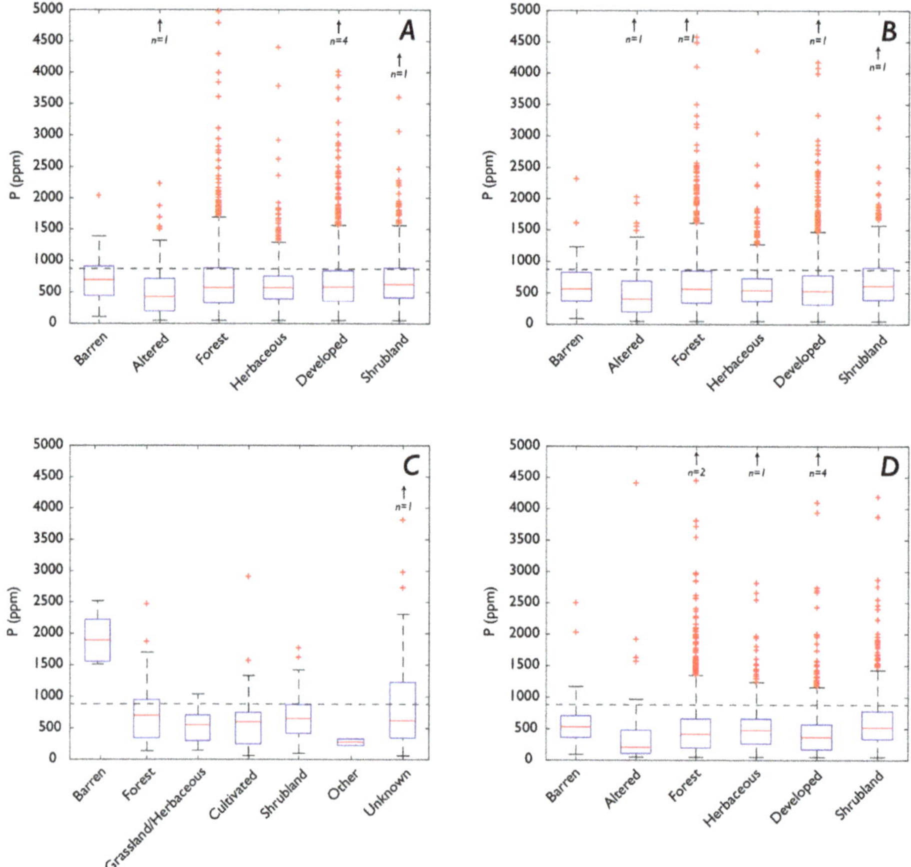

Figure 2. P concentrations in all horizons, binned by vegetation type. Dashed gray line in all is the crustal average P (870 ppm) [77]. Arrows indicate off-plot values. (**A**) P in the Top 5 cm, where Barren (no vegetation) and Altered soils have the lowest range of P concentrations. (**B**) P in A horizons, with similar distributions to Top 5 cm. (**C**) P in B horizons, with much smaller bin sizes than other horizons. (**D**) P in C horizons. Bin sizes for (**A,B,D**): Barren (68), Altered (209), Forest (1252), Herbaceous (815), Developed (1568), Shrubland (945). Bin sizes for (**C**): Forest (43), Grassland/Herbaceous (19), Shrubland (50), Barren (5), Cultivated (59), Unknown (74).

3.1.2. Fe

The crustal composition of total Fe (Fe_{tot}) is more variable than that of P because while P is sourced primarily by apatite-group minerals, Fe_{tot} can be present in a wide range of minerals and lithologies. The Phanerozoic upper continental crust is estimated to have 3.5 wt% Fe [77]. Fe_{tot} averages in the soils studied here were 2.1 wt% for Top 5 cm, 1.6 wt% for A horizon, 2.6 wt% for C horizon, and 4 wt% for our dataset (primarily B horizons). Density-normalized Fe_{tot} varied slightly by soil order

(Supplemental Table S3), showing the expected trend of modest loss shifting to modest accumulation during the Alfisol-Ultisol transition (Figure 1B), but otherwise was relatively consistent. Oxisols had the highest values and range in Fe, with some variability among the other soil orders but generally within consistent ranges (Figure 1B).

Fe_{tot} is generally consistent both within a given horizon and between different vegetation cover types (Figure 3). While most soil orders and most horizons had <3 wt% Fe_{tot}, values ranged up to >15% (Figure 3; Supplemental Figure S2). The only notable variability in Fe concentrations among vegetation is in Cultivated soil B horizons (Figure 4C), which are depleted, and 'Other' soil B horizons that are substantially enriched relative to the other vegetation cover groups. 'Other' vegetation contains mostly basalt-parented soils from a limited geographical area (primarily Iceland), so this result is likely an artifact of our sampling. There was no strong correlation between Fe_{tot} in a soil and mean annual precipitation (R^2: 0.2; Supplemental Figure S3B).

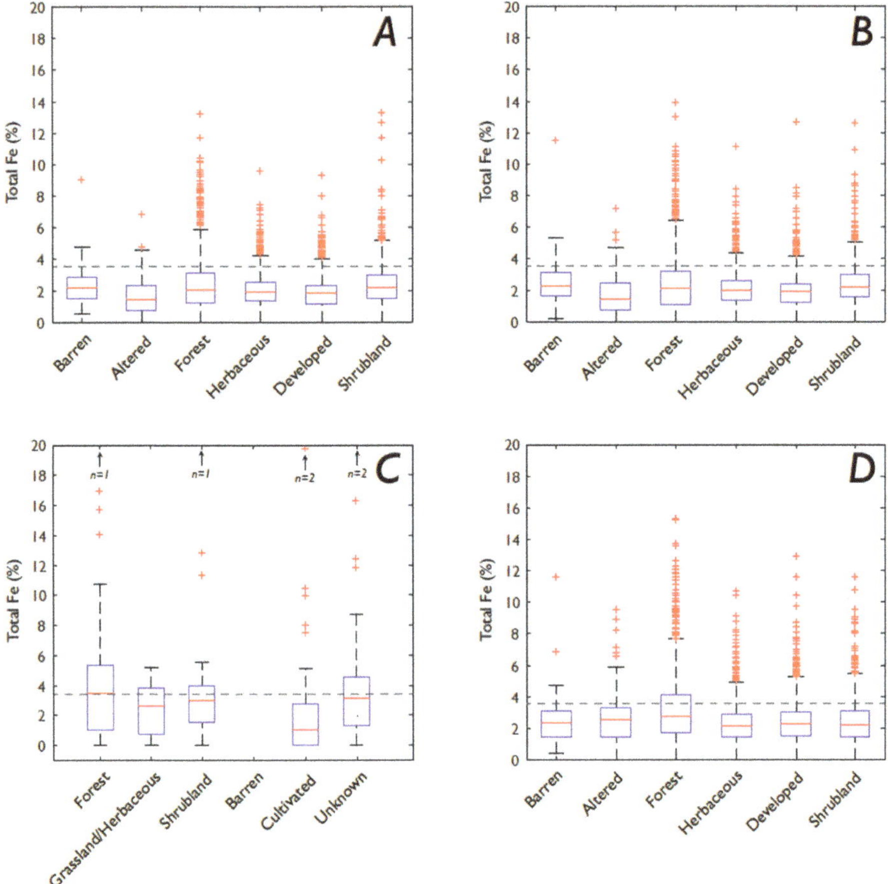

Figure 3. Fe concentrations in all horizons, binned by vegetation type. Dashed gray line in all is the crustal average (3.5%) [77]. Arrows indicate off-plot values. (**A**) Fe in the Top 5 cm, where Barren and Altered soils have the lowest maximum Fe concentrations. (**B**) Fe in A horizons, with similar trends to Top 5 cm. (**C**) Fe in B horizons. There were no data for Barren B horizons' Fe. (**D**) Fe in C horizons. Bin sizes for (**A,B,D**): Barren (68), Altered (209), Forest (1252), Herbaceous (815), Developed (1568), Shrubland (945). Bin sizes for (**C**): Forest (43), Grassland/Herbaceous (19), Shrubland (50), Barren (0), Cultivated (59), Unknown (74).

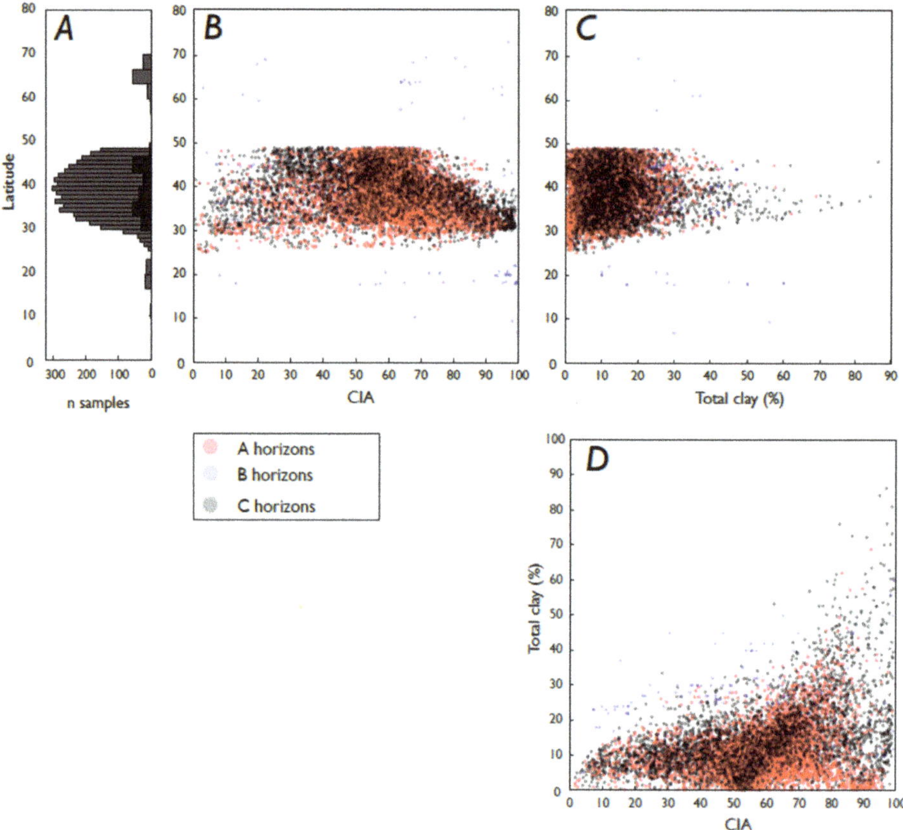

Figure 4. (**A**) Latitudinal distribution of soils used in this study, where smaller bins represent the USGS dataset [64] and the wider bins are our soils. The range of latitudes covered here is 7° to 73° N, but most of the samples fall between 20 and 50° N. (**B**) Latitudinal trends in Chemical Index of Alteration (CIA). (**C**) Latitudinal trends in clay content. (**D**) Correlation between CIA and clay content (R^2: 0.05, <0.01, and 0.19 for A, B, and C horizons, respectively). Overall, these results support common assumptions about the relationships between latitude and weathering.

3.2. Fe and P: Latitude, Weathering, Soil Order, and Clay Content

Latitude, weathering intensity, and clay content should each be associated with each other as well as with Fe (e.g., [24]). Some of the trends described here could be influenced by a latitudinal sampling bias, with most of the samples coming from the mid-latitudes (Supplemental Figure S1; Figure 4A). In the soils analyzed here, maximum weathering (as measured by CIA) decreases as latitude increases (Figure 4B), while clay content (a weathering product) does not show a strong trend with latitude (Figure 4C), with the exception of a weak mid-latitude (35–40° N) peak in clay content in some C horizons. Weathering trends between soil orders behave as expected, with CIA values increasing from Entisols to Ultisols/Oxisols (Figure 5). Clay content and weathering show a well-behaved and expected pattern of increase that matches the theoretical understanding of that metric [24,75,78,79], with B horizons specifically showing the highest clay content until the highest CIA values are reached (Figure 4D, green points). Clay content and geochemical data were not available for the Top 5 cm subset of the USGS dataset, so that horizon is excluded from those comparisons.

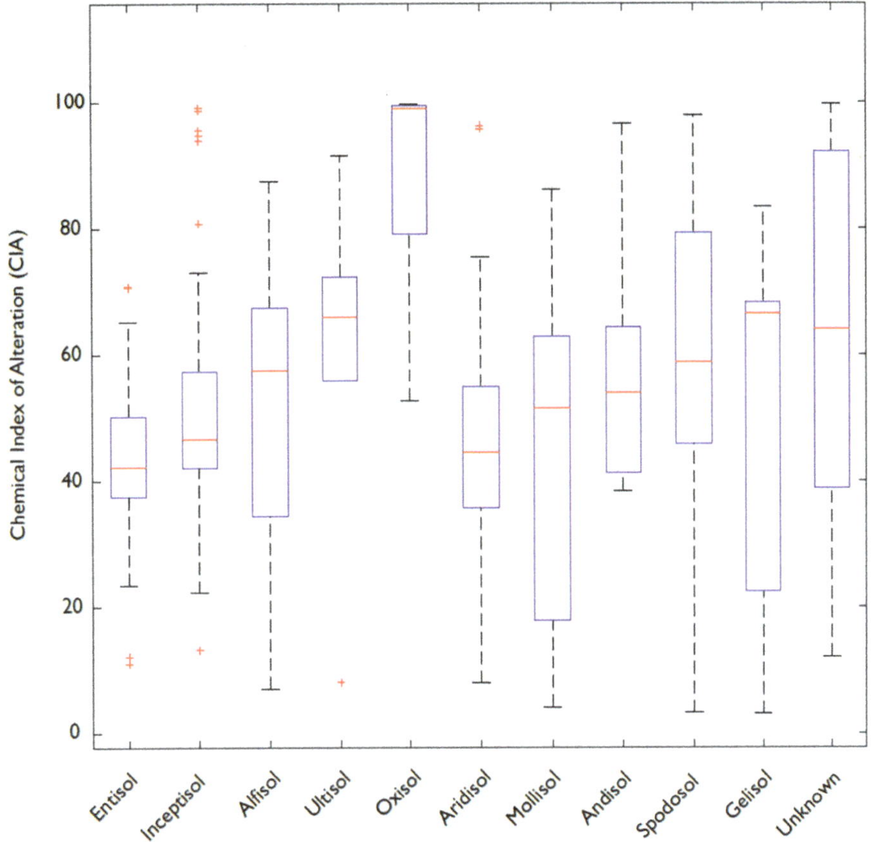

Figure 5. Chemical Index of Alteration (CIA) binned by soil order. CIA increases through the Entisol-Ultisol soil development progression, as expected, and Oxisols have the highest CIA with a median near 100. Aridisols' median CIA falls between Entisols and Inceptisols. Bin sizes: Entisols (44), Inceptisols (75), Alfisols (42), Ultisols (6), Oxisols (8), Aridisols (76), Mollisols (60), Andisols (15), Spodosols (9), Gelisols (30), Unknown (39).

Fe_{tot} content increases moderately with higher latitudes (Figure 6A), higher weathering (Figure 6C), and clay content (Figure 6E). P in soils behaves less linearly with respect to these variables: P increases moderately with latitude (Figure 6B), but rather than decreasing with weathering, as might be expected based on soil age-P relationships, average P concentrations in all horizons vary little, though there is a general increase in range at moderate weathering intensities (CIA~40–70) (Figure 6D). P concentrations are weakly negatively correlated with clay content (Figure 6F). Fe_{tot} and P showed positive correlations with one another in all horizons, as expected (Supplemental Figure S4).

3.3. Weathering, Clay Content, and Vegetation

Between vegetation types, there is low variability in clay content, with Forests showing slightly lower values in B horizons than the other groups (Supplemental Figure S5). Forests have the highest weathering in natural (unaltered/not developed) soils, followed by Grasslands and Shrublands (Figure 7).

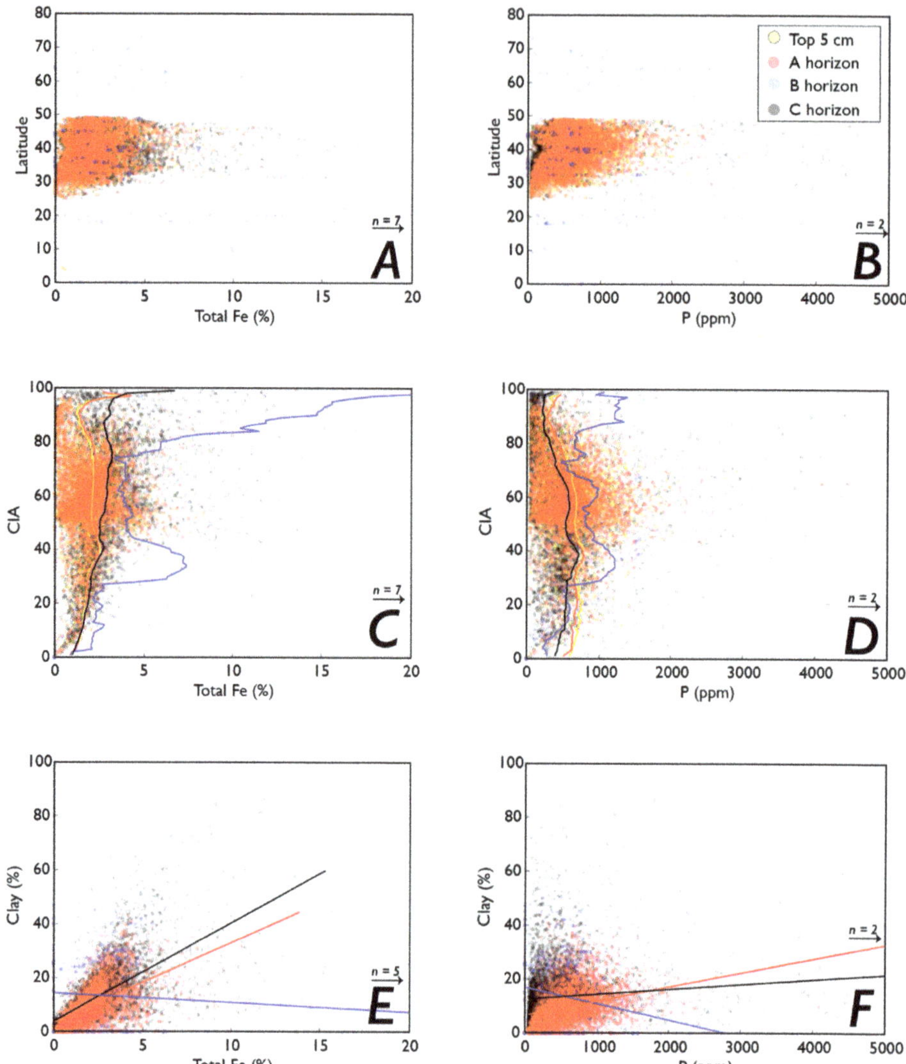

Figure 6. Trends between latitude, weathering, and clay content and total Fe and P concentrations. Solid lines in (**C,D**) are running averages for Top 5 (yellow), A (red), B (blue), and C (black) horizons. Solid lines in (**E,F**) are linear least-squares regressions for A (red), B (blue) and C (black) horizons. (**A**) Fe in soils by latitude. (**B**) P in soils by latitude. (**C**) Fe$_{tot}$ and weathering intensity. Solid lines are the running averages (10-CIA-unit window) for each horizon; colors correspond to data colors. The B horizon average (blue line) is highest at very high CIAs due in part to lower sample density at very high CIA values, especially between 98–100. (**D**) P concentrations and weathering intensity. (**E**) Fe and clay content (R^2 is 0.34 and 0.37 for A and C horizons; 0.01 for B horizons, much smaller dataset; $p = 0.7$ for B horizon, <<0.01 for other horizons). Solid lines are linear least-squares regressions; the B horizon (blue line) is skewed due to high-Fe, low-clay points off plot. (**F**) P and clay content. Solid lines are linear least-squares regressions (R^2 is 0.11, 0.08, and 0.01 for A, B, and C horizons, respectively; p << 0.01 for all horizons).

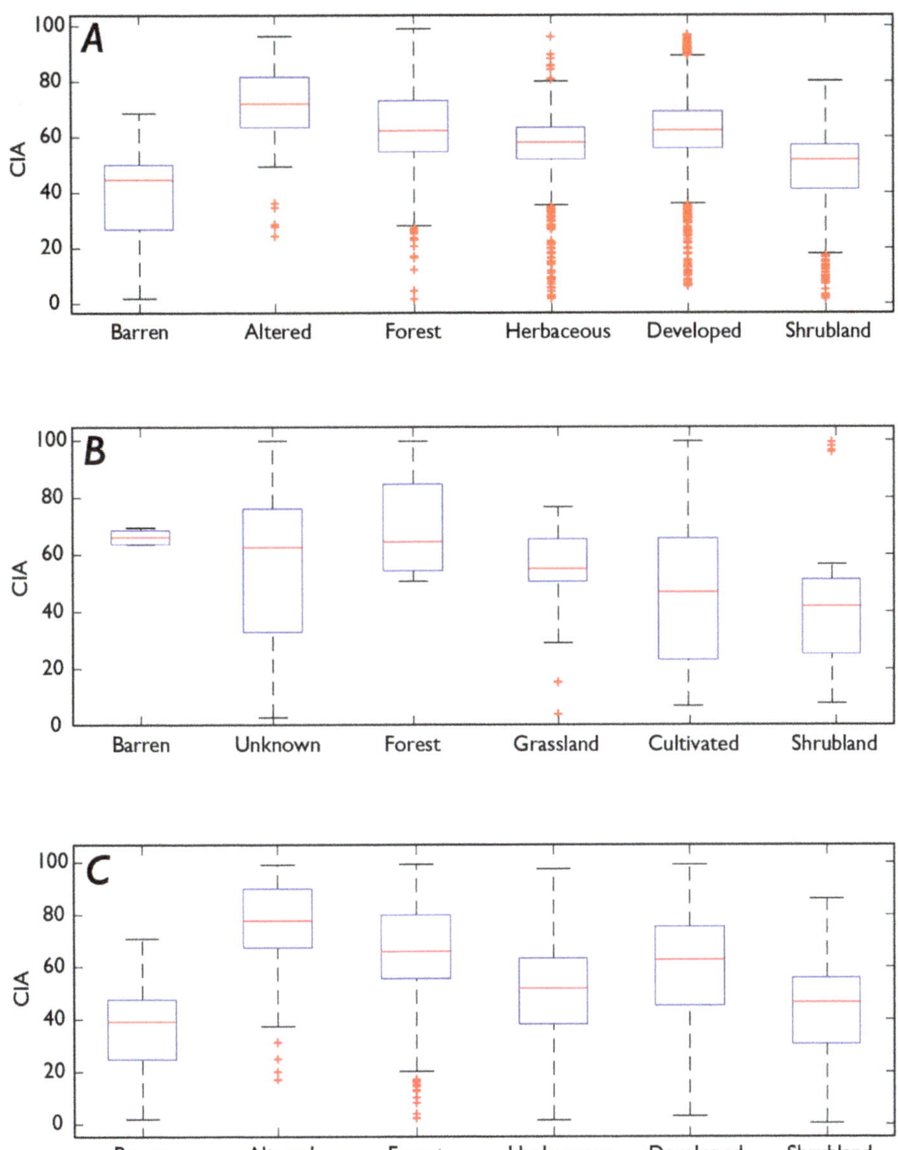

Figure 7. Chemical Index of Alteration (CIA) binned by vegetation type, for horizons (**A–C**) (bulk chemistry data for CIA were not available for the Top 5 cm horizon). (**A**) Altered soils have the highest CIA values overall, and Forest soils have the highest CIA for non-anthropogenically-affected soils, in A horizons. (**B**) Forests have the highest CIA values in B horizons, followed by Grasslands and Shrublands (Barren soils' small bin size, n = 5, precludes it from analysis here). (**C**) Altered soils also have the highest CIA in C horizons, followed by Forests.

3.4. Fe and P: Drainage and Soil Moisture

Moderately poorly-drained soils had higher Fe and P concentrations than the other degrees of drainage. Perudic moisture regimes had the highest Fe and P values (Figure 8); however, these samples were dominated by basalt-parented soils in Iceland, which may skew the results. Aside from that, Aquic and Xeric moisture regimes had high Fe, and Ustic and Udic soils had high P (Figure 8). Some trends emerge with relatively high Fe/low P (Aquic) and vice versa (Udic).

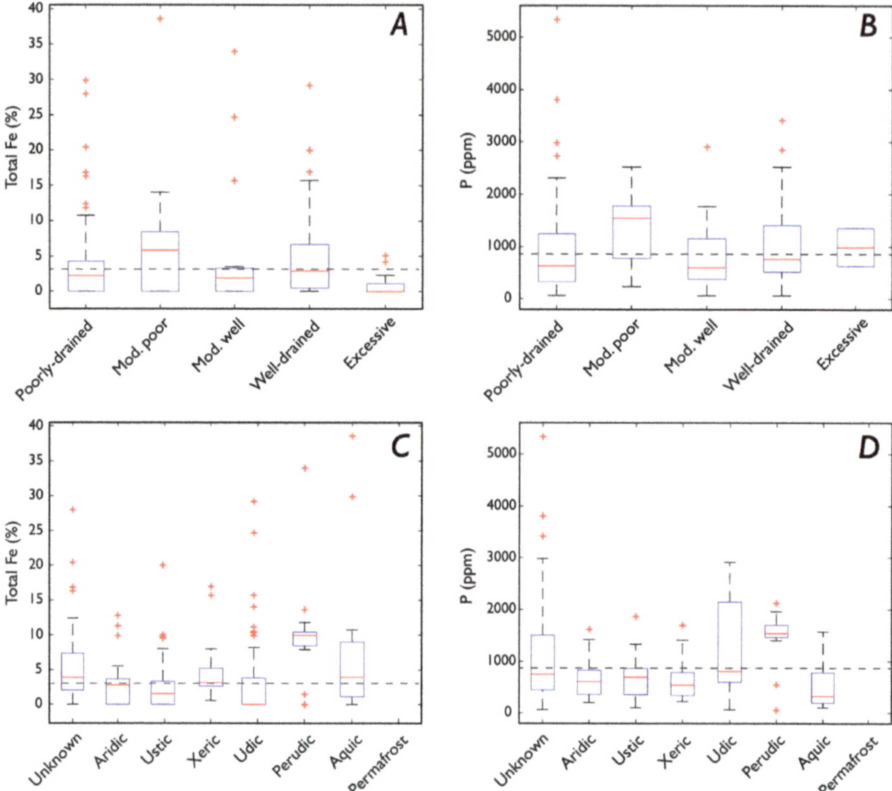

Figure 8. Variability in Fe concentrations (**A**,**C**) and P concentrations (**B**,**D**) in different soil drainage and moistures regimes. Dashed gray lines represent crustal averages for Fe and P in respective plots. (**A**) Fe binned by drainage. (**B**) P binned by drainage. (**C**) Fe binned by moisture regime. (**D**) P binned by moisture regime. Some patterns emerge when comparing Fe and P between soil moisture regimes, e.g., Udic soils having low Fe/high P and Aquic soils having high Fe/low P. Perudic soils are dominated by basalt-parented soils, influencing their Fe and P concentrations and highlighting the importance of bedrock parent material. No data for Permafrost Fe or P concentrations. Bin sizes for (**A**,**B**): Poorly-drained (166), Moderately poorly drained (25), Moderately well-drained (16), Well-drained (185), Excessively well-drained (12). Bin sizes for (**C**,**D**): Unknown (111), Aridic (73), Aquic (15), Udic (82), Perudic (27), Ustic (54), Xeric (21), Permafrost (N.D.).

3.5. Fe Speciation

3.5.1. Fe Speciation and P

Pyrite/Fe in sulfides, normalized to average soil order density, were above detection by titration in only 22 tested non-wetland soils out of 63 from around the continental U.S. (Supplemental Table S5), so that test was excluded from subsequent analyses. Density-normalized pyrite yields from all wetland sediments analyzed (n = 15) were measurable/above detection limit (>0.1 wt%), as opposed to non-wetland soils (Alfisols, Aridisols, and Mollisols), which typically fell below this limit after being density-normalized. In a non-perennially-waterlogged soil (e.g., not wetlands), it is reasonable to assume that Fe contents in sulfides are negligible.

Contrary to expectations (see Section 4.3 below), Fe_{dith} and Fe_{ox} pools did not display stronger or more robust correlations with P (Figure 9) than the other Fe pools. There were no clear correlations between P and any Fe species (for all, $R^2 < 0.01$). See Supplemental Table S4 for all Fe species results and Fe^{3+}/Fe^{2+} ratios.

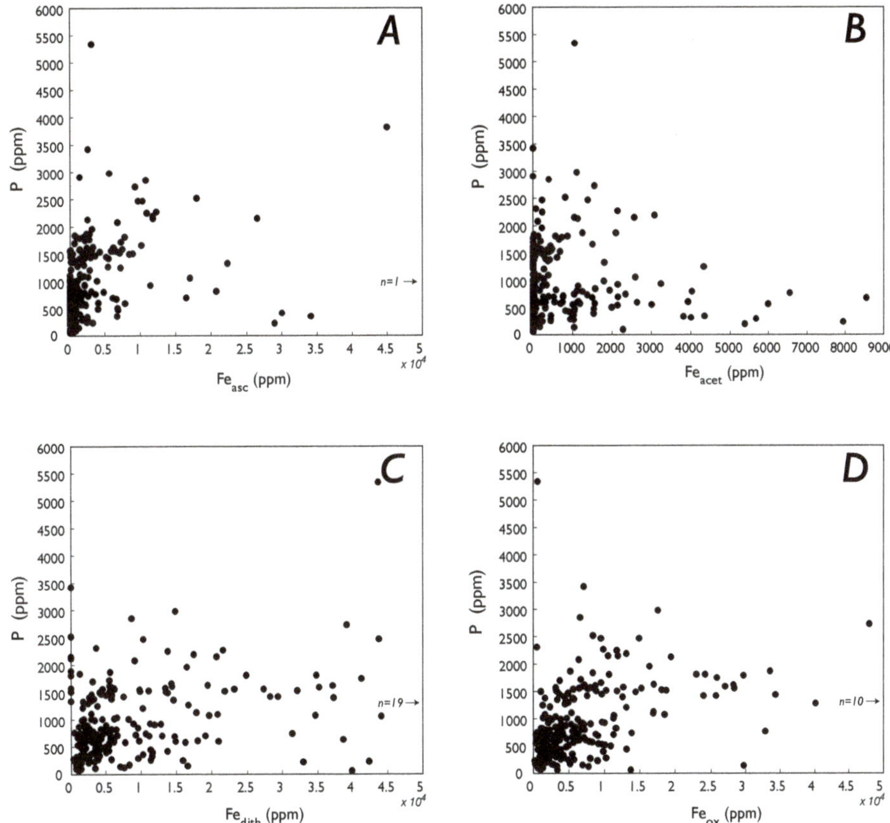

Figure 9. Relationships between Fe species and P concentrations, all with R^2 values < 0.1. Arrows indicate off-plot values. (**A**) No strong correlation between Fe_{asc} and P ($p = 10^{-6}$). (**B**) No significant correlation between Fe_{acet} and P ($p = 0.2$). Note the different *x*-axis scale. (**C**) No strong correlation between Fe_{dith} and P ($p = 0.003$), where a positive correlation had been expected due to P's affinity to sorb to Fe (oxyhydr)oxides. (**D**) No strong correlation is present between Fe_{ox} and P ($p = 10^{-9}$), again where a strong positive correlation might have been expected.

3.5.2. Fe Speciation, Precipitation, Soil Moisture, and Drainage

Contrary to expectations (see Section 4.2.2 below), Fe speciation pools showed no strong predictive relationships (all $R^2 < 0.2$) with mean annual precipitation (Figure 10), but there were differences between soil moisture regimes. Permafrost soils from the North Slopes of Alaska, high in organic matter and Fe, were dominated by labile Fe (Fe_{asc}). Aquic and Udic soils had high Fe in carbonates (Fe_{acet}), and Perudic soils (primarily from Iceland, with basalt parent material) had high Fe_{ox} (Figure 11). Mean Fe_{dith} was consistent between soil moisture regimes but has an extremely high range in Aridic soils.

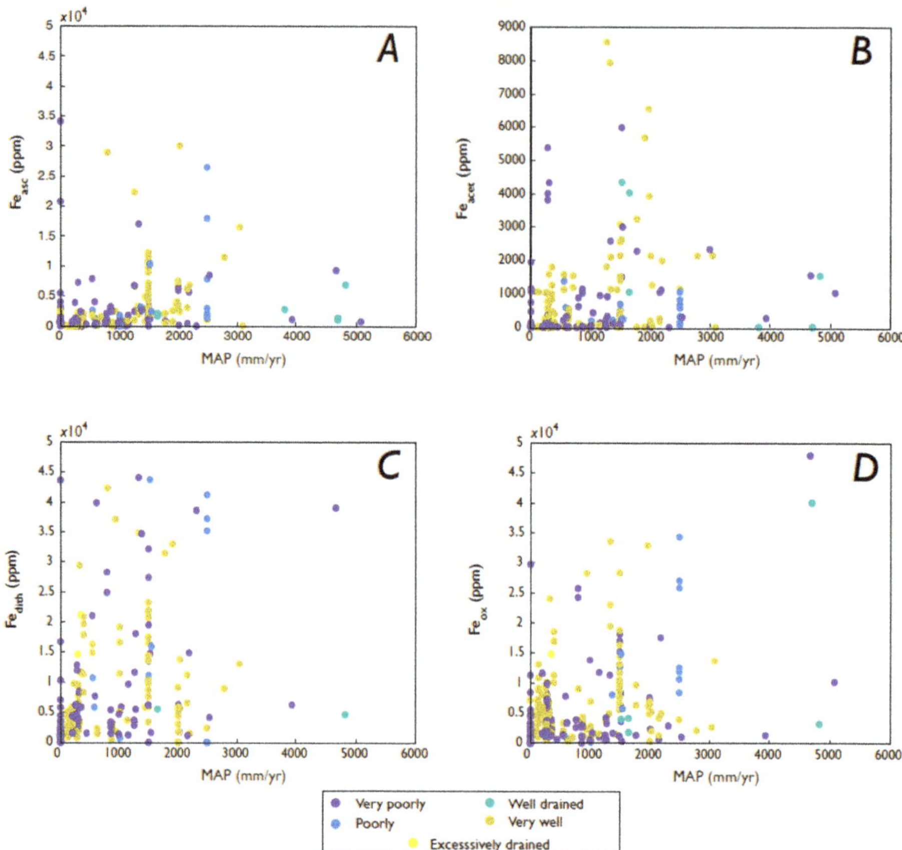

Figure 10. Relationships between mean annual precipitation (MAP; mm yr^{-1}) and Fe species, colored by soil drainage. Arrows indicate off-plot values. (**A**) No correlation between MAP and Fe$_{asc}$ (R^2 < 0.1, $p = 10^{-9}$). (**B**) No strong correlation between MAP and Fe$_{acet}$ (R^2 < 0.01; $p = 0.0005$); there could be a maximum MAP value beyond which Fe$_{acet}$ is less kinetically favorable. Note the different y-axis scale. (**C**) No significant correlation between MAP and Fe$_{dith}$ is present (R^2 < 0.01, $p = 0.7$). (**D**) Modest significant correlation between MAP and Fe$_{ox}$, (R^2 = 0.16, $p = 10^{-17}$). For all Fe species, there are no strong predictive trends between Fe species concentration and MAP, and no patterns in drainage.

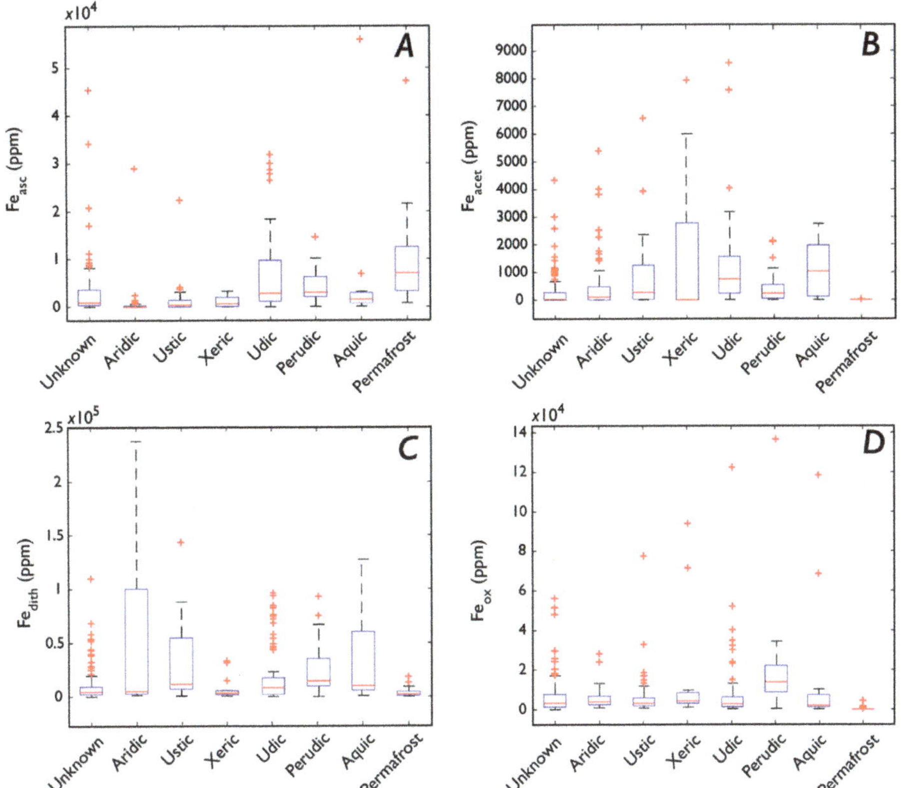

Figure 11. Fe species binned by soil moisture regime. (**A**) Udic, Perudic, and Permafrost soils have the highest median Fe_{asc}. (**B**) Fe_{acet} is quite low in most soils. Note the different *y*-axis scale. (**C**) Aridic soils have high Fe_{dith}, but other moisture regimes show little variability. (**D**) Little variability can be seen in Fe_{ox} across soil moisture regimes. Bin sizes (slight variability between Fe species is due to extraction yields): (**A**) Unknown (106), Aridic (73), Aquic (15), Udic (82), Perudic (26), Ustic (53), Xeric (21), Permafrost (21). (**B**) Unknown (111), Aridic (73), Aquic (15), Udic (82), Perudic (27), Ustic (54), Xeric (21), Permafrost (18). (**C**) Unknown (111), Aridic (61), Aquic (14), Udic (78), Perudic (26), Ustic (48), Xeric (15), Permafrost (21). (**D**) Unknown (106), Aridic (73), Aquic (15), Udic (82), Perudic (27), Ustic (54), Xeric (21), Permafrost (21).

Poorly-drained soils had a high range of Fe_{asc} and Fe_{dith}, while well-drained soils had high amounts of Fe_{acet} (Figure 12A). The other Fe species and drainages were consistent. The role of slope was considered, but local, small-scale topographic relief was not readily determinable for most samples.

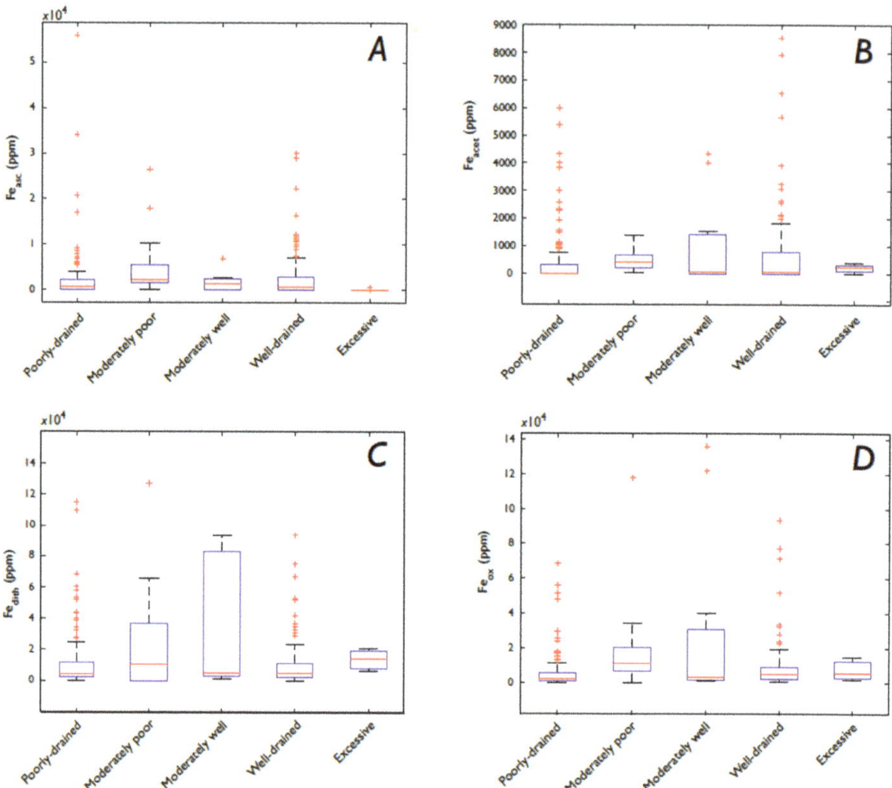

Figure 12. Fe species binned by soil drainage. (**A**) Excessively well-drained soils have very low Fe_{asc}. (**B**) Note the different y-axis scale. Fe_{acet} is highest in well-drained soils, although poorly-drained soils show an unexpectedly high range as well. (**C**) Excessive soils show the lowest maximum Fe_{dith}, with the other drainages showing similar medians and ranges. (**D**) Fe_{ox} has similar patterns to Fe_{dith}, with three very high values for moderate drainages. Bin sizes: (**A**) Poorly-drained (104), Moderately poorly drained (16), Moderately well-drained (11), Well-drained (140), Excessively well-drained (3). (**B**) Poorly-drained (109), Moderately poorly drained (16), Moderately well-drained (11), Well-drained (142), Excessively well-drained (3). (**C**) Poorly-drained (101), Moderately poorly drained (15), Moderately well-drained (9), Well-drained (123), Excessively well-drained (3). (**D**) Poorly-drained (109), Moderately poorly drained (16), Moderately well-drained (11), Well-drained (142), Excessively well-drained (3).

3.5.3. Fe Speciation, Vegetation, and Soil Order

Forest soils had higher Fe_{acet} (Figure 13B) than the other vegetation groups, but Fe species concentrations were consistent otherwise. Oxisols had higher Fe_{dith} and Fe_{ox} than other soil orders (Figure 14C), but Fe species had little variability between soil orders otherwise.

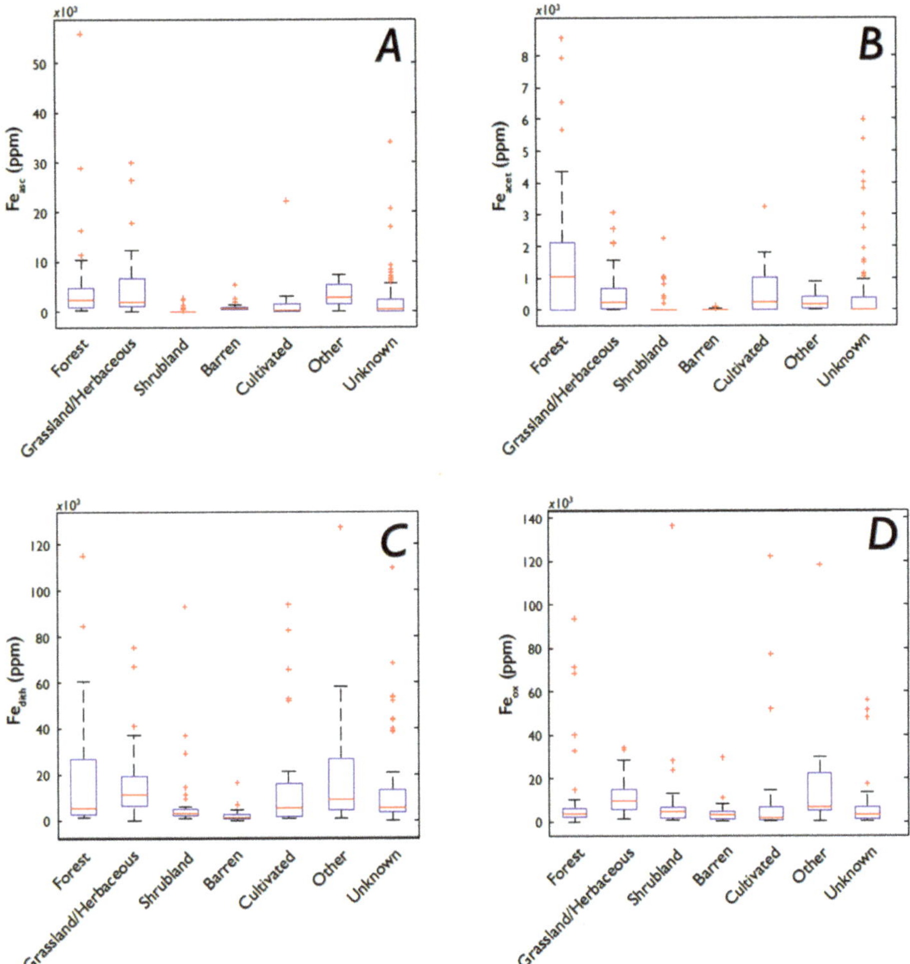

Figure 13. Fe species binned by vegetation type. (**A**) Forests and Grasslands have higher Fe_{asc} than other vegetations (except for Unknown). (**B**) Forests have the highest Fe_{acet} values (median and range). (**C**) No pattern seen in Fe_{dith} and vegetation, except that Barren soils have low Fe_{dith}. (**D**) No patterns in Fe_{ox} and vegetation.

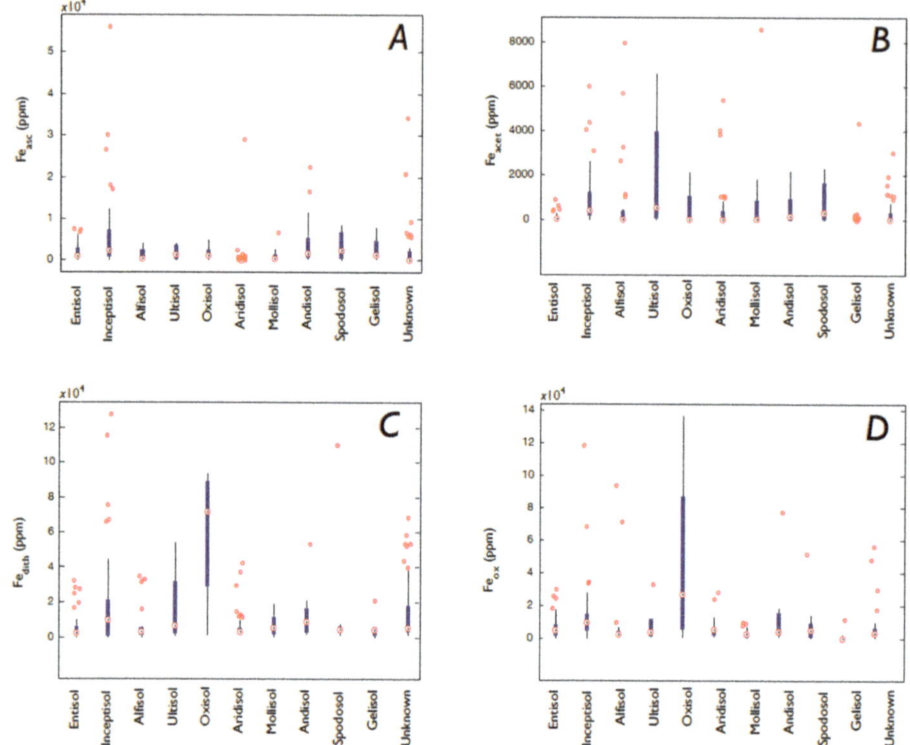

Figure 14. Boxplots of Fe species binned by soil orders, where centered red circles are the medians, the dark blue boxes are the 25th inner quartiles, the whiskers are the outer 25th quartiles, and the red circles are the statistical outliers (outside 3σ). (**A**) No differences in Fe_{asc} between soil orders. (**B**) Ultisols have the highest Fe_{acet}, although Alfisols have a high range of Fe_{acet} as well. (**C**) Oxisols have the highest Fe_{dith}, as expected. (**D**) Oxisols have the highest Fe_{ox}, as expected.

3.6. Organic Carbon

Organic carbon (C_{org}) in A horizons ranged from 0 to 61 wt%, and in C horizons, ranged from 0 to 43 wt% [64]. The average C_{org}:P ratio for A horizons was 17.9:1, and for C horizons, 18.3:1. Relationships between C_{org} and CIA, P, Fe, and clay content were weak (Supplemental Figures S6 and S7). There were no strong correlations between inorganic C and Fe_{tot}, or between C_{tot} and Fe_{tot} (Supplemental Figure S7).

3.7. Parent Material

There is slight variability between Fe_{tot} and P concentrations and parent material, with igneous bedrock and ash/volcanics having the highest median values for Fe, with high P values as well (Supplemental Figure S8). While limestone parent materials show high Fe, there are only two samples in this group and they are not included in this discussion. The parent material does not appear to be a predictive control in either Fe_{tot} or P (Supplemental Figure S8). However, this could be due to the mixed provenance of many modern soils' parent material (i.e., alluvium/colluvium, glacial deposits, etc. as opposed to compositionally homogeneous bedrock). A more targeted exploration into bedrock-parented soils, Fe_{tot}, and P could well show more well-behaved relationships.

3.8. Principle Components Analysis

Results from the principal components analyses (PCA) for the USGS data are shown in Supplemental Figure S10. Principle components 1 and 2 explain ~60% of the variance at a maximum. However, all eigenvalues are very low (<0.6) and data are essentially clustered around the origin, so while there are associations, they are very weak and should be interpreted conservatively. Latitude, P, Fe, and Al were associated with PC2 in the Top 5cm and A horizons, and vegetation and CIA were associated with PC1. Groupings changed in the C horizons, with Fe and Al associated with PC2, but P and latitude associated with C_{org} along PC1. In all horizons, CIA and vegetation vectors are opposite to each other, and were orthogonal to subparallel to most other vectors.

4. Discussion

In this section, we explore our results with a series of questions focused on constraining controls on Fe and total P in soils, defining their relationships on continental scales, and making inferences about how those relationships may impact terrestrial P fluxes. They are centered around key soil-forming factors, such as climate, vegetation, and weathering intensity, as well as soil redox-specific factors (i.e., precipitation, moisture and drainage, and Fe species). Because a majority of samples come from between 20 and 50°N, the interpretations made here are most robust for those latitudes.

4.1. How do Latitude, Weathering, and Clay Content Associate with P and Fe Concentrations?

Most of the expected relationships between soil order, Fe_{tot} and P, weathering, and latitude were supported. As expected, weathering generally decreases with latitude, dropping from a maximum CIA of >95 at 30–35° N to a CIA ~ 75 at 50° N (the northern limit of the large USGS dataset). Farther north, some B horizons deviate from that pattern, with CIA values higher than might be expected based on their climatic regime [80,81]. Interpreting the latitudinal trends here should include a caveat for latitudinal sampling bias and limitations of the dataset to, mostly, between 20° and 50° N. While the PCA results (Supplemental Figure S10) do not support a relationship between P and weathering, we interpret this discrepancy as being due to the complexities within the P-weathering relationship rather than negating the observations made between latitude, weathering, and P concentrations because the PCA eigenvalues are so low (most < ±0.2), and essentially, are clustered on the origin, which indicates a not-predictive value. P's relationship with weathering is complex and due to a variety of factors (climate, time, slope/erosion rate, etc.), and by looking only at the end product—which is what is left in the geologic record for analysis—we must inherently work around those limitations and draw the most robust conclusions possible from limited data on these large scales.

Fe_{tot} behaves as expected, accumulating in B horizons as soil weathering increases, while P accumulation in all horizons peaks at moderately-weathered soils (CIA ~ 60). The latter relationship is expected because a soil that has experienced moderate weathering and has developed somewhat (e.g., Inceptisol, Alfisol) has had sufficient time to have P mobilized from the bedrock/substrate and biotically cycled, but not so long or with such intense weathering that P is depleted from the substrate and removed via erosion and/or loss of biomass. An older and/or more intensely weathered soil (e.g., Ultisol) will have a substrate more depleted of P and will have lost more P. In terms of terrestrial P transport, a moderately-weathered soil is the most likely to have a large pool of potential P to transport. An interesting next step, building off this work, could be to test these expected correlations by mapping soil P with soil age and collecting fluvial P loads.

P concentrations are, on average, far below the crustal average of 870 ppm P. Even accounting for density differences between a typical soil (ρ_{soil} ~ 1.62 ± 0.2 g cm^{-3}; n = 659) [72] and continental crust (e.g., granite; ρ_{crust} ~ 2.7 g cm^{-3}), all of the soil horizons are depleted in P relative to the crustal abundance. There is some variability between soil orders, but all but a small subset were below the crustal abundance (i.e., recent basalt-parented soils). When considering bulk density, the total Fe concentrations show the opposite, with the soil Fe average of 4.4% being greater than the crustal average of 3.5%, again with some variability between soil orders but overall greater than the crustal average (Supplemental Table S3). This has implications for biogeochemical modeling (see Section 4.7.3). While P concentrations could be expected to generally decrease with increased weathering intensity, a moderately-weathered soil is more likely to be at a mid-developmental stage (e.g., Inceptisol, Alfisol) and supporting vegetation that can cycle P without being P-depleted (e.g., [19]). This mid-CIA accumulation could be reflective of the vegetation's impact on soil nutrients (e.g., mycorrhizal P mobilization, P recycling through biomass/organic matter decomposition), pointing to a key balancing point in a soil's lifespan where the bedrock is being weathered enough to maximize fertility without yet being depleted. Forest and Altered/Cultivated soils have the highest CIA values (Figure 7), but P concentrations were relatively invariant between types of vegetation, suggesting that weathering intensity exerts a greater control on soil P than the type of vegetation.

Clay content (both percent of grain-size fraction and abundance of clay minerals) should increase with the degree of weathering a soil has undergone, and that is reflected in the results (Figure 4D). Consequently, there are secondary trends associated with clay content, i.e., clay content peaks at lower latitudes where CIA values are the highest. Fe_{tot} and clay content show a strong positive correlation, as expected (Figure 6E), but P shows only a weak potential decrease as clay content increases, supporting traditional thinking that P is depleted as a soil is increasingly weathered and/or older (Figure 6F). While it is kinetically favorable for P to sorb to Fe and Al oxides (which accumulate in B horizons as a soil ages), clays (often with surficial Fe-Al oxides) may be equally as important for immobilizing P in soils [82].

Interestingly, our results indicate no positive correlations between Fe oxides (Fe_{dith} and Fe_{ox}) and clay content grain-size fraction (Supplemental Figure S9), as would be expected based on soil development patterns (i.e., higher weathering intensity leads to accumulation of clay minerals and Fe oxides), suggesting that while particle size may mediate Fe speciation due to varying degrees of reactivity, it is not a primary control. While the USGS database's clay content is based on mineralogy (determined by XRD), soils used in the Fe species work had clay content determined by settling, which would include both clay- and non-clay minerals that were clay-sized particles. This could explain at least partially the lack of the expected trend. Further exploration of clay mineralogy-specific correlations (or lack thereof) with Fe speciation pools, in the context of soil redox state, is warranted to explain this deviation from our expectations. Therefore, degree of weathering remains a primary control on soil P concentrations (e.g., [19]).

Although dust deposition is a source of P, it is typically secondary to bedrock weathering, not really applicable on anthropogenic-change timescales, and is not rapid in most of our study area (primarily the conterminous United States). Because it is not separately examined or included in this work, anyone using our results of soil P should take regional dust deposition and subsequent P replenishment into account when possible.

4.2. How Do Precipitation, Soil Moisture, and Soil Drainage Affect P and Fe Concentrations, and Fe Species?

4.2.1. How Does Soil Redox Affect P?

The relationships between soil redox factors (precipitation, soil moisture, and soil drainage) and soil P contents suggest that moisture is not a dominant control on soil P. While there is a weak possible correlation between MAP and P (Supplemental Figure S3A), the range of P concentrations for a given MAP is large (ca. 1500 ppm) and is not particularly predictive. Adding more climate data (e.g., correlating continental climate data with the USGS sample locations) could improve this potential relationship. As with other possible relationships in soils, however, it could be mediated by an additional soil-forming factor (i.e., precipitation is related to weathering intensity, and possibly to the presence of vegetation).

There exist only weak patterns between P and soil moisture or drainage. Udic soils have higher mean and ranges of P than other moisture regimes (Figure 8D), and poorly-drained soils can have higher P than well- or excessively-drained soils (Figure 8B). However, the bin sizes for these soil moisture regime types are relatively small; more data could clarify these trends. A relationship between soil P concentration and moisture regime would not necessarily be expected because soil moisture regimes are defined by the number of saturated days per year [83], but not necessarily if those days are linked by season (i.e., they do not need to be consecutive). While soil redox state would affect the (short-term) phase of P adsorption sites (such as Fe oxides) and possibly the accumulation of organic matter, which can mediate P-oxide interactions [8,9], as well as sequester P in organic compounds, only precipitation and its links to weathering intensity would be expected to exert long-term control on P concentrations.

4.2.2. How Does Soil Redox Affect Fe Concentrations and Speciation?

Our primary hypothesis was that precipitation would exert control on soil moisture and therefore on redox conditions in the soil porewaters, controlling Fe speciation. However, mean annual precipitation did not correlate with total Fe and showed none of the expected correlations with the four Fe species. There was perhaps a maximum precipitation (ca. 2000 mm yr^{-1}) for most Fe in carbonates (Figure 10B), but little else. Turning to a more direct approach of constraining soil moisture—soil moisture regime—revealed one interesting correlation: although variability in total Fe was low across soil moisture regimes (Figure 9C,D), there are some patterns of trade-off between species within a moisture regime (Aquic, Udic). Similarly, variability between soil drainage was low, but a general pattern of moderately-drained soils having higher Fe in oxides and carbonate, whereas labile Fe (reduced) showed no trend. Based on these data, soil moisture, redox state, and associated chemistry are highly heterogeneous even within soil orders, moisture regimes, and drainage ability, and broad generalizations cannot be made.

Iron oxide mineralogy and soil magnetic properties have been linked to regional precipitation [84–87] and soil moisture [42,43,88,89], but relatively few studies explore the links between soil iron oxide mineralogy and soil order [90,91]. Cervi et al. [90] found parent material and climate to be primary controls on soil magnetic susceptibility. A recent survey of Australian soils explored 471 unaltered topsoils' magnetic properties [92], which was the only study we could find of a similar scale to the work presented here. Hu et al. [92] found parent material to be the strongest control on soil iron mineralogy, with climate (precipitation, temperature) showing weaker influence and vegetation effects on soil iron being regional.

In addition to using the large soil datasets that are increasingly available, more targeted studies on the influence of soil moisture and drainage on redox conditions—using larger datasets and including more detail of soil texture—could strengthen these trends and reveal additional patterns. Many of the soils included in this study were well- or very-well drained, with a smaller subset being poorly- or very poorly-drained. It is possible that relationships between Fe species and moisture are stronger in soil drainage regimes for which we have effectively too small datasets in this study. Additionally, highly localized slopes (along with the context of grain size composition) likely exert strong controls on soil moisture and redox state, but we were not able to expand this study to include those data. Doing so would help clarify potential relationships between soil moisture and Fe species. Using high-resolution slope data (e.g., fine-scale LiDAR/digital elevation models) could be useful for determining these relationships between local soil redox conditions and geochemistry.

4.3. How Do Fe Species Associate with P?

We expected that Fe oxides, having been shown to efficiently sorb P onto their surfaces in other soils (e.g., [8,9,44,93]), would have a more robust correlation than between either labile Fe and P (in reducing microenvironments) or Fe in carbonates and P. However, no species showed a strong correlation with P (Figure 9). Overall, Fe$_{tot}$ and P in all soils showed a positive correlation, suggesting that Fe species is not a dominant control on P concentration, and that no single species is more likely to be associated with P, at least on the regional scales examined in this work. Because of the weak relationships between

Fe species and assorted soil moisture proxies, as well as a surprising lack of correlation with clay content, there is no simple rule connecting Fe species, soil redox state, and P. However, an increase in any Fe-bearing species should correspond to an increase in P due to the general positive relationship (Supplemental Figure S4). Fe (and Al) oxides in the clay fraction specifically may be most important for terrestrial P transport (both in soils and fluvial systems) (e.g., [39,44,94]). We expect that repeated analyses within an individual profile or on a small local scale (i.e., meters) would show stronger expected relationships, but they cannot be generalized to regional or continental scales.

4.4. How Does Vegetation Affect P and Fe Concentrations?

Biology—here, generalized as dominant vegetation on the soil surface (not considering microbial life and mesoscale organisms)—plays an important role in terrestrial P cycling. We found that Barren soils' B horizons had more P accumulation than other, vegetated soils' B horizons, reflecting plants' effectiveness at moving P from regolith to higher in the soil profile ([15,16]; see Section 4.7.4). The other vegetation types showed surprisingly little variability, but P does tend to increase up-profile, suggesting that once a landscape is vegetated—regardless of specific ecosystem type—P is efficiently mobilized by plants to the upper horizons and recycled and stored there. Additionally, vegetated soils had far higher ranges/maximum P concentrations than unvegetated soils, supporting the importance of plants' role in P mobilization and accumulation. Barren landscapes lack the rooting systems to mobilize and transport P, hence the increased accumulation in B horizons. Halsted and Lynch [95] found essentially no discrimination in P uptake between C_3 and C_4 plants, which our results further support.

Aside from 'Other' vegetation (which was biased by Fe-rich parent material), Cultivated soils' B horizons were the only ones that showed depletion, suggesting that ecosystem variability is less important than parent material and soil mineral accumulation rates for Fe in soils. Deep tilling in agricultural practice could be an additional factor in lower-soil-profile nutrient depletion, but it is less common than it used to be. Similarly, vegetation type does not appear to be a direct control on Fe species within soils; Forests did have greater Fe_{acet}, which is likely linked to the defined property of Alfisols of high cation exchange capacity and cation mobility, as well as their tendency to be well-drained.

Overall, the presence of vegetation—rather than differences in the type of vegetation—exerts the most control on P accumulation in soils (Figure 2). Although the mean P values were similar between vegetation types within a soil horizon (with the exception of B horizons in Barren soils), non-Barren (vegetated) soils had higher ranges of P than Barren soils. Forests also have higher CIA values than grasslands and shrublands (Figure 7), suggesting that the evolution and spread of different ecosystems could have changed the distribution of weathering intensity. This has interesting implications for the ongoing debate on land plants' influence on continental P fluxes (see Section 4.7.4).

Vegetation changes with latitude; the soils and ecosystems represented in this work span tropical and temperate forests, continental grasslands, dry shrublands, and high-latitude, little-vegetated landscapes. This inherently contributes to the observed latitudinal trend in weathering because different plants have different interactions with soil and bedrock. While there are ground-level differences between plants within a plant type (e.g., grasses), the functional difference between plant functional types (e.g., grasses vs. trees) are far larger and more meaningful on continental scales. While at local or regional scales, high-resolution, species-specific studies on weathering and plant-soil relationships can be done, that level of analysis is neither possible nor necessary for continental-scale analysis of vegetation-weathering relationships. Global climate, biogeochemical, and environmental change models rely on continent-scale data; therefore, while locally, species variability within a plant functional type may lead to nuanced plant-soil geochemical relationships, the continental scale and umbrella vegetation type is appropriate for this style of inquiry.

4.5. Is Soil Order Predictive of P and Fe Concentrations?

Soil order is not quantitatively predictive of P and Fe concentrations. Soil order, Fe, and P are linked indirectly through the correlation between Fe/P and CIA; this relationship is stronger for Fe, which shows a clear Fe accumulation increasing with weathering (Figure 6A,C), which is also linked to the Inceptisol to Ultisol soil progression common in chronosequences. Soil order and weathering show a very well-behaved and expected relationship, with CIA values increasing from Entisols to Ultisols/Oxisols (Figure 5). An exception to this trend is high-latitude Gelisols, which have longer formation times and lower weathering rates. These results align with existing soil formation frameworks and support proxies that rely upon their use, such as the CIA-based mean annual precipitation proxy, the paleosol weathering index for mean annual temperature, and Bt thickness ([79] and refs. therein). However, because soil order depends on a variety of factors (i.e., weathering stage, vegetation), quantitative relationships between soil order and either Fe_{tot} or P concentrations cannot be generalized, and samples within one soil order should not necessarily be used to predict Fe or P concentrations quantitatively for other samples of the same order. This is important to remember when using limited numbers of samples of fossil soils (see Section 4.7.1).

4.6. Implications for P in Modern Soils, Climate Change, and Soil Fertility/Food Security

4.6.1. Soil P, Erosion and Transport, and Human Activity

Aside from the geologically-recent uptick in anthropogenic influence on landscapes, continental scale-vegetation was relatively stable over the past millennium in North America (e.g., [96]). Concern for changing P fluxes in terrestrial ecosystems largely stems from the occurrence of harmful algal blooms in rivers, lakes, and coastal areas, often linked to mass-production agriculture and nutrient-rich runoff. The implications for these fluxes from this work center on land use change and soil degradation, as well as climate change-driven shifts in vegetation (e.g., grassland to barren). P would also be lost in that shift because vegetation loss exacerbates erosion and landscape degradation [97].

Human-induced changes (e.g., industrial agriculture, extensive monoculture, unsustainable farming practices, etc.) can lead to natural P being lost through subsidence and erosion, P being added through fertilizer, and reduced plant P retention if natural, diverse ecosystems are replaced with monoculture (e.g., [98]). Land-use changes typically result in the removal of the upper portion of a soil profile (various, e.g., [99]). Because P tends to accumulate in the upper horizons (O,A), our results suggest that this could lead to a noticeable loss in P and in agricultural soils, increased need for fertilizer (see Section 4.6.3). Land-use changes in vegetated soils could increase the P flux from land to lakes and coastal waters, leading to eutrophication. The spread of Aridisols through desertification could mean more regions experience erosion due to vegetation loss, but perhaps in turn lower weathering due to that loss of vegetation and to drier climatic conditions [100]. As desert regions expand, global dust volumes and distributions will change, which would change which areas see P removal (through increased erosion and P loss through dust) and which see P accumulation through dust deposition. Changes in atmospheric and soil CO_2 concentrations are likely to affect both plant and soil productivity (as well as mycorrhizal efficiency) under climate change, although the extent and duration of changes to productivity are debated [100–103].

4.6.2. Weathering, Climate Change, and Soil P

In addition to changes in vegetation (ecosystem), climate change will impact weathering intensity and erosion, affecting global soil P reservoirs. As a consequence, soil fertility and food security risk will shift and vary depending on how different regions and biomes respond to climate change [104]. Currently, there is a general trend of intense, rapid weathering at lower latitudes as a consequence of high heat and precipitation. Higher latitudes tend to be cooler and drier, and therefore, weathering intensity and rates are lower [14,24,105]. While low-latitude areas are expected to undergo desertification (decreased precipitation, loss of vegetation), higher latitudes may experience an increase in precipitation and temperature. Therefore, the style and intensity of weathering will change regionally. Based on our results, where the range in P concentrations peaks at mid-intensity weathering (i.e., CIA~60; Figure 5) and decreases in older and/or more strongly-weathered soils (as expected), the climate change expected for higher-latitude regions may result in decreased soil fertility (limiting agricultural production, see Section 4.6.3) and increased risk of eutrophication in surrounding waters. The former effect could potentially be mediated by vegetation, which holds P and slows erosion.

Except for the increase in range at mid-intensity weathering, P concentrations were relatively consistent across weathering intensities, which was unexpected. We had hypothesized that as older soils tend to be depleted in P, a higher weathering value would correspond to lower P. This was not demonstrated by our results. We interpret the lack of correlation as highlighting the significance of soil age—rather than weathering intensity (climate)—in controlling soil P, as some previous literature has suggested (e.g., [19,106]). A nuance that this large dataset could be missing is P speciation, which could still show P phase-specific variability with weathering intensity. The implication for P transport from soils remains unchanged from previous studies, where older soils would likely have smaller pools of P for mobilization. Mapping soil P with soil age on a large scale (such as the USGS dataset used here) would be a valuable new addition to the field and could elucidate the P-soil age-weathering puzzle. Additionally, exploring relationships between soils and weathering with species-level variability (on smaller, local scales) could help refine interpretations based on dominant regional plant functional type.

The difference in weathering intensity/CIA values between Forests and other natural vegetation types (Grasslands, Shrublands), as well as the high CIA values in Altered/Cultivated soils suggests that if more land is converted to farming (cultivated) from natural grasslands (e.g., the Great Plains) or if agricultural intensity increases, weathering rates would increase, depleting the soils' nutrients more quickly and perhaps increasing P fluxes. To our knowledge, this is the first data-based demonstration of the principle on the continental scale rather than as a model result. Understanding how human activity and agricultural degradation of soils affects weathering and resulting P is crucial for predicting changes in terrestrial P.

4.6.3. Soil P, Plants, and Agriculture

Climate change is also likely to affect plant physiology and biological processes, such as P uptake (mediated by mycorrhizal fungi) and C storage, as a result of changing atmospheric carbon dioxide concentrations (pCO$_2$). Plants rely heavily on symbiotic relationships with mycorrhizae to mobilize and uptake recalcitrant mineral P and fix nitrogen [102,107–109]. Mycorrhizal activity and P uptake can also be affected by soil moisture [110,111], which depends on a number of factors (discussed in Section 2.2) that could change in response to altered land use and climate change. P uptake efficiency also varies between species of symbiont [112], which may respond differently to climate stressors such as pCO$_2$ and temperature [110,113,114]. Some work suggests that increased atmospheric pCO$_2$ will be advantageous for plant growth, leading to greater terrestrial biomass and C storage [115,116], though potentially with an upper limit on pCO$_2$ vs. growth returns [117]. However, observations have shown that increases in pCO$_2$ during and after industrialization did not always increase plant C richness or biomass growth [118–120], and the biomass increase effect can still be mediated by P (and N) limitation in soils [121–123] and mycorrhizal species [102,114].

Ecosystem-specific C:N:P (Redfield) ratios vary, but throughout natural marine and terrestrial systems, P is the limiting nutrient [4,5]. For example, while the original marine phytoplankton-based Redfield ratio is 106:16:1, in soils, it is naturally higher at 186:13:1 [124]. The C:P ratios in both A and C horizons (16.7:1 and 14:1, respectively) were far lower than what is observed for O horizons (186:1 [124]) or inland streams/rivers (167:1 [125]), and are closer to ratios for microbial biomass in soils (~50:1 [124]). However, this ratio was derived with total P rather than organic P, and is therefore likely an overestimation. In these samples, the C$_{org}$:P$_{org}$ may be closer to the expected value for soils, but further speciation work is needed. Additionally, it is reasonable to expect that mineral soil horizons (rather than organic-rich horizons, like O) would have lower C$_{org}$ and nutrient concentrations. For developed soils (mean C$_{org}$:P = 15.25), this discrepancy could be linked to deep tilling removing nutrients from subsurface horizons and potentially increasing nutrient fluxes as compared to undisturbed or shallow-tilled soils. For natural soils with the potential for cultivation/deep tilling (e.g., grasslands, mean C_{org}:P = 20.5), the lower C:P ratio in subsurface mineral horizons could be inferred as diminishing returns for deeper tilling. There were no strong correlations between C_{org}, C_{inorg}, or C_{tot} and variables of interest (CIA, clay content, P, Fe) (Supplemental Figures S6 and S7).

Complicating this response is the fact that increases in atmospheric $p\mathrm{CO_2}$ could increase the rate of bedrock weathering (e.g., [126]), which mediates P availability and biolimitation. Moderate increases in $p\mathrm{CO_2}$ and weathering could lead to higher P availability and greater plant biomass, but if weathering increases too much, P in soils could be depleted while plant growth continues to increase. If high weathering rates lead to faster or premature depletion of P, especially exacerbated by anthropogenic activity (agriculture), crop stresses and regional food shortages could occur. Additionally, demand for P as fertilizer could increase; fertilizer P is ultimately sourced from bedrock, which, on human timescales, is a nonrenewable material [127]. Without developing efficient P recycling methods or adding sources such as bone char (e.g., [128,129]), P could run out in a matter of decades [130–132]. Therefore, when considering how climate change will affect agricultural yields, it is important to consider plant physiology and P uptake mechanisms together with mineral/geochemical changes in the soil driven by natural and anthropogenic forces.

Overall, this work aids in predicting changes in terrestrial P fluxes primarily by linking soil P, weathering intensity, and the presence of vegetation. However, in this study, "vegetation" remains something of a black box; the mechanism by which the presence of vegetation affects soil P could be through slowing erosion, increasing biotic P cycling, retaining P via plant/microbial biomass, increasing bedrock apatite weathering, or—most likely—a combination of those factors. Exploring each of these factors at large spatial scales, and to the extent possible pairing those analyses with smaller-scale species-specific soil-plant relationships, is a critical next step in elucidating the potential for P fluxes in different terrestrial regions.

4.7. Implications for P in the Fossil Record of Soils and Its Geologic Use

4.7.1. Heterogeneity and Paleosol Representativeness

Fossilized soils (paleosols), which are present in the rock record as far back as at least 3.0 billion years ago, serve as important windows into Earth's terrestrial past [133]. They have been used to reconstruct ancient atmospheres (e.g., [134–138]), climates ([79] and refs. therein), and terrestrial biology and biogeochemical cycling (e.g., [139,140]). The geochemical composition of fossil soils is used to reconstruct a range of climatic and environmental changes; however, these tools are primarily based on small modern soil datasets. This work provides ranges of reasonable values for soil chemistries under known environmental and atmospheric conditions, providing critical background information to improve our paleoclimate and paleoenvironment reconstructions. Based on the highly variable geochemical results in this work, we urge researchers using paleosols for these types of reconstructions to be cautious in assuming a single paleosol profile to be representative of an entire landscape or basin. More work on how representative a paleosol profile is, chemically and climatically, should be done to incorporate landscape-scale variability in soils (e.g., [141]).

4.7.2. Paleosol Fe, Atmospheric Oxygen Reconstructions, and Microbial Life

Because paleosols form on the Earth's surface and in direct contact with the atmosphere, their Fe chemistry has been used to reconstruct atmospheric oxygen levels [134,138,142]. Across a range of soil orders, environments, and climates, non-wetland soils in this work reflect oxidizing conditions (based on Fe^{3+}/Fe^{2+} ratios; Supplemental Table S4). While Fe^{3+}/Fe^{2+} ratios are no longer the primary tool for reconstructing paleo-oxygen levels, these results indicate that that tool is qualitatively robust in the absence of other observations such Cr or S isotopes.

A concern with using fossil soils for reconstructing atmospheric oxygen is that these soils likely hosted microbial life, which could have affected the redox signal ('biosignature') left behind. Indeed, redox-sensitive metals have been proposed as biosignatures because they can be mobilized by biologic processes and redeposited within a soil profile [143]. However, if it is suspected that the soils have variable redox states due either to high moisture input or poor drainage, how reliably might they have recorded the atmospheric oxygen signal as opposed to microscale oxic/reducing conditions due to microbial activity? Modern biological soil crusts (BSC; or, cryptogamic soils) are symbiotic communities of microbes (predominantly cyanobacteria), algae, and fungi; these communities have been proposed as analogues for early life on Earth (and Mars) [143]. Here, we found that the soil redox state being recorded by BSCs was oxic (again, using Fe^{3+}/Fe^{2+} ratios and dominant Fe speciation; Supplemental Table S4), despite BSCs' ability to retain water and temporarily shift to reducing conditions. It is unlikely that the microbial communities would have 'overprinted' the atmospheric oxygen signal even if they temporarily held water to cycle nutrients and leave behind biosignatures.

4.7.3. Continental Weathering, Nutrient Fluxes, and the Atmosphere

Geologists are interested in constraining many of the processes described above in ancient soils and ecosystems. Weathered P, transported to the oceans via fluvial networks and continental drainage, is typically invoked as a first-order control on marine productivity, which is associated with organic C and organic-associated and inorganic carbonate sedimentation, atmospheric CO_2 drawdown, and oxygen production [56]. Continental weathering, and therefore P flux, is a central control in many biogeochemical models of Earth's past. However, the continent-to-ocean transport of P is usually prescribed and not based on actual changes in fossil soil (paleosol) P values or weathering intensity data. Rather, it has been generalized based on crustal P values and fluxes controlled by large-magnitude changes, such as global glaciations [144,145] or limited modern observations [61,62].

Based on our results, continental weathering intensity and resulting elemental fluxes should vary with paleogeography (latitude effect), in addition to changes in climate and landmass area. Clay content through time could be taken into account and paired with CIA to reconstruct actual weathering intensity where those data are available in the rock record [146]. Forests also have higher CIA values than grasslands and shrublands (Figure 7), suggesting that the evolution and spread of different ecosystems could have changed the distribution of weathering intensity. Biogeochemical models should take these differences into account as opposed to assuming uniform behavior spatially.

Continental sulfide weathering in the Precambrian (pre-540 Ma) has been invoked as a major source of sulfate to the oceans, affecting organic matter respiration rates, marine alkalinity and oxidation state, and ultimately the concentration of atmospheric oxygen [147–150]. At the Great Oxygenation Event (GOE) 2.45 billion years ago, pyrite burial and increased ocean sulfate concentrations are thought to have decreased atmospheric methane through microbial sulfate reduction, lowering the greenhouse effect and leading to global glaciations [151,152]. Our results (Supplemental Table S5) constrain the number of Fe-bound sulfides in modern soils, forming under an oxic atmosphere, and suggest the potential for a 'false positive' for a reducing atmosphere if a paleosol is not correctly identified as waterlogged (and therefore reducing) or histic. Although it should be noted that using the mere presence of minerals, such as pyrite or uraninite, in the rock record as reducing indicators [153] is simplistic and no longer widely-used in the field.

4.7.4. Vegetation and P in the Phanerozoic (542 Ma Onwards)

The rise of land plants during the Devonian period (419–349 Ma) has been suggested as a major driver of change to the Phanerozoic P and C cycles, which could have affected atmospheric oxygen [15,56,63,154]. While B horizons in Barren vegetation had increased P retention, the lack of variability between other vegetation domains suggests a somewhat binary response—either soils are vegetated and mobilize/accumulate P, or they are not vegetated and are relatively depleted in P. Only soils with vegetation accumulated high ranges of P, supporting the need for biological mediation of P in terrestrial settings.

5. Conclusions

From this large-scale analysis of modern soils' physical and chemical properties, we draw conclusions about the relationships between modern soil biogeochemistry and climate, and link those to terrestrial P transport. We also describe implications about past terrestrial biogeochemistry and interpreting the terrestrial rock record.

In modern soils, some of the most well-defined (and expected) relationships were between soil composition and weathering. Latitude, clay content, and weathering correlate as expected, with higher latitudes corresponding with lower weathering intensity and clay content. Contrary to our expectations, average P concentrations are relatively consistent across weathering intensities but show a slight peak at middle weathering intensities, which supports commonly applied models of soil age and P loss. Specifically, it helps to clarify the conflation of time and weathering intensity that can occur, pointing to the importance of soil age as a control on soil P depletion. Maximum terrestrial P transport, then, would still likely occur at a midpoint in the weathering-P spectrum, where a soil is mature enough to have mobilized P from apatite and be linked to high erosion rates, but before soil P is too depleted due to age (which would be a lower P flux overall). Contrary to expectations, P in soils was not strongly associated with Fe (oxyhydr)oxides. The scale of such relationships may be much smaller than continental (i.e., locally or per profile). Additionally, no strong, predictive relationships were present between Fe species and precipitation, soil moisture, or drainage, so predicting P transport based on climate or soil redox properties is not possible. Finally, the presence of vegetation (but not plant functional type or generalized ecosystem type) is important for P accumulation in soils, which has implications for how the rise of terrestrial plants may have changed P cycling through geologic history.

Based on modern relationships between soil P, weathering, and vegetation, the spread of land plants in the early Phanerozoic likely increased P accumulation in soils and on continents, rather than increasing the flux of P to the oceans. This would limit the role of terrestrial-marine P transport in marine productivity. Because of latitudinal variability in weathering intensity, biogeochemical models should take paleolatitude into account when using weathering intensity as a driver for P fluxes. Additionally, models should use density-normalized values for terrestrial P rather than bulk crustal values and should use the range of P in modern soils as a quantitative constraint. Finally, while we did not find the relationships between Fe species and climate that we had expected, the overall Fe signature in soils (including microbially-colonized) was oxic despite analyzing a range of soil moistures and drainages, which supports the use of paleosol Fe/redox geochemistry for paleo-atmosphere reconstruction.

Supplementary Materials: The following are available online at http://www.mdpi.com/2571-8789/4/4/73/s1, Figure S1: Map of soils used in this work; Figure S2: Maps of P_2O_5 and Fe concentrations by horizon from USGS database; Figure S3: Scatterplots of mean annual precipitation with P_2O_5 (a) and Fe (b); Figure S4: Scatterplot of Fe and P_2O_5, for all samples; Figure S5: Boxplots of the CIA and clay content, sorted by vegetation type; Figure S6: Scatterplots of organic carbon with the CIA, clay content, and P_2O_5.; Figure S7: Scatterplots of organic, inorganic, and total carbon with total Fe; Figure S8: Boxplot of Fe and P_2O_5, sorted by parent material; Figure S9: Scatterplot of Fe species and clay content; Figure S10: Principle components analysis biplot; Table S1: Description of the datasets and which has what horizons, geochemical info, etc.; Table S2: Sample locations and soil details for soils compiled in this work (not in the USGS database); Table S3: Soil geochemistry for soils compiled in this work (not USGS data); Table S4: Fe species and Fe3/2 ratios, including BSC samples; Table S5: Pyrite in soils, normalized to density.

Author Contributions: Conceptualization, R.M.D. and N.D.S.; methodology, R.M.D. and N.D.S.; formal analysis, R.M.D.; investigation, R.M.D.; resources, N.D.S.; data curation, R.M.D. and N.D.S.; writing—original draft preparation, R.M.D. and N.D.S.; writing—review and editing, R.M.D. and N.D.S.; visualization, R.M.D.; supervision, N.D.S.; project administration, N.D.S.; funding acquisition, R.M.D. and N.D.S. All authors have read and agreed to the published version of the manuscript.

Funding: This research was funded by Lewis and Clark Astrobiology to R.M.D; Turner Award from the University of Michigan Department of Earth and Environmental Sciences to R.M.D and the NSF Award #1812949.

Acknowledgments: NRCS KSSL for soil sample storage, preparation, and shipping; USGS Southwest Biological Science Center for assistance in soil crust sampling; soils from Adrianna Trusiak/Rose Cory lab; Catherine Seguin for her thesis work on biological soil crusts and lab assistance; Sonya Vogel, Bianca Gallina, and Jordan Tyo for assistance in the lab; Emily Beverly for assistance in the field; and Kathryn Rico and Nikolas Midttun for helpful discussions. Our thanks to our anonymous reviewers for their helpful comments and suggestions to improve our manuscript.

Conflicts of Interest: The authors declare no conflict of interest. The funders had no role in the design of the study; in the collection, analyses, or interpretation of data; in the writing of the manuscript, or in the decision to publish the results.

References

1. Froelich, P.N.; Bender, M.L.; Luedtke, N.A.; Heath, G.R.; Devries, T. The marine phosphorus cycle. *Am. J. Sci.* **1982**, *282*, 474–511. [CrossRef]
2. Du, E.; Terrer, C.; Pellegrini, A.F.A.; Ahlström, A.; Van Lissa, C.J.; Zhao, X.; Xia, N.; Wu, X.; Jackson, R.B. Global patterns of terrestrial nitrogen and phosphorus limitation. *Nat. Geosci.* **2020**, *13*, 221–226. [CrossRef]
3. Correll, D.L. The Role of Phosphorus in the Eutrophication of Receiving Waters: A Review. *J. Environ. Qual.* **1998**, *27*, 261–266. [CrossRef]
4. Filippelli, G.M. The Global Phosphorus Cycle. *Rev. Miner. Geochem.* **2002**, *48*, 391–425. [CrossRef]
5. Filippelli, G.M. The Global Phosphorus Cycle: Past, Present, and Future. *Elements* **2008**, *4*, 89–95. [CrossRef]
6. Peretyazhko, T.; Sposito, G. Iron(III) reduction and phosphorus solubilzation in humid tropical soils. *Geochim. Cosmochim. Acta* **2005**, *69*, 3643–3652. [CrossRef]
7. Chacon, N.; Silver, W.L.; Dubinsky, E.A.; Cusack, D.F. Iron Reduction and Soil Phosphorus Solubilization in Humid Tropical Forests Soils: The Roles of Labile Carbon Pools and an Electron Shuttle Compound. *Biogeochemistry* **2006**, *78*, 67–84. [CrossRef]
8. Fink, J.R.; Inda, A.V.; Tiecher, T.; Barrón, V. Iron oxides and organic matter on soil phosphorus availability. *Ciência Agrotec.* **2016**, *40*, 369–379. [CrossRef]
9. Herndon, E.M.; Kinsman-Costello, L.; Duroe, K.A.; Mills, J.; Kane, E.S.; Sebestyen, S.D.; Thompson, A.A.; Wullschleger, S.D. Iron (oxyhydr) oxides serve as phosphate traps in tundra and boreal peat soils. *J. Geophys. Res. Biogeosci.* **2019**, *124*, 227–246. [CrossRef]
10. Reinhard, C.T.; Planavsky, N.J.; Gill, B.C.; Ozaki, K.; Robbins, L.J.; Lyons, T.W.; Fischer, W.W.; Wang, C.; Cole, N.J.P.D.B.; Konhauser, L.J.R.K.O. Evolution of the global phosphorus cycle. *Nat. Cell Biol.* **2017**, *541*, 386–389. [CrossRef]
11. Lenton, T.M.; Daines, S.J.; Mills, B.J. COPSE reloaded: An improved model of biogeochemical cycling over Phanerozoic time. *Earth Sci. Rev.* **2018**, *178*, 1–28. [CrossRef]
12. Guidry, M.W.; Mackenzie, F.T. Apatite weathering and the Phanerozoic phosphorus cycle. *Geology* **2000**, *28*, 631–634. [CrossRef]
13. D'Antonio, M.P.; Ibarra, D.E.; Boyce, C.K. Land plant evolution decreased, rather than increased, weathering rates. *Geology* **2019**, *48*, 29–33. [CrossRef]
14. Jenny, H.J. *Factors in Soil Formation*; McGraw-Hill: New York, NY, USA, 1941.
15. Quirk, J.; Leake, J.R.; Johnson, D.A.; Taylor, L.L.; Saccone, L.; Beerling, D.J. Constraining the role of early land plants in Palaeozoic weathering and global cooling. *Proc. R. Soc. B Boil. Sci.* **2015**, *282*, 20151115. [CrossRef]
16. Leake, J.R.; Read, D.J. Mycorrhizal symbioses and pedogenesis throughout Earth's history. In *Mycorrhizal Mediation of Soil: Fertility, Structure, and Carbon Storage*; Johnson, N.C., Gehring, C., Jansa, J., Eds.; Elsevier: New York, NY, USA, 2017; pp. 9–33.
17. Chadwick, O.; Derry, L.; Vitousek, P.M.; Huebert, B.J.; Hedin, L. Changing sources of nutrients during four million years of ecosystem development. *Nat. Cell Biol.* **1999**, *397*, 491–497. [CrossRef]
18. Okin, G.S.; Mahowald, N.; Chadwick, O.A.; Artaxo, P. Impact of desert dust on the biogeochemistry of phosphorus in terrestrial ecosystems. *Glob. Biogeochem. Cycles* **2004**, *18*, 2005. [CrossRef]
19. Walker, T.; Syers, J. The fate of phosphorus during pedogenesis. *Geoderma* **1976**, *15*, 1–19. [CrossRef]
20. Wardle, D.A.; Walker, L.R.; Bardgett, R.D. Ecosystem Properties and Forest Decline in Contrasting Long-Term Chronosequences. *Science* **2004**, *305*, 509–513. [CrossRef]
21. Arvin, L.J.; Riebe, C.S.; Aciego, S.; Blakowski, M.A. Global patterns of dust and bedrock nutrient supply to montane ecosystems. *Sci. Adv.* **2017**, *3*, eaao1588. [CrossRef]
22. Gu, C.; Hart, S.C.; Turner, B.L.; Hu, Y.; Meng, Y.; Zhu, M. Aeolian dust deposition and the perturbation of phosphorus transformations during long-term ecosystem development in a cool, semi-arid environment. *Geochim. Cosmochim. Acta* **2019**, *246*, 498–514. [CrossRef]

23. Sherman, G.D. The genesis and morphology of the alumina-rich laterite clays. In *Problems of Clay and Laterite Genesis*; Fredericksen, A.F., Ed.; American Institute of Mining and Metallurgical Engineering: New York, NY, USA, 1952; pp. 154–161.

24. Jobbágy, E.G.; Jackson, R.B. The distribution of soil nutrients with depth: Global patterns and the imprint of plants. *Biogeochemistry* **2001**, *53*, 51–77. [CrossRef]

25. Riebe, C.S.; Kirchner, J.W.; Finkel, R.C. Erosional and climatic effects on long-term chemical weathering rates in granitic landscapes spanning diverse climate regimes. *Earth Planet. Sci. Lett.* **2004**, *224*, 547–562. [CrossRef]

26. Dixon, J.L.; Heimsath, A.M.; Amundson, R. The critical role of climate and saprolite weathering in landscape evolution. *Earth Surf. Process. Landf.* **2009**, *34*, 1507–1521. [CrossRef]

27. Gabet, E.J.; Mudd, S.M. A theoretical model coupling chemical weathering rates with denudation rates. *Geology* **2009**, *37*, 151–154. [CrossRef]

28. Dixon, J.L.; Hartshorn, A.S.; Heimsath, A.M.; DiBiase, R.A.; Whipple, K.X. Chemical weathering response to tectonic forcing: A soils perspective from the San Gabriel Mountains, California. *Earth Planet. Sci. Lett.* **2012**, 40–49. [CrossRef]

29. Hewawasam, T.; Von Blanckenburg, F.; Bouchez, J.; Dixon, J.L.; Schuessler, J.A.; Maekeler, R. Slow advance of the weathering front during deep, supply-limited saprolite formation in the tropical Highlands of Sri Lanka. *Geochim. Cosmochim. Acta* **2013**, *118*, 202–230. [CrossRef]

30. Uhlig, D.; Schuessler, J.A.; Bouchez, J.; Dixon, J.L.; Von Blanckenburg, F. Quantifying nutrient uptake as driver of rock weathering in forest ecosystems by magnesium stable isotopes. *Biogeosciences* **2017**, *14*, 3111–3128. [CrossRef]

31. Eger, A.; Yoo, K.; Almond, P.C.; Boitt, G.; Larsen, I.J.; Condron, L.M.; Wang, X.; Mudd, S.M. Does soil erosion rejuvenate the soil phosphorus inventory? *Geoderma* **2018**, *332*, 45–59. [CrossRef]

32. Kurtz, A.C.; Derry, L.A.; Chadwick, O.A. Accretion of Asian dust to Hawaii soils: Isotopic, elemental, and mineral mass balances. *Geochim. Cosmochim. Acta* **2001**, *65*, 1971–1983. [CrossRef]

33. Porder, S.; Hilley, G.E.; Chadwick, O.A. Chemical weathering, mass loss, and dust inputs across a climate by time matrix in the Hawaiian Islands. *Earth Planet. Sci. Lett.* **2007**, *258*, 414–427. [CrossRef]

34. Swap, R.; Garstang, M.; Greco, S.; Talbot, R.; Kållberg, P. Saharan dust in the Amazon Basin. *Tellus B* **1992**, *44*, 133–149. [CrossRef]

35. Bristow, C.S.; Hudson-Edwards, K.A.; Chappell, A. Fertilizing the Amazon and equatorial Atlantic with West African dust. *Geophys. Res. Lett.* **2010**, *37*, 14807. [CrossRef]

36. Yu, H.; Chin, M.; Yuan, T.; Bian, H.; Remer, L.A.; Prospero, J.M.; Omar, A.; Winker, D.M.; Yang, Y.; Zhang, Y.; et al. The fertilizing role of African dust in the Amazon rainforest: A first multiyear assessment based on data from Cloud-Aerosol Lidar and Infrared Pathfinder Satellite Observations. *Geophys. Res. Lett.* **2015**, *42*, 1984–1991. [CrossRef]

37. Bullard, J.E.; White, K. Dust production and the release of iron oxides resulting from the aeolian abrasion of natural dune sands. *Earth Surf. Process. Landforms* **2005**, *30*, 95–106. [CrossRef]

38. Lafon, S.; Rajot, J.L.; Alfaro, S.C.; Gaudichet, A. Quantification of iron oxides in desert aerosol. *Atmos. Environ.* **2004**, *38*, 1211–1218. [CrossRef]

39. Froelich, P.N. Kinetic control of dissolved phosphate in natural rivers and estuaries: A primer on the phosphate buffer mechanism. *Limnol. Oceanogr.* **1988**, *33*, 649–668. [CrossRef]

40. Karathanasis, A.D. Phosphate Mineralogy and Equilibria in Two Kentucky Alfisols Derived from Ordovician Limestones. *Soil Sci. Soc. Am. J.* **1991**, *55*, 1774–1782. [CrossRef]

41. Turner, B.L.; Cade-Menun, B.J.; Westermann, D.T. Organic Phosphorus Composition and Potential Bioavailability in Semi-Arid Arable Soils of the Western United States. *Soil Sci. Soc. Am. J.* **2003**, *67*, 1168–1179. [CrossRef]

42. Herndon, E.; Albashaireh, A.; Singer, D.; Chowdhury, T.R.; Gu, B.; Graham, D. Influence of iron redox cycling on organo-mineral associations in Arctic tundra soil. *Geochim. Cosmochim. Acta* **2017**, *207*, 210–231. [CrossRef]

43. Kuo, S.; Mikkelsen, D.S. Distribution of Iron and Phosphorus in Flooded and Unflooded Soil Profiles and Their Relation to Phosphorus Adsorption. *Soil Sci.* **1979**, *127*, 18–25. [CrossRef]

44. Peña, F.; Torrent, J. Relationships between phosphate sorption and iron oxides in Alfisols from a river terrace sequence of Mediterranean Spain. *Geoderma* **1984**, *33*, 283–296. [CrossRef]

45. Torrent, J. Fast and Slow Phosphate Sorption by Goethite-Rich Natural Materials. *Clays Clay Miner.* **1992**, *40*, 14–21. [CrossRef]

46. Jaisi, D.; Kukkadapu, R.K.; Stout, L.M.; Varga, T.; Blake, R.E. Biotic and Abiotic Pathways of Phosphorus Cycling in Minerals and Sediments: Insights from Oxygen Isotope Ratios in Phosphate. *Environ. Sci. Technol.* **2011**, *45*, 6254–6261. [CrossRef] [PubMed]

47. Prietzel, J.; Klysubun, W.; Werner, F. Speciation of phosphorus in temperate zone forest soils as assessed by combined wet-chemical fractionation and XANES spectroscopy. *J. Plant Nutr. Soil Sci.* **2016**, *179*, 168–185. [CrossRef]

48. Coronato, F.R.; Bertiller, M.B. Climatic controls of soil moisture dynamics in an arid steppe of northern Patagonia, Argentina. *Arid. Soil Res. Rehabil.* **1997**, *11*, 277–288. [CrossRef]

49. Sandvig, R.M.; Phillips, F.M. Ecohydrological controls on soil moisture fluxes in arid to semiarid vadose zones. *Water Resour. Res.* **2006**, *42*, W08422. [CrossRef]

50. Vivoni, E.R.; Rinehard, A.J.; Mendez-Barroso, L.A.; Aragon, C.A.; Bisht, G.; Bayani Cardenas, M.; Engle, E.; Forman, B.A.; Frisbee, M.D.; Gutierrez-Jurado, H.A.; et al. Wyckoff, Vegetation controls on soil moisture distribution in the Valles Caldera, New Mexico, during the North American Monsoon. *Ecohydrology* **2008**, *1*, 225–238. [CrossRef]

51. Gaur, N.; Mohanty, B.P. Evolution of physical controls for soil moisture in humid and subhumid watersheds. *Water Resour. Res.* **2013**, *49*, 1244–1258. [CrossRef]

52. Fatichi, S.; Katul, G.G.; Ivanov, V.Y.; Pappas, C.; Paschalis, A.; Consolo, A.; Kim, J.; Burlando, P. Abiotic and biotic controls of soil moisture spatiotemporal variability and the occurrence of hysteresis. *Water Resour. Res.* **2015**, *51*, 3505–3524. [CrossRef]

53. Wang, T.; Wedin, D.; Franz, T.E.; Hiller, J. Effect of vegetation on the temporal stability of soil moisture in grass-stabilized semi-arid sand dunes. *J. Hydrol.* **2015**, *521*, 447–459. [CrossRef]

54. Surridge, B.W.J.; Heathwaite, A.L.; Baird, A.J. The Release of Phosphorus to Porewater and Surface Water from River Riparian Sediments. *J. Environ. Qual.* **2007**, *36*, 1534–1544. [CrossRef] [PubMed]

55. KerrMichele, J.G.; Burford, M.A.; Olley, J.; Udy, J.W. Phosphorus sorption in soils and sediments: Implications for phosphate supply to a subtropical river in southeast Queensland, Australia. *Biogeochemistry* **2011**, *102*, 73–85. [CrossRef]

56. Berner, R.A. The carbon cycle and carbon dioxide over Phanerozoic time: The role of land plants. *Philos. Trans. R. Soc. B Biol. Sci.* **1998**, *353*, 75–82. [CrossRef]

57. Martin, J.H.; Fitzwater, S.E. Iron deficiency limits phytoplankton growth in the north-east Pacific subarctic. *Nat. Cell Biol.* **1988**, *331*, 341–343. [CrossRef]

58. Martin, J.H.; Coale, K.H.; Johnson, K.S.; Fitzwater, S.E.; Gordon, R.M.; Tanner, S.J.; Hunter, C.N.; Elrod, V.A.; Nowicki, J.L.; Coley, T.L.; et al. Testing the iron hypothesis in ecosystems of the equatorial Pacific Ocean. *Nature* **1994**, *371*, 123–129. [CrossRef]

59. Coale, K.H. Effects of iron, manganese, copper, and zinc enrichments on productivity and biomass in the subarctic Pacific. *Limnol. Oceanogr.* **1991**, *36*, 1851–1864. [CrossRef]

60. Boyd, P.W.; Watson, A.J.; Law, C.S.; Abraham, E.R.; Trull, T.; Murdoch, R.; Bakker, D.C.E.; Bowie, A.R.; Buesseler, K.O.; Chang, H.; et al. A mesoscale phytoplankton bloom in the polar Southern Ocean stimulated by iron fertilization. *Nature* **2000**, *407*, 695–702. [CrossRef]

61. Berner, R.A.; Rao, J.-L. Phosphorus in sediments of the Amazon River and estuary: Implications for the global flux of phosphorus to the sea. *Geochim. Cosmochim. Acta* **1994**, *58*, 2333–2339. [CrossRef]

62. Moulton, K.L.; Berner, R.A. Quantification of the effect of plants on weathering: Studies in Iceland. *Geology* **1998**, *26*, 895. [CrossRef]

63. Lenton, T.M.; Dahl, T.W.; Daines, S.J.; Mills, B.J.W.; Ozaki, K.; Saltzman, M.R.; Porada, P. Earliest land plants created modern levels of atmospheric oxygen. *Proc. Natl. Acad. Sci. USA* **2016**, *113*, 9704–9709. [CrossRef]

64. Smith, D.B.; Cannon, W.F.; Woodruff, L.G.; Solano, F.; Ellefsen, K.J. *Geochemical and Mineralogical Maps for Soils of the Conterminous United States*; U.S. Geological Survey Open-File Report; United States Geological Survey: Denver, CO, USA, 2014; 386p, ISSN 2331-1258. [CrossRef]

65. Box, E.O. Plant functional types and climate at the global scale. *J. Veg. Sci.* **1996**, *7*, 309–320. [CrossRef]

66. Diaz, S.; Cabido, M. Plant functional types and ecosystem function in relation to global change. *J. Veg. Sci.* **2009**, *8*, 463–474. [CrossRef]

67. DiMichele, W.A.; Montañez, I.P.; Poulsen, C.J.; Tabor, N.J. Climate and vegetational regime shifts in the late Paleozoic ice age earth. *Geobiology* **2009**, *7*, 200–226. [CrossRef] [PubMed]

68. Diefendorf, A.F.; Mueller, K.E.; Wing, S.L.; Koch, P.L.; Freeman, K.H. Global patterns in leaf 13C discrimination and implications for studies of past and future climate. *Proc. Natl. Acad. Sci. USA* **2010**, *107*, 5738–5743. [CrossRef]

69. Poulton, S.W.; Canfield, D.E. Development of a sequential extraction procedure for iron: Implications for iron partitioning in continentally derived particulates. *Chem. Geol.* **2005**, *214*, 209–221. [CrossRef]

70. Raiswell, R.; Vu, H.P.; Brinza, L.; Benning, L.G. The determination of labile Fe in ferrihydrite by ascorbic acid extraction: Methodology, dissolution kinetics and loss of solubility with age and de-watering. *Chem. Geol.* **2010**, *278*, 70–79. [CrossRef]

71. Canfield, D.E.; Raiswell, R.; Westrich, J.T.; Reaves, C.M.; Berner, R.A. The use of chromium reduction in the analysis of reduced inorganic sulfur in sediments and shales. *Chem. Geol.* **1986**, *54*, 149–155. [CrossRef]

72. Sheldon, N.D.; Retallack, G.J. Equation for compaction of paleosols due to burial. *Geology* **2001**, *29*, 247–250. [CrossRef]

73. Raiswell, R.; Hardisty, D.S.; Lyons, T.W.; Canfield, D.E.; Owens, J.D.; Planavsky, N.J.; Poulton, S.W.; Reinhard, C.T. The iron paleoredox proxies: A guide to the pitfalls, problems and proper practice. *Am. J. Sci.* **2018**, *318*, 491–526. [CrossRef]

74. Algeo, T.J.; Liu, J. A re-assessment of elemental proxies for paleoredox analysis. *Chem. Geol.* **2020**, *540*, 119549. [CrossRef]

75. Nesbitt, H.W.; Young, G.M. Early Proterozoic climates and plate motions inferred from major element chemistry of lutites. *Nat. Cell Biol.* **1982**, *299*, 715–717. [CrossRef]

76. Li, C.; Yang, S. Is chemical index of alteration (CIA) a reliable proxy for chemical weathering in global drainage basins? *Am. J. Sci.* **2010**, *310*, 111–127. [CrossRef]

77. Taylor, S.R.; McLennan, S.M. The geochemical evolution of the continental crust. *Rev. Geophys.* **1995**, *33*, 241–265. [CrossRef]

78. Maynard, J.B. Chemistry of Modern Soils as a Guide to Interpreting Precambrian Paleosols. *J. Geol.* **1992**, *100*, 279–289. [CrossRef]

79. Sheldon, N.D.; Tabor, N.J. Quantitative paleoclimatic and paleoenvironmental reconstruction using paleosols. *Earth Sci. Rev.* **2009**, *95*, 1–52. [CrossRef]

80. Gallagher, T.M.; Sheldon, N.D. A new paleothermometer for forest paleosols and its implications for Cenozoic climate. *Geology* **2013**, *41*, 647–650. [CrossRef]

81. Stinchcomb, G.E.; Nordt, L.C.; Driese, S.G.; Lukens, W.E.; Williamson, F.C.; Tubbs, J.D. A data-driven spline model designed to predict paleoclimate using paleosol geochemistry. *Am. J. Sci.* **2016**, *316*, 746–777. [CrossRef]

82. Gérard, F. Clay minerals, iron/aluminum oxides, and their contribution to phosphate sorption in soils—A myth revisited. *Geoderma* **2016**, *262*, 213–226. [CrossRef]

83. Soil Survey Staff. *Keys to Soil Taxonomy*, 12th ed.; USDA-NRCS: Washington, DC, USA, 2014.

84. Maher, B.A. Magnetic properties of modern soils and Quaternary loessic paleosols: Paleoclimatic implications. *Palaeogeogr. Palaeoclim. Palaeoecol.* **1998**, *137*, 25–54. [CrossRef]

85. Maher, B.A.; Possolo, A. Statistical models for use of palaeosol magnetic properties as proxies of palaeorainfall. *Glob. Planet. Chang.* **2013**, *111*, 280–287. [CrossRef]

86. Hyland, E.G.; Badgley, C.; Abrajevitch, A.; Sheldon, N.D.; Van Der Voo, R. A new paleoprecipitation proxy based on soil magnetic properties: Implications for expanding paleoclimate reconstructions. *GSA Bull.* **2015**, *127*, B31207.1. [CrossRef]

87. Maxbauer, D.P.; Feinberg, J.; Fox, D.L. Magnetic mineral assemblages in soils and paleosols as the basis for paleoprecipitation proxies: A review of magnetic methods and challenges. *Earth Sci. Rev.* **2016**, *155*, 28–48. [CrossRef]

88. Prem, M.; Hansen, H.C.B.; Wenzel, W.W.; Heiberg, L.; Sørensen, H.; Borggaard, O.K. High Spatial and Fast Changes of Iron Redox State and Phosphorus Solubility in a Seasonally Flooded Temperate Wetland Soil. *Wetlands* **2014**, *35*, 237–246. [CrossRef]

89. Valaee, M.; Ayoubi, S.; Khormali, F.; Lu, S.G.; Karimzadeh, H.R. Using magnetic susceptibility to discriminate between soil moisture regimes in selected loess and loess-like soils in northern Iran. *J. Appl. Geophys.* **2016**, *127*, 23–30. [CrossRef]

90. Cervi, E.C.; Maher, B.; Poliseli, P.C.; Junior, I.G.D.S.; Da Costa, A.C.S. Magnetic susceptibility as a pedogenic proxy for grouping of geochemical transects in landscapes. *J. Appl. Geophys.* **2019**, *169*, 109–117. [CrossRef]

91. Chaparro, M.A.E.; Moralejo, M.d.P.; Bohnel, H.N.; Acebal, S.G. Iron oxide mineralogy in Mollisols, Aridisols and Entisols from the southwestern Pampean region (Argentina) by environmental magnetism approach. *Catena* **2020**, *190*, 104534. [CrossRef]

92. Hu, P.; Heslop, D.; Rossel, R.A.V.; Roberts, A.P.; Zhao, X. Continental-scale magnetic properties of surficial Australian soils. *Earth Sci. Rev.* **2020**, *203*, 103028. [CrossRef]

93. Borggaard, O.K. The influence of iron oxides on phosphate adsorption by soil. *Eur. J. Soil Sci.* **1983**, *34*, 333–341. [CrossRef]

94. Stone, M.; Mudroch, A. The effect of particle size, chemistry and mineralogy of river sediments on phosphate adsorption. *Environ. Technol. Lett.* **1989**, *10*, 501–510. [CrossRef]

95. Halsted, M.; Lynch, J. Phosphorus responses of C3and C4species. *J. Exp. Bot.* **1996**, *47*, 497–505. [CrossRef]

96. Griffith, D.M.; Cotton, J.M.; Powell, R.L.; Sheldon, N.D.; Still, C.J. Multi-century stasis in C3 and C4 grass distributions across the contiguous United States since the industrial revolution. *J. Biogeogr.* **2017**, *44*, 2564–2574. [CrossRef]

97. Katra, I.; Gross, A.; Swet, N.; Tanner, S.; Krasnov, H.; Angert, A. Substantial dust loss of bioavailable P from agricultural soils. *Sci. Rep.* **2016**, *6*, 24736. [CrossRef] [PubMed]

98. Ewel, B.; Mazzarino, M.J.; Berish, C.W. Tropical Soil Fertility Changes under Monocultures and Successional Communities of Different Structure. *Ecol. Appl.* **1991**, *1*, 289–302. [CrossRef] [PubMed]

99. Gregorich, E.G.; Anderson, D. Effects of cultivation and erosion on soils of four toposequences in the Canadian prairies. *Geoderma* **1985**, *36*, 343–354. [CrossRef]

100. Cotton, J.M.; Jeffery, M.L.; Sheldon, N.D. Climate controls on soil respired CO_2 in the United States: Implications for 21st century chemical weathering rates in temperate and arid ecosystems. *Chem. Geol.* **2013**, *358*, 37–45. [CrossRef]

101. Leakey, A.D.B.; Ainsworth, E.A.; Bernacchi, C.J.; Rogers, A.; Long, S.P.; Ort, D.R. Elevated CO_2 effects on plant carbon, nitrogen, and water relations: Six important lessons from FACE. *J. Exp. Bot.* **2009**, *60*, 2859–2876. [CrossRef]

102. Terrer, C.; Vicca, S.; Hungate, B.A.; Phillips, R.P.; Prentice, I.C. Mycorrhizal association as a primary control of the CO_2 fertilization effect. *Science* **2016**, *353*, 72–74. [CrossRef]

103. Obermeier, W.A.; Lehnert, L.W.; Kammann, C.I.; Müller, C.; Grünhage, L.; Luterbacher, J.; Erbs, M.; Moser, G.; Seibert, R.; Yuan, N.; et al. Reduced CO_2 fertilization effect in temperate C3 grasslands under more extreme weather conditions. *Nat. Clim. Chang.* **2016**, *7*, 137–141. [CrossRef]

104. Mearns, L.O.; Gutowski, W.; Jones, R.; Leung, R.; McGinnis, S.; Nunes, A.; Qian, Y. A Regional Climate Change Assessment Program for North America. *Eos* **2009**, *90*, 311. [CrossRef]

105. Bluth, G.J.S.; Kump, L.R. Phanerozoic Paleogeology. *Am. J. Sci.* **1991**, *291*, 284–308. [CrossRef]

106. Crews, T.E.; Kitayama, K.; Fownes, J.H.; Riley, R.H.; Herbert, D.A.; Mueller-Dombois, D.; Vitousek, P.M. Changes in Soil Phosphorus Fractions and Ecosystem Dynamics across a Long Chronosequence in Hawaii. *Ecology* **1995**, *76*, 1407–1424. [CrossRef]

107. Schachtman, D.P.; Reid, R.J.; Ayling, S.M. Phosphorus Uptake by Plants: From Soil to Cell. *Plant Physiol.* **1998**, *116*, 447–453. [CrossRef] [PubMed]

108. Lynch, J.P.; Brown, K.M. Root strategies for phosphorus acquisition. In *The Ecophysiology of Plant-Phosphorus Interactions*; White, P.J., Hammond, J.P., Eds.; Springer: Dordecht, Germany, 2008.

109. Smith, S.E.; Jakobsen, I.; Grønlund, M.; Smith, F.A. Roles of Arbuscular Mycorrhizas in Plant Phosphorus Nutrition: Interactions between Pathways of Phosphorus Uptake in Arbuscular Mycorrhizal Roots Have Important Implications for Understanding and Manipulating Plant Phosphorus Acquisition. *Plant Physiol.* **2011**, *156*, 1050–1057. [CrossRef] [PubMed]

110. Fitter, A.H.; Heinemeyer, A.; Husband, R.; Olsen, E.; Ridgway, K.; Staddon, P.L. Global environmental change and the biology of arbuscular mycorrhizas: Gaps and challenges. *Can. J. Bot.* **2004**, *82*, 1133–1139. [CrossRef]

111. Deepika, S.; Kothamasi, D. Soil moisture—A regulator of arbuscular mycorrhizal fungal community assembly and symbiotic phosphorus uptake. *Mycorrhiza* **2015**, *25*, 67–75. [CrossRef]

112. Pearson, J.N.; Jakobsen, I. The relative contribution of hyphae and roots to phosphorus uptake by arbuscular mycorrhizal plants, measured by dual labelling with 32P and 33P. *New Phytol.* **1993**, *124*, 489–494. [CrossRef]

113. Gavito, M.E.; Schweiger, P.; Jakobsen, I. P uptake by arbuscular mycorrhizal hyphae: Effect of soil temperature and atmospheric CO_2 enrichment. *Glob. Chang. Biol.* **2003**, *9*, 106–116. [CrossRef]

114. Treseder, K.K. A meta-analysis of mycorrhizal responses to nitrogen, phosphorus, and atmospheric CO_2 in field studies. *New Phytol.* **2004**, *164*, 347–355. [CrossRef]

115. Kimball, B.A. Carbon Dioxide and Agricultural Yield: An Assemblage and Analysis of 430 Prior Observations1. *Agron. J.* **1907**, *75*, 779–788. [CrossRef]

116. Solomon, A.M.; Cramer, W. Biospheric implications of global environmental change. In *Vegetation Dynamics and Global Change*; Solomon, A.M., Shugart, H.H., Eds.; Chapman and Hall: New York, NY, USA, 1993; pp. 25–52.

117. Caseldine, C. Book Review: Encyclopedia of global environmental change. Volume two—The Earth system: Biological and ecological dimensions of global environmental change. *Holocene* **2003**, *13*, 147. [CrossRef]

118. Girardin, M.P.; Bouriaud, O.; Hogg, E.H.; Kurz, W.; Zimmermann, N.E.; Metsaranta, J.M.; De Jong, R.; Frank, D.C.; Esper, J.; Büntgen, U.; et al. No growth stimulation of Canada's boreal forest under half-century of combined warming and CO_2 fertilization. *Proc. Natl. Acad. Sci. USA* **2016**, *113*, E8406–E8414. [CrossRef] [PubMed]

119. Stein, R.A.; Sheldon, N.D.; Smith, S.Y. Rapid response to anthropogenic climate change by Thuja occidentalis: Implications for past climate reconstructions and future climate predictions. *PeerJ* **2019**, *7*, e7378. [CrossRef] [PubMed]

120. Sheldon, N.D.; Smith, S.Y.; Stein, R.; Ng, M. Carbon isotope ecology of gymnosperms and implications for paleoclimatic and paleoecological studies. *Glob. Planet. Chang.* **2020**, *184*, 103060. [CrossRef]

121. Norby, R.; Warren, J.; Iversen, C.; Garten, C.; Medlyn, B.; McMurtrie, R. CO_2 Enhancement of Forest Productivity Constrained by Limited Nitrogen Availability. *Nat. Précéd.* **2009**, *107*, 19368–19373. [CrossRef]

122. Ellsworth, D.S.; Anderson, I.C.; Crous, K.Y.; Cooke, J.; Drake, J.E.; Gherlenda, A.N.; Gimeno, T.E.; Macdonald, C.A.; Medlyn, B.E.; Powell, J.R.; et al. Elevated CO_2 does not increase eucalypt forest productivity on a low-phosphorus soil. *Nat. Clim. Chang.* **2017**, *7*, 279–282. [CrossRef]

123. Terrer, C.; Jackson, R.B.; Prentice, I.C.; Keenan, T.; Kaiser, C.; Vicca, S.; Fisher, J.B.; Reich, P.B.; Stocker, B.D.; Hungate, B.; et al. Nitrogen and phosphorus constrain the CO_2 fertilization of global plant biomass. *Nat. Clim. Chang.* **2019**, *9*, 684–689. [CrossRef]

124. Cleveland, C.C.; Liptzin, D. C:N:P stoichiometry in soils: Is there a "Redfield ratio" for the microbial biomass? *Biogeochemistry* **2007**, *85*, 235–252. [CrossRef]

125. Maranger, R.; Jones, S.E.; Cotner, J.B. Stoichiometry of carbon, nitrogen, and phosphorus through the freshwater pipe. *Limnol. Oceanogr. Lett.* **2018**, *3*, 89–101. [CrossRef]

126. Berner, R.A. Weathering, plants, and the long-term carbon cycle. *Geochim. Cosmochim. Acta* **1992**, *56*, 3225–3231. [CrossRef]

127. Samreen, S.; Kausar, S. Phosphorus fertilizer: The original and commercial sources. In *Phosphorus–Recovery and Recycling*; Zhang, T., Ed.; IntechOpen: London, UK, 2019. [CrossRef]

128. Siebers, N.; Godlinski, F.; Leinweber, P. The phosphorus fertilizer value of bone char for potatoes, wheat, and onions: First results. *Agric. For. Res.* **2012**, *62*, 59–64.

129. Zwetsloot, M.J.; Lehmann, J.; Bauerle, T.; Vanek, S.; Hestrin, R.; Nigussie, A. Phosphorus availability from bone char in a P-fixing soil influenced by root-mycorrhizae-biochar interactions. *Plant Soil* **2016**, *408*, 95–105. [CrossRef]

130. Cordell, D.; Drangert, J.-O.; White, S. The story of phosphorus: Global food security and food for thought. *Glob. Environ. Chang.* **2009**, *19*, 292–305. [CrossRef]

131. Dawson, C.; Hilton, J. Fertiliser availability in a resource-limited world: Production and recycling of nitrogen and phosphorus. *Food Policy* **2011**, *36*, S14–S22. [CrossRef]

132. Cordell, D.; White, S. Sustainable Phosphorus Measures: Strategies and Technologies for Achieving Phosphorus Security. *Agronomy* **2013**, *3*, 86–116. [CrossRef]

133. Retallack, G.J. *Soils of the Past: An Introduction to Paleopedology*, 2nd ed.; Blackwell Science Ltd.: Oxford, UK, 2001.

134. Rye, R.; Holland, H.D. Paleosols and the evolution of atmospheric oxygen; a critical review. *Am. J. Sci.* **1998**, *298*, 621–672. [CrossRef]

135. Cerling, T.E. Carbon dioxide in the atmosphere; evidence from Cenozoic and Mesozoic Paleosols. *Am. J. Sci.* **1991**, *291*, 377–400. [CrossRef]

136. Sheldon, N.D. Using paleosols of the Picture Gorge Basalt to reconstruct the Middle Miocene Climatic Optimum. *PaleoBios* **2006**, *26*, 27–36.

137. Driese, S.G.; Jirsa, M.A.; Ren, M.; Brantley, S.L.; Sheldon, N.D.; Parker, D.F.; Schmitz, M.D. Neoarchean paleoweathering of tonalite and metabasalt: Implications for reconstructions of 2.69Ga early terrestrial ecosystems and paleoatmospheric chemistry. *Precambrian Res.* **2011**, *189*, 1–17. [CrossRef]

138. Murakami, T.; Sreenivas, B.; Das Sharma, S.; Sugimori, H. Quantification of atmospheric oxygen levels during the Paleoproterozoic using paleosol compositions and iron oxidation kinetics. *Geochim. Cosmochim. Acta* **2011**, *75*, 3982–4004. [CrossRef]

139. Driese, S.G. Pedogenic Translocation of Fe in Modern and Ancient Vertisols and Implications for Interpretations of the Hekpoort Paleosol (2.25 Ga). *J. Geol.* **2004**, *112*, 543–560. [CrossRef]

140. Retallack, G.J.; Noffke, N.; Chafetz, H. Criteria for Distinguishing Microbial Mats and Earths. *Soc. Econ. Paleont. Mineral. Spec. Pap.* **2012**, *101*, 136–152.

141. Hyland, E.G.; Sheldon, N.D. Examining the spatial consistency of palaesol proxies: Implications for palaeoclimatic and palaeoenvironmental reconstructions in terrestrial sedimentary basins. *Sedimentology* **2016**, *63*, 959–971. [CrossRef]

142. Beukes, N.J.; Dorland, H.; Gutzmer, J.; Nedachi, M.; Ohmoto, H. Tropical laterites, life on land, and the history of atmospheric oxygen in the Paleoproterozoic. *Geology* **2002**, *30*, 491. [CrossRef]

143. Beraldi-Campesi, H.; Hartnett, H.E.; Anbar, A.; Gordon, G.W.; Garcia-Pichel, F. Effect of biological soil crusts on soil elemental concentrations: Implications for biogeochemistry and as traceable biosignatures of ancient life on land. *Geobiology* **2009**, *7*, 348–359. [CrossRef]

144. Donnadieu, Y.; Goddéris, Y.; Ramstein, G.; Nédélec, A.; Meert, J. A 'snowball Earth' climate triggered by continental break-up through changes in runoff. *Nature* **2004**, *541*, 303–306. [CrossRef]

145. Mills, B.; Watson, A.J.; Goldblatt, C.; Boyle, R.; Lenton, T.M. Timing of Neoproterozoic glaciations linked to transport-limited global weathering. *Nat. Geosci.* **2011**, *4*, 861–864. [CrossRef]

146. Dzombak, R.M.; Sheldon, N.D. Three billion years of continental weathering and implications for marine biogeochemistry. In Proceedings of the Goldschmidt Annual Geochemical Conference, Honolulu, HI, USA, 21–26 June 2020. Abstract Number 635.

147. Canfield, D. A new model for Proterozoic ocean chemistry. *Nat. Cell Biol.* **1998**, *396*, 450–453. [CrossRef]

148. Habicht, K.S.; Gade, M.; Thamdrup, B.; Berg, P.; Canfield, D.E. Calibration of Sulfate Levels in the Archean Ocean. *Science* **2002**, *298*, 2372–2374. [CrossRef]

149. Halevy, I.; Peters, S.E.; Fischer, W.W. Sulfate Burial Constraints on the Phanerozoic Sulfur Cycle. *Science* **2012**, *337*, 331–334. [CrossRef]

150. Fakhraee, M.; Hancisse, O.; Canfield, D.E.; Crowe, S.A.; Katsev, S. Proterozoic seawater sulfate scarcity and the evolution of ocean–atmosphere chemistry. *Nat. Geosci.* **2019**, *12*, 375–380. [CrossRef]

151. Cameron, E.M. Sulphate and sulphate reduction in early Precambrian oceans. *Nat. Cell Biol.* **1982**, *296*, 145–148. [CrossRef]

152. Hoffman, P.F.; Kaufman, A.J.; Halverson, G.P.; Schrag, D.P. A Neoproterozoic Snowball Earth. *Science* **1998**, *281*, 1342–1346. [CrossRef] [PubMed]

153. Prasad, N.; Roscoe, S. Evidence of anoxic to oxic atmospheric change during 2.45-2.22 Ga from lower and upper sub-Huronian paleosols, Canada. *Catena* **1996**, *27*, 105–121. [CrossRef]

154. Algeo, T.J.; Scheckler, S.E. Terrestrial-marine teleconnections in the Devonian: Links between the evolution of land plants, weathering processes, and marine anoxic events. *Philos. Trans. R. Soc. B Biol. Sci.* **1998**, *353*, 113–130. [CrossRef]

Publisher's Note: MDPI stays neutral with regard to jurisdictional claims in published maps and institutional affiliations.

 *soil systems*

Review

Accessing Legacy Phosphorus in Soils

Sarah Doydora [1], Luciano Gatiboni [1], Khara Grieger [2], Dean Hesterberg [1], Jacob L. Jones [3], Eric S. McLamore [4], Rachel Peters [5], Rosangela Sozzani [5], Lisa Van den Broeck [5] and Owen W. Duckworth [1,*]

[1] Department of Crop and Soil Sciences, North Carolina State University, Raleigh, NC 27695, USA; sadoydor@ncsu.edu (S.D.); luke_gatiboni@ncsu.edu (L.G.); sscdlh@ncsu.edu (D.H.)

[2] Department of Applied Ecology, North Carolina State University, Raleigh, NC 27695, USA; kdgriege@ncsu.edu

[3] Department of Materials Science and Engineering, North Carolina State University, Raleigh, NC 27695, USA; JacobJones@ncsu.edu

[4] Department of Agricultural Sciences, Clemson University, Clemson, SC 29634, USA; emclamo@clemson.edu

[5] Department of Plant and Microbial Biology, North Carolina State University, Raleigh, NC 27695, USA; rsozzan@ncsu.edu (R.S.); rmpeter4@ncsu.edu (R.P.); lfvanden@ncsu.edu (L.V.d.B.)

* Correspondence: owduckwo@ncsu.edu

Received: 30 October 2020; Accepted: 17 December 2020; Published: 18 December 2020

Abstract: Repeated applications of phosphorus (P) fertilizers result in the buildup of P in soil (commonly known as legacy P), a large fraction of which is not immediately available for plant use. Long-term applications and accumulations of soil P is an inefficient use of dwindling P supplies and can result in nutrient runoff, often leading to eutrophication of water bodies. Although soil legacy P is problematic in some regards, it conversely may serve as a source of P for crop use and could potentially decrease dependence on external P fertilizer inputs. This paper reviews the (1) current knowledge on the occurrence and bioaccessibility of different chemical forms of P in soil, (2) legacy P transformations with mineral and organic fertilizer applications in relation to their potential bioaccessibility, and (3) approaches and associated challenges for accessing native soil P that could be used to harness soil legacy P for crop production. We highlight how the occurrence and potential bioaccessibility of different forms of soil inorganic and organic P vary depending on soil properties, such as soil pH and organic matter content. We also found that accumulation of inorganic legacy P forms changes more than organic P species with fertilizer applications and cessations. We also discuss progress and challenges with current approaches for accessing native soil P that could be used for accessing legacy P, including natural and genetically modified plant-based strategies, the use of P-solubilizing microorganisms, and immobilized organic P-hydrolyzing enzymes. It is foreseeable that accessing legacy P will require multidisciplinary approaches to address these limitations.

Keywords: legacy phosphorus; speciation; transformation; accessibility

1. Introduction

Phosphorus (P) is essential to life on Earth. It plays critical roles in core biological systems associated with energy storage, cell replication, and protein synthesis [1]. Among essential macronutrients, it is often the concentration of bioaccessible P, or dissolved inorganic P, in soil that limits plant growth [2]. This fact makes the utilization of P, a finite resource that is mined from specific locations worldwide [3], a critical component of the global food system that aims to feed a growing population [4]. When inorganic fertilizers are added to soil to ameliorate deficiencies, P undergoes sorption, precipitation, and organic matter complexation reactions that render it unavailable for plant uptake [5]. Consequently, large quantities of fertilizers (globally, 16.5 million metric tons P y^{-1} [4]) are added to maintain soil solution P levels that are optimal for plant growth [6]. With only 10–36% of added P taken up by most crops in the year of application [7,8], long-term P fertilization has led to the buildup of residual P in soil (at a rate of ≈10 million metric tons P y^{-1} globally [9]) that is not immediately accessible to plants, and is commonly known as legacy P. Legacy P can be categorized as inorganic and organic legacy P, referring to excess, unassimilated inorganic and organic P, respectively, from added inorganic and/or organic fertilizers in the year of application.

Long-term accumulation of soil P is undesirable from agricultural, economic, and environmental perspectives. Apart from simply being an inefficient use of a finite resource, the buildup of legacy P, such as in European and U.S. soils under long-term P fertilizer applications [10], also presents environmental challenges. Excessive legacy soil P can result in its loss from soil by leaching and erosion to surrounding water bodies. For instance, it is estimated that freshwater ecosystems have had a 75% increase in total P as compared to their pre-industrial revolution state [9]; estuaries and other waterbodies may also be impacted [11]. These increased loadings of P can lead to elevated concentrations and widespread eutrophication of water bodies [12]. For example, 40% of U.S. lakes contain excess P, and ≈80% of states reported the annual occurrence of harmful algal blooms in fresh waterbodies [13,14]. At the same time, legacy P has drawn interest as a potential resource that may be harnessed to reduce use of P fertilizers. Worldwide, soil P is estimated to be five times greater than minable P [15], represented graphically in Figure 1, making the enhancement of legacy P for plant use potentially transformative to the global food system.

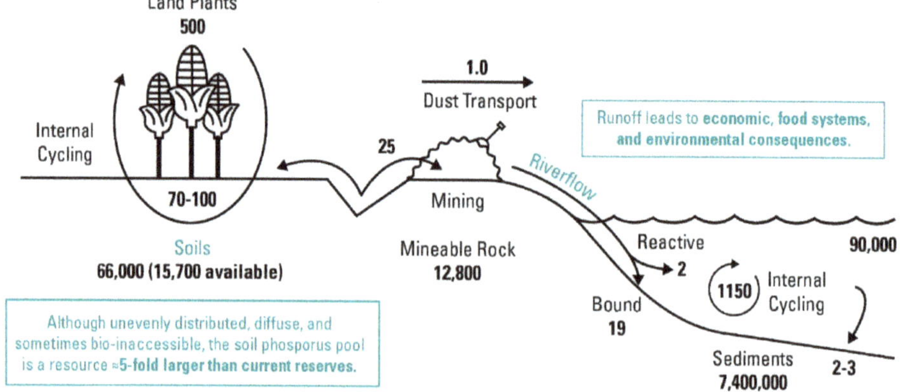

Figure 1. Global phosphorus flows with flux shown in units of 10^{12} g P/year. Recreated from [15].

Currently, there are significant technological and societal challenges associated with accessing legacy soil P as a resource for agriculture. First, legacy P naturally becomes available for plant use. For example, highly P-enriched soils in North Carolina, USA, are estimated to support 50–250 years of crop growth without P application [16]. However, actual crop recoveries of this residual P are highly variable across soils, with 4 to >100 years after last fertilizer application needed to recover up to 80% of added P [7,17,18]. Secondly, controlling and increasing soil P solubilization of inorganic legacy P or mineralization of organic legacy P for plant use requires a detailed understanding of the occurrence and reactivity of different soil chemical P forms. Most soil P is bound in bioinaccessible forms with iron, aluminum, or calcium minerals, and organic matter [10,19]. Therefore, with more than 20,000 different soil types mapped across the U.S. alone [20], determining the relative importance of these soil P chemical forms in relation to edaphic factors is a challenge for developing effective management strategies. Third, understanding the dynamic processes associated with P transformation from mineral and animal waste fertilizers into these soil P chemical forms is necessary to maximize the contributions of legacy P as a complement or substitute for externally added P fertilizers. Finally, technologies and strategies for potentially enhancing the capacity to dissolve inorganic legacy P or hydrolyze organic legacy P, herein referred to as bioaccessibility, that are feasible on a large scale are not yet available.

In light of these challenges, this paper aims to review (1) the chemical forms of P in soil and their variation with chemically relevant soil properties (Section 2), (2) the transformation of P into different proportions of various chemical species after mineral and organic P fertilizer applications (Section 3), and (3) different plant- and microbial-based approaches for accessing native soil P and potentially legacy P and associated challenges (Section 4). We conclude with an overall outlook on legacy P as a currently untapped resource that could potentially decrease the dependence on external fertilizer applications in the future as research and technology advances, particularly with the generation of plant cultivars with more efficient P mobilization and utilization processes/capabilities, and with the use of P-solubilizing microorganisms (Section 5).

2. Occurrence and Bioaccessibility of Different Chemical Forms of Soil Legacy P

Inorganic P typically accounts for the bulk of total P in mineral soils [21]. Phosphorus is predominantly taken up by plant roots via dissolved inorganic forms, *viz.* soluble orthophosphate and its protonated forms. However, concentrations of soluble or bioaccessible phosphate in soils are typically too low for optimal plant growth (ranging from 0.1 to 1 μmol P L^{-1} soil solution [22]). This low dissolved concentration stems from the strong association of inorganic P with a myriad of minerals. Kizewski et al. [19] compiled a list of the most commonly identified inorganic P species in soils by spectroscopic techniques, which include the calcium phosphate minerals hydroxyapatite and octacalcium phosphate, and phosphate adsorbed on Fe- or Al-oxide minerals.

Unlike inorganic P wherein solubilization equals bioaccessibility, organic P becomes bioaccessible only after mineralization, a biological degradation process catalyzed by specific phosphatase enzymes. Orthophosphate monoesters and diesters are the most common classes of organic P, with monoesters typically comprising 50–70% of organic soil P [23,24]. Among monoester compounds, either the inositol hexakisphosates (IHPs) or humic P constitute the predominant portion of this class [25–29]. These species have differing behaviors in soils, and thus understanding organic speciation of soil P is also critical.

Building off of this knowledge, the following section focuses on the relationship between different P forms and bioaccessibility as influenced by relevant soil properties—a topic that is critical to understanding legacy P accumulation and the potential for legacy P utilization.

2.1. Inorganic P Forms and Bioaccessibility

Among edaphic factors, soil pH largely governs the speciation and bioaccessibility of inorganic P species. In general, acidic pH (pH < 7) favors P association with Al and Fe whereas alkaline pH (pH > 7) favors association with Ca [5]. For instance, in slightly acidic agricultural soils (pH 5.5–6.0), bulk sample P-XANES (X-ray absorption near edge structure) spectroscopy analysis showed 46–56% of total P adsorbed to Fe, 31–42% as Al phosphate mineral, 8–15% as apatite, and 0–12% as organic P [30]. In alkaline, calcareous soils (pH 7.6–7.9), 54–74% of total P existed as hydroxyapatite and/or dicalcium phosphate dihydrate, 25–35% adsorbed to Fe mineral (goethite), and 0–19% organic P [31].

Under acidic soil conditions, Fe- or Al-adsorbed and precipitated P forms control soil solution P (Figure 2a–c). For adsorbed species, composition and phase impact their bioaccessibility. For example, P adsorbed to goethite (α-FeOOH) desorbs similarly to gibbsite (Al(OH)$_3$) and alumina (Al$_2$O$_3$) (within 10% difference) at pH $\approx$6 under similar experimental conditions [32]. In fact, plant P uptake by ryegrass has been demonstrated with goethite- and poorly crystalline Al(OH)$_3$-adsorbed P at near sorption capacity as lone P sources, demonstrating the potential bioaccessibility of adsorbed P species [33]. However, specific soil mineralogies can strongly impact P extractability and potentially bioaccessibility. Phosphorus adsorbed to amorphous Fe (hydr)oxide phases (i.e., ferrihydrite and Fe(OH)$_3$) is 400–500 times less desorbable than P adsorbed to goethite and amorphous Al(OH)$_3$, with negligible desorption when Fe to Al ratios are increased in amorphous Fe/Al(OH)$_3$ mixtures [34]. It is worth noting that current analytical capabilities, including spectroscopic techniques, such as P K-edge XANES, are limited in their ability to distinguish between P species lacking in distinct spectral features [35], emphasizing that identifying the key adsorbent species in soils is not always straightforward. Thus, functionally defined extractions have been designed to estimate the concentration of poorly crystalline Fe and Al minerals and the P associated with them [36]. Derived from these measurements, the degree of P saturation, defined as the ratio of oxalate-extractable P to the sum of oxalate extractable Fe and Al [37], is an estimate of how much of the P binding capacity of these minerals is occupied, and may serve as a useful tool in assessing the relatively bioaccessible P stocks of legacy P, particularly for acidic soils [38].

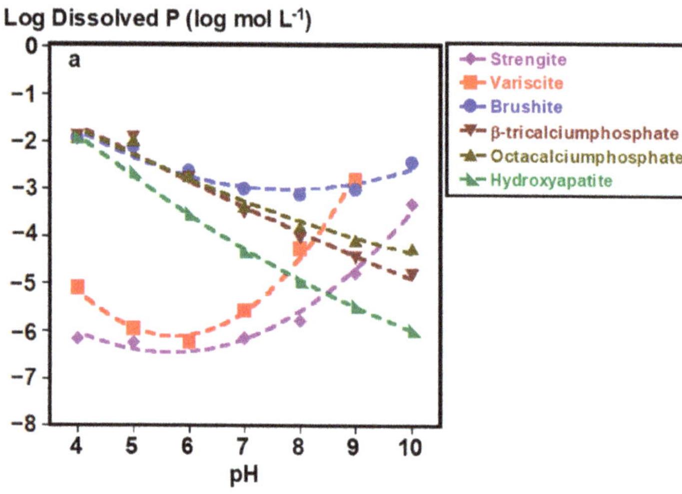

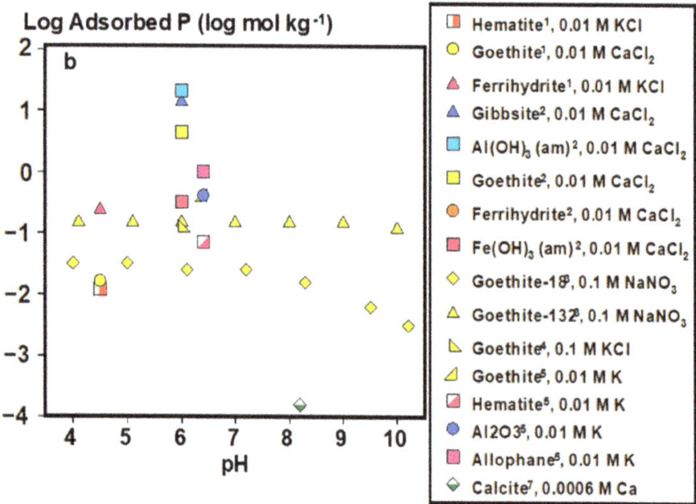

Figure 2. *Cont.*

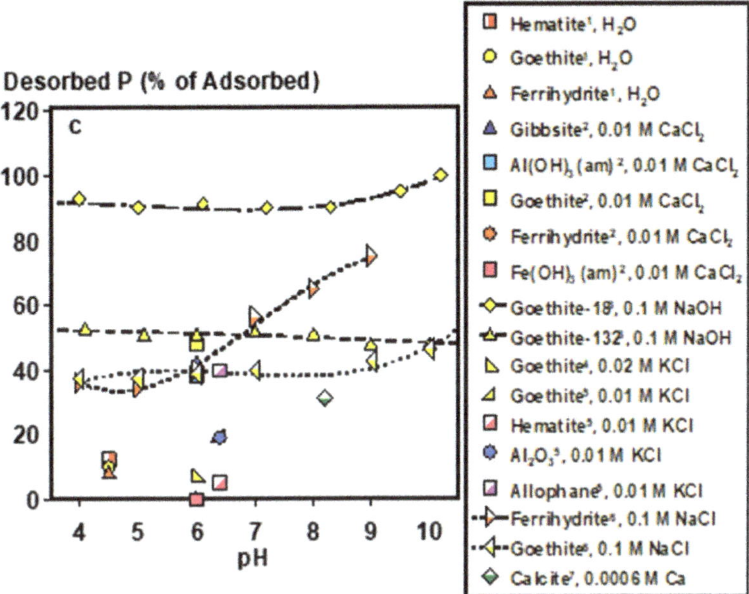

Figure 2. Soil solution pH controls: (a) dissolved phosphate concentrations in equilibrium with major phosphate minerals, (b) adsorbed P concentrations on to various soil minerals, and (c) desorbed P from different soil minerals as fractions of adsorbed concentrations. *Plotted data points were calculated by Visual MINTEQ Ver. 3.1 at 0.001 KCl and excess concentrations of each mineral (2 g L^{-1}). Adsorption and desorption data were computed from the following works: [1] [39], [2] [34], [3] [40], [4] [41], [5] [32], [6] [42], [7] [43]. Lines are meant to guide the eye and do not represent model fits.

In alkaline soils, bioaccessibility of inorganic precipitated P is controlled by the following Ca phosphate minerals of decreasing solubility: brushite ($CaHPO_4 \bullet 2H_2O$) > β-tricalcium phosphate ($Ca_3(PO_4)_2$) > octacalcium phosphate ($Ca_8H_2(PO_4)_6 \bullet 5H_2O$) > hydroxyapatite ($Ca_5(PO4)_3(OH)$) (Figure 2a). In addition to brushite, calcite-adsorbed P can be another source of labile P, particularly for calcareous alkaline soils (Figure 2). Although P adsorption to Fe (hydr)oxide minerals still occurs under alkaline pH [40], fractions of desorbed P could be similar or greater from these Fe adsorbents than from calcite at pH 8.2 (Figure 2b,c). This suggests that Fe adsorbents could still be controlling more of solution P even in alkaline conditions. However, there is a lack of data on the plant uptake of P from Fe or Al adsorbents under alkaline conditions.

In soils, P is often categorized into labile or non-labile pools, but this concept is rather arbitrary with unclear molecular boundaries [44]. For instance, the use of Hedley sequential extraction method could lead to chemical P redistributions in the course of the extraction process, overestimating Ca-bound P species (i.e., hydroxyapatite), particularly for soils high in exchangeable Ca [45]. Similarly, agronomic soil P tests using various types of chemical solutions (e.g., Bray 1 [46], Mehlich-3 [47], Olsen [48], or Morgan [49] extractants), measure labile P, but the extractable P does not equate to P molecules that plants access in soil. Hence, these tests are agronomically useful only when correlated and calibrated with actual crop response [50]. The use of anion exchange resin membranes may offer a more direct alternative index for bioaccessible fractions of soil legacy P by simulating root surfaces in removing dissolved P at native soil pH, thus providing a better approach for predicting bioavailable P [51–53]. However, use of resins can be laborious, time-consuming, and likely incompatible with commercial soil test laboratories where the farmers require quick soil diagnostic tests. Moreover, when well calibrated, soil test extraction methods (e.g., Mehlich-3, Bray, Olsen, and other tests [54]) can determine the fertilizer needs for crops. Consequently, despite being a better method to measure soil bioavailable P, the current use of the anion exchange method is largely limited to research purposes and has low adoption by commercial soil test laboratories. It is worth noting that neither the soil test nor other commonly used labile P extractions, anion resin P measurements, or any field sensing tools can distinguish the individual P species contributing to dissolved inorganic P in soil solution.

2.2. Organic P Forms and Bioaccessibility

Soil pH has been shown to influence occurrence of total organic P and specific organic P species differently. At higher pH in tropical rainforest soils, total organic P concentrations were found to be greater (pH 3.3–7; $R^2 = 0.89$), but this relationship appears to be reversed in temperate agricultural grasslands (pH 5.3–6.7; $R^2 = -0.60$) [55,56]. A more comprehensive regional study by Hou et al. [57] showed generally uniform concentrations of total organic P with pH in tropical, subtropical, and temperate forest soils. This agrees with total organic P concentrations remaining constant across soil pH gradients (pH 3.7–7.8) along the arable Hoosfield strip at Rothamsted Research Station, UK [58]. In terms of specific organic P species, Turner and Blackwell [58] found that non-IHP phosphomonoesters constituted the bulk of soil organic P (51–67%) and did not vary considerably across the Hoosfield pH gradient. However, concentrations of IHPs (myo- and scyllo-isomers), DNA, and phosphonates increased for pH ≤ 5 from relatively uniform concentrations (for IHP) or undetected presence (for DNA and phosphonates) at higher pH (pH 5–8) [58]. The degradation of IHP may be limited by lower enzymatic hydrolysis (breakdown) by phytase enzyme at higher pH, as its activity is mostly optimum between pH 2 and 6.5 with fewer characterized phytases showing optimum activities between pH 7 and 8 [59–63]. In contrast, phosphodiesterase activities decrease at lower soil pH [64]. Additionally, greater accumulation of IHPs and DNAs at highly acidic pH may arise from their stronger associations with clay surfaces at low pH [65,66], with presumably decreased microbial activity under these conditions.

Soil organic P is positively related with soil organic C in natural ecosystems and cultivated grassland soils [57,67–69]. No relationships have been drawn between IHPs and organic C, but mixed relationships have been reported between DNA and organic C in non-agricultural and cropland soils [67,70]. In arable soils, humic P has been highly correlated with soil organic C [70]. Positive associations with organic C, however, do not necessitate equal rates of turnover for organic P. In a meta-analysis of 80 topsoils in different countries, Spohn [71] found no changes in organic P concentrations in bulk soils or clay-size fractions in comparison to the significant reductions in organic C concentrations due to land-use change (i.e., native woodlands to croplands). Although different organic P species were not distinguished in the meta-analysis, this analysis suggests that organic P forms are likely more persistent than organic C, or their degradation is limited by factors independent of those of bulk organic C.

Among commonly identifiable organic P compounds, simple non-IHP phosphomonesters enzymatically degrade the fastest, followed by phosphodiester, with IHP being most resistant [72]. The relatively rapid degradation of non-IHP monoesters could either be due to greater enzymatic efficiencies of non-IHP phosphomonesterases [73,74] or to the longer persistence of these enzymes in soils relative to phosphodiesterases or phytases [75,76]. Although sustained activities have been reported for adsorbed non-IHP phosphomonoesterase [77] and adsorbed phytase [78], it is not clear how much of these immobilized enzyme activities constitute in their activities in field soils. Moreover, although many laboratory experiments imply limited IHP mineralization, when added in the field, IHP has been reported to degrade rapidly, with 12–18% remaining in calcareous soils after 13 weeks [79,80]. However, it has also been recognized that freshly added IHP may be different from residual IHP that has aged in field soils under much longer periods [78]. Dissolved organic P (<0.2 μm, molybdate-unreactive) in soil water extracts hydrolyzed between 0 and 61% of molybdate-unreactive P in the presence of added enzymes with highly variable fractions of hydrolyzed P class or species [81,82]. Similarly, organic P in soil leachate from soils receiving inorganic fertilizer and dairy effluent hydrolyzed 36–54% of malachite-green unreactive P (another measure for organic P) using enzyme additions [83,84]. However, Toor et al. [84] reported 10–21% of unreactive P inherently hydrolyzed in the leachate in the absence of phosphatase enzymes. These studies suggest that organic P could mineralize during transport in water, but their field hydrolysis may be less than laboratory measurements considering the optimized conditions used in added phosphatase assays [85].

As for humic P, information is currently limited regarding its potential mineralization. Jarosch et al. [86] reported highly significant correlation between enzyme-stable P and high molecular weight unhydrolyzed organic P, consistent with humic P bearing monoester P linkages [87]. Collectively, these studies not only show a gradient chemical stability of organic P but also suggest that humic P might be the most recalcitrant form of organic P, thereby presenting a challenge towards mineralizing this pool of soil organic legacy P for plant use.

3. Legacy P Transformations with Mineral and Organic Fertilizer Applications

As noted above, legacy P accumulates in soil with repeated additions of excess P fertilizers. In the USA, inorganic P fertilizer is applied approximately 10 times more than organic fertilizers (by mass; Table 1). This section reviews particularly how legacy inorganic P transforms into various species with mineral and organic fertilizer applications as it relates to its overall speciation and potential bioaccessibility in soils. We also review studies on impacts of fertilizer application and cessation on implied transformations of specific organic legacy P. This section focuses on soil P species transformation studies before and after fertilizer applications as probed by spectroscopic techniques (i.e., P-XANES for inorganic P and solution P-NMR (nuclear magnetic resonance) for organic P). Considering the analytical limitations of these techniques in distinguishing between fertilizer-derived versus native parent material-derived inorganic P in soil, we consider existing soil inorganic P largely as fertilizer-derived or legacy inorganic P throughout this section.

Table 1. Types and annual usage of phosphate fertilizers in the USA (the year 2015). Data are from the United States Department of Agriculture Economic Research Service (https://www.ers.usda.gov/data-products/fertilizer-use-and-price.aspx).

Inorganic P Fertilizers			
Superphosphates	Diammonium phosphate (DAP) (18-46-0)	Monoammonium phosphate (MAP) (11-(51-55)-0)	Other nitrogen–phosphate grades
		metric tons of material	
651,162	2,236,864	2,458,331	1,723,509
	Organic P Fertilizers		
Compost	Dried manure	Sewage sludge	Other organic materials
		metric tons of material	
101,062	86,900	192,101	196,800

3.1. Soil Inorganic Legacy P Transformations

Different types of P fertilizers have been shown to transform inorganic legacy P species of contacted soils depending on soil properties, fertilizer placement, and incubation time. For instance, addition of granular monoammonium phosphate (MAP; $(NH_4)H_2PO_4$) or liquid ammonium polyphosphate (APP) changes P speciation from predominantly Al-adsorbed P (>90%) to Fe-adsorbed (>70%) within 7.5 mm around the point of application. These shifts towards Fe-adsorbed species also yielded the greatest resin P extractability with MAP in acidic Oxisol (pH 3.9) and APP in Andisol (pH 5.9) [88]. Khatiwada et al. [51] also reported predominantly Fe- and Al-adsorbed P (63%) and greater resin-extractable P from liquid MAP than granular MAP (≈50%) when both fertilizers were deep-banded (at 10 cm below the soil surface) and the soils were sampled at 7.5–10 cm depth. Incorporating single superphosphate (monocalcium phosphate) with or without hog manure in bulk soil (0–20 cm) in an acidic soil (pH 5.87–6.13), increased the proportion of Fe-adsorbed P (at 43 or 47%, respectively), and produced Ca-adsorbed P (31.5%) while consuming 100% of apatite in an unfertilized control after 21 years of fertilization [89]. For the same soils, NPK (nitrogen–phosphorus–potassium) addition more than tripled Olsen-extractable P to 10.4

and 38.2 mg kg^{-1} for surface and subsurface layers, respectively; however, NPK application coupled with hog manure increased extractable P up to nine times relative to unfertilized control soils (11.4 to 99.4 and 9.6 to 43.4 mg kg^{-1} for surface and subsurface soils, respectively) [89]. Positive association suggested by these studies between soluble P and adsorbed P forms from additions of synthetic and organic P fertilizers may indicate greater contribution of adsorbed P forms over precipitated P minerals in supplying dissolved P in soils. However, although liquid MAP remained mostly adsorbed ($\approx$80%) even after 6 months of field application, resin-P extractability decreased by as much as two thirds [51]. This highlights the need to couple molecular-scale P speciation studies with macroscale measurements evaluating soluble P (i.e., resin-P extractability), as well as experiments that address in-field kinetics of P solubilization from identified P forms, ideally throughout the growing season and in the long term (i.e., between years of production), when developing P fertility regimes.

Transformations of legacy P have also been shown to differ in alkaline soils upon additions of inorganic P fertilizers. For example, Lou et al. [89] observed a slight increase in hydroxyapatite (71–78% from 67–72%) with concomitant decreases in proportions of P adsorbed to goethite and alumina from NPK additions to a calcareous soil (pH 8.1–8.4). However, to a non-calcareous alkaline soil (pH 7.4–8), the same NPK applications (i.e., P as diammonium phosphate) shifted P speciation from predominantly hydroxyapatite-P in unfertilized soil (52–57%) to brushite-P (44–63%), with a slight increase in goethite- or alumina-bound P [89]. Inorganic P fertilizer applications generally increased Olsen-P from 11 to 19% for calcareous or 9 to 41% for the non-calcareous surface soils, respectively [89]. On the other hand, Kar et al. [90] reported that most soil P (55–90% of total P) precipitated as hydroxyapatite after the addition of MAP coupled with urea, with the rest being adsorbed in a non-calcareous alkaline soil (pH 7.9) from 0–20 cm away from the point of application. However, 27 years of urea-based fertilization with or without P led to >50% reduction of hydroxyapatite species and enrichment of FePO$_4$ for an acidified alkaline Mollisol (originally pH 7.6 to 5.7) [91]. Together, these studies demonstrate how both soil and fertilizer types influence predominant soil legacy P forms in the short and the long term.

Contrary to purely mineral fertilizations, adding manure alone or in combination with inorganic fertilizers tends to show more consistent outcomes in diminishing proportions of more stable legacy P species in alkaline soils. For example, applying hog manure with NPK promoted transformation of hydroxyapatite into brushite (39–50%) on both non-calcareous and calcareous surface soils examined by Lou et al. [89]. This agrees with the findings of Kar et al. [90] who showed that addition of solid cattle manure to a calcareous alkaline soil (pH 7.8) also led to dominant precipitation of brushite (60%), particularly at the point of application (i.e., center of the band). However, for both MAP- and manure-fertilized soils, resin-extractable P decreased with increasing distance away from the point of subsurface band application. Nevertheless, resin-P fractions were 10 times greater in manured compared to MAP-fertilized soils [90]. Under laboratory conditions, Ajiboye et al. [92] also reported decreased hydroxyapatite precipitation and altered distributions among more soluble P species (i.e., β-tricalcium phosphate or calcite-adsorbed P) when different organic amendments were incorporated with calcareous alkaline soils. Increased fractions of soluble P observed in manured alkaline soils has been attributed to organic acids inhibiting precipitation of sparingly soluble Ca phosphate minerals and to the lower Ca to P ratios in manure-amended relative to MAP-amended soils [90]. Together these speciation studies suggest that manure enhances transformation of inorganic legacy P into more soluble chemical forms, confirming their lability in resin P extractions. Management strategies or technological advances that catalyze the solubilization from these pools would be necessary for legacy P to either substitute, if not eliminate, mineral P fertilizers for crop needs.

Overall, results of reviewed literature in this section show that inorganic legacy P could convert to more soluble forms with appropriate fertilizer management (i.e., fertilizer type and placement), but these effects are soil-specific and vary with space and time. For example, formation of insoluble soil P species was decreased in an Andisol (pH 5.9) with a liquid synthetic APP fertilizer, but decreased with a granular fertilizer (MAP) in an Oxisol (pH 3.9) [88]. Additionally, although subsurface banding of some inorganic and manure fertilizers enhanced more soluble inorganic legacy P speciation (e.g., brushite and adsorbed P forms) along with increased resin-extractable P, these effects were diminished farther away from the point and time of application [51]. These results suggest that, depending on long-term P application rates, placement, and soil type, fertilization-enhanced solubilization of inorganic legacy P may be time-sensitive and likely microscopic in scale, which may not necessarily impact the bulk soils in the rooting zone. This observation is consistent with the findings of Weyers et al. [31] and Koch et al. [30], who reported insignificant effects of 3 to 16 years of mineral or manure fertilizer applications on inorganic legacy P speciation of bulk soils. Moreover, the reviewed studies suggest that manure applications may offer a benefit for enhancing the solubilization of soil legacy P (i.e., hydroxyapatite), particularly for alkaline soils, and thus combining inorganic and organic P fertilizer applications may be a simple management strategy for decreasing dependence for mineral P fertilizers. However, it should also be mentioned that hydroxyapatite has also been detected as the main P species in acid soils (pH 4.2–5.9) with a long history of receiving higher rates of poultry litter [93,94]. Therefore, the relationships between application, solubilization, and ultimately crop utilization of manure-driven legacy P requires additional long-term evaluation both for acid and alkaline soils. To further enrich our understanding on legacy P transformations, we review organic legacy P transformations with fertilizer applications and cessations in the succeeding section.

3.2. Soil Organic P Transformations

Existing soil organic P does not seem to be impacted by P fertilizer applications [95], suggesting that organic legacy P may not be actively accumulating in organically P-fertilized soils. However, unlike those of inorganic soil legacy P, many P fertilization studies evaluating organic legacy P transformations are limited in investigating other specific fertilization variables (i.e., changes in organic legacy P speciation as affected by fertilizer application in space and time). For example, in a field study on bulk soils (0–20 cm) by Annaheim et al. [96], no significant changes were observed in concentrations of different organic P forms receiving organic fertilizers relative to unfertilized controls, despite stark differences in contents of organic P species in applied organic fertilizers (i.e., dairy manure, compost, dry sewage sludge). By comparing the expected concentrations of what could have been accumulated after 62 years of additions, they also demonstrated the limited accumulation for all forms of added organic P species, including IHPs, nucleic acids, and unidentified monoesters. Similarly, Dou et al. [29] reported negligible IHP accumulation in farm soils receiving different types of manure for 8–10 years, despite an estimated 30 kg P ha^{-1} annual additions from manure-derived IHPs. These long-term studies suggested organic P mineralization that may have released and supplied inorganic P for plant uptake, although they did not exclude transport losses (i.e., leaching and runoff). Although the fate of organic P species was not determined, these studies suggest that organic fertilizers are not likely to build up organic legacy P in mineral soils in a matter of years or decades. Moreover, their results indicate that inherent or native soil organic P species in bulk soils are relatively more stable against mineralization (and/or transport) in comparison to newly added organic P forms that appeared to be rapidly transformed to inorganic P in the soil.

Indeed, native organic P has been found stable in soils even after cessation of external P additions. For inorganically P-fertilized soils in Canada without history of manure additions, halting chemical P fertilization for 15 years increased organic P and orthophosphate diesters and decreased most of the IHP isomers including myo-IHP, although total IHP remained unchanged relative to continuously P-fertilized soils [23]. Build-up of organic P when inorganic P decreased has been attributed to drawdown or crop utilization of inorganic legacy P [95] from 28 years of prior inorganic fertilizations before stopping P inputs [23]. Although Liu et al. [23] demonstrated crop utilization of inorganic legacy P from soils, mineralization and consequently crop utilization of organic P appeared to be limited in these soils. In another experiment in Northern Ireland, Cade-Menum et al. [97] reported similar findings when orthophosphate decreased, but neither total organic P nor specific organic P forms changed between various P-fertilized treatments from zero-P controls 5 years after P fertilizations had stopped. Liu et al. [91] also found similar levels of accumulated organic P for both P-fertilized and non-P-fertilized soils in China after 27 years of cropping relative to baseline soils, except that phosphodiesters considerably accumulated more in P-fertilized soils. Liu et al. [91] suggested that P deficits induced degradation of phosphodiesters in soils not receiving P. Taken together, these studies suggest that native organic P forms are not likely to mineralize or contribute to bioaccessible P for soils with prolonged inorganic P fertilization history when soil test P is still sufficient.

Based on the reviewed studies above, organic P may be constituting a minor portion of legacy P. Moreover, existing non-fertilizer-derived soil organic P may contribute less to plant uptake than inorganic P in soils highly enriched with inorganic legacy P (i.e., soil test P above optimum) (Figure 3). Organic P mineralization by microorganisms or plant roots may not be promoted when sufficient levels of bioaccessible inorganic P exists. Conversely, mineralization of organic P into inorganic P may occur to a greater extent than desorption and solubilization of inorganic P when bioaccessible inorganic P is deficient (i.e., soil test P is below optimum). This is consistent with the work of Recena et al. [98], who observed greater contribution of organic P to actual plant P uptake more in low-P than in high-P soils. Mineralization of organic P forms under deficient soil P conditions could arise from stimulation of microbial growth when C and N levels are not limiting [99,100], meaning when sufficient C supply is present and soil C/N ratios are lower. Hence, balanced nutrient and organic matter management may be required for releasing bound orthophosphates from natural organic P reserves in P-deficient soils. More work is needed to better track the transformations of added organic P compounds from organic P fertilizers under different fertilizer management scenarios and field conditions over different periods of time. Building off this knowledge, the following section reviews the current literature on potential plant, microbial, and biochemical strategies for accessing soil legacy P.

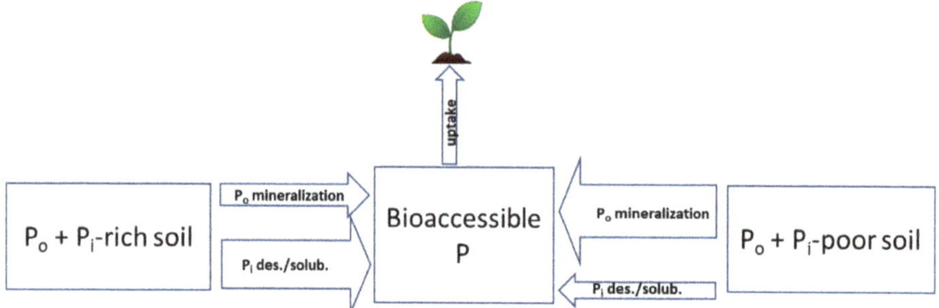

Figure 3. Conceptual representation of organic and inorganic P transformations towards releasing dissolved P and their expected magnitude of occurrence when legacy P is higher and largely dominated by bioaccessible inorganic P (P_o + P_i-rich soil) versus when legacy P is lower and deficient of bioaccessible inorganic P (P_o + P_i-poor soil), assuming equal concentrations of organic P. Arrow size indicates the relative magnitude of fluxes. Legend: P_o, soil organic P; P_i, soil inorganic P; des/solub, desorption and dissolution.

4. Approaches for Accessing Native Soil P and Associated Challenges

4.1. Plant-Based Strategies

Plants can possibly utilize soil legacy P through naturally evolved strategies, such as induction of root architectural changes and exudation of organic acids and enzymes, as well as possibly through scientific pursuits for improving internal P use efficiency (i.e., plant yield per unit P uptake) [101,102]. Another way that plants can thrive on legacy P is through facilitation of symbiotic and beneficial fungi and other soil microbes. However, the role of microorganisms will be discussed separately in the ensuing section.

Root architectural response is a well-recognized plant mechanism for alleviating soil conditions of low bioaccessible soil P [103], making it likely adaptable for extracting legacy P. Examples of efficient root configurational changes include highly branched root systems, increased production and length of lateral roots, proliferation of root hairs, and formation of cluster roots among various crops [101,104–109]. However, these adaptations vary among crop species and genetic varieties [110–113]. Moreover, most of root architectural studies have been performed on seedlings, which do not necessary correspond to root architecture of mature plants grown in the field [114,115]. With limited field measurements in this area [116], soil factors, particularly those associated with low soil P availability (i.e., soil pH), could potentially restrict efficient root architectural responses in field settings. For instance, foraging traits of maize cultivars characterized by greater root surface correlated well with improved P uptake in neutral field soil, but not in alkaline soil [117]. Furthermore, root architecture has been found to vary depending on the chemical forms and supply patterns of P [118], highlighting the importance for understanding soil legacy P speciation and likely distributions across agricultural soils. Interestingly, genotypes of common bean bearing more adaptive root architectures effectively grew in an unfertilized low-P Ultisol, showing comparable P uptake with P-fertilized commercial varieties [111]. Together these studies emphasize that, while efficient root systems can potentially mine soil legacy P, there is a need for a field-based approach for identifying and matching efficient root systems with soil legacy P and other soil conditions. Moreover, soil properties need to be considered in breeding programs aiming for plants with high P uptake efficiencies.

Another plant adaptive means for potentially exploiting soil legacy P is root exudation of organic acids and phosphatases. Organic acids are known to increase dissolved P by rhizosphere acidification, competition for P adsorption sites, complexation of P-precipitating metals, and provision of C source for P-mobilizing microorganisms [119–123]. However, organic acid production is highly specific to certain plant species [121]. For instance, legumes are known to produce more root exudates than grasses [2,22,124,125], which could be advantageous for P acquisition when legumes are intercropped with cereals. However, increasing evidence exists regarding the minor role of organic acids on actual plant P uptake compared to phosphatase activities or root architectural changes. For instance, legacy P uptake was positively correlated with rhizosphere acid phosphatases, but not with organic anions, with wheat (*Triticum aestivum*), oat (*Avena sativa*), potato (*Solanum tuberosum*), and canola (*Brassica napus*) in low bioaccessible P soils [126]. Using statistical redundancy analysis, Sun et al. [127] also found that P uptake of field-grown maize and alfalfa was explained more by root architectural changes (58–87%) than total organic anion concentrations (<0.2–24%). This observation emphasizes the importance of root architecture in acquiring P as compared to other mechanisms (e.g., strategic placement of organic acids [128]), although experimental variables also present challenges in understanding organic anions effects [120]. Nevertheless, these studies suggest that enhanced root exudation could effectively allow plants to utilize soil legacy P when combined with highly adaptive root systems.

Apart from natural plant adaptations, genetic engineering can also be used to increase plant uptake capacity and internal use efficiency of soil legacy P. Efforts related to this field have included increasing P uptake by overexpressing P transporters responsible for root P uptake and P transport to shoots, driving crop resilience to P deficiency [129]. For example, overexpression of transporter TaPht1.4 in wheat promoted growth and P accumulation [130]. Similarly, modulating the expression of OsPHT1 transporters in rice has shown to directly impact P uptake [131–136]. Phosphorus transporters are regulated by genes and transcription factors, which are altogether induced under P-deficient conditions [129,137]. Phosphorus-related genes have already been identified in major agronomic crops, such as rice (*Oryza sativa*), maize (*Zea mays*), and soybean (*Glycine max*) [138–141]. Candidate gene overexpression is a powerful approach to enhance nutrient uptake under conditions of low P availability. It is also foreseeable that a combination of both increased uptake capacity and internal P-use efficiency will be equally desirable for crops that have adapted and developed better P uptake and efficiency use strategies.

Whereas natural plant adaptations (i.e., root architectural changes and root exudation) to mobilize legacy P can be highly specific to plant species or varieties as noted above, genetic engineering can increase legacy P acquisition and internal use efficiency for any desired crops. However, this work continues to be a long-term research trajectory. Consequently, much is still unknown as to whether genetically engineered plants may successfully utilize soil legacy P and reduce, if not eliminate, dependence of mineral P fertilizers. As with natural adaptive plant mechanisms, there is an added uncertainty that these plants could address other soil stresses associated with low bioaccessible P (i.e., Al toxicity in highly acidic soils). To this end, material science, microelectronics, and nanotechnology should develop and deploy several sensor platforms (chemical, electromagnetic, optical, and genetic) for detecting real-time below-ground root development and plasticity response to bioaccessible legacy P. Genetic and genomics approaches should complement this sensor approach and will hone in on the identification of the genes regulating P uptake, use efficiency, and tolerance to other soil conditions. For example, the systems level understanding of the many genes, and their regulation, associated with phosphate deficiency once unveiled would allow for predictive and prescriptive interventions. It would be foreseeable to use genome editing technologies, such as CRISPR (clustered regularly interspaced short palindromic repeats) technologies [142,143], to genetically test the functional relevance of known (e.g., OsPSTOL1, AVP1, PHO1 and OsPHT1;6) [141,144–146] and unknown genes identified in molecular studies. At the same time, it will be important to consider and incorporate stakeholder perceptions and socio-economic considerations of genetically engineered crops for increased legacy P mobilization and crop utilization to ensure adoption by end-users.

4.2. Phosphate-Solubilizing Microorganisms

A diverse genera of microorganisms including bacteria, fungi, actinomycetes, and cyanobacteria have been reported to solubilize native soil P [147], making them potentially capable of solubilizing legacy P. Phosphate-solubilizing microorganisms (PSM) dissolve or mineralize soil P largely through the production of organic acids and phosphatases [148], and recent reviews have reported generally increased crop growth and yield from the use of PSMs [147,149–151]. In general, most studies in this field have focused on effects of PSMs on crop performance, but information is lacking on how PSMs transform legacy P species during or after plant growth, Hence, in the following studies reviewed below, we discuss PSM effects mostly on plant P uptake and growth and suggest research directions for ascertaining the roles of PSMs in impacting legacy P.

Mycorrhizal fungi are perhaps the longest known microorganisms capable of accessing native soil P owing to their capacity for symbiotic associations with 72% of vascular plant roots that began 450 million years ago [152] and could likely be used to access legacy P. Arbuscular mycorrhiza (AM) scavenge greater soil volumes than plant roots for soluble P and rapidly transport acquired P to roots via arbuscules, thereby minimizing the impact of naturally slow replenishment of dissolved P in soil solution [153]. However, AM fungi are diverse; different taxa do not produce the same response from the same plant species, and that the same fungus does not yield the same effects across different plant species or varieties [154]. These effects could range from positive to negative with unknown reasons for these inconsistencies, although elevated C expense by the plant to maintain the symbiotic relationship has been suggested as being detrimental [126,155]. This suggests that compatibility between the plant host and colonizing AM are required for a beneficial outcome of the symbiosis. However, interspecific compatibility between the host plant and the fungi may also vary depending on soil chemical conditions such as soil pH and associated occurrence of different forms of legacy P. Research is therefore needed in understanding t legacy P speciation impacts symbiotic mycorrhizal associations, and vice-versa.

Inoculation of other PSMs to soils has also shown inconsistent effects on plant P uptake and production. For instance, maize increased P uptake by 10% without P addition after 28 d on acidic soil (pH 5.6) with *Enterobacter radicincitans* DSM 16,656 and *Pseudomonas fluorescens* DR54 inoculations [156]. However, the observed increase in P uptake had no effect on plant growth. In contrast, *Bacillus mucilaginosus* inoculation to acidic soil (pH 5.2) did not improve shoot P uptake in maize [157], but *Aspergillus* FS9 and *Bacillus* FS-3 increased strawberry (*Fragaria x ananasa* cv. Fern) yield by 7% and 30%, respectively, in a calcareous soil (pH 7.6) without added P [158]. Although PSM inoculations have been shown to increase water- and resin-extractable P [156], future research needs to focus on how PSMs transform actual soil legacy P species. Direct information on solubilization or mineralization of soil legacy P forms could serve as a tool in matching appropriate PSMs for the a given soil. Moreover, these data are needed both in the short- and the long-term to explore sustainability issues with the use of PSMs in the context of potentially minimizing mineral P fertilizer applications.

PSMs may not thrive on soil legacy P alone and thus may not completely eliminate dependence to external P inputs. The need for supplemental P with PSMs is expected, considering that microorganisms also need bioaccessible P and other essential nutrients for their metabolic processes to establish and sustain microbial activity in soil. Supplementing PSMs with P fertilizers has improved shoot production [156], increased P uptake [157], or exceeded yield of plants from uninoculated P-fertilized control soils by 54–71% [158]. However, inoculating PSM with co-added P not improving plant P uptake and yield has also been reported. This has been observed with *Pseudomonas fluorescens* Pf153 inoculations to soils (pH 4.8 to pH 6.6) after 8 weeks of maize growth compared to uninoculated controls, despite supplemental rock phosphate [159]. In this respect, future research needs to evaluate compatibilities between supplemental P fertilizers and individual or combinations of various PSM species for a given soil. Considering how mineral and organic fertilizers transform soil legacy P (Section 3.1), co-adding mineral and manure with PSM inoculations may enhance utilization of existing P in soils. On one hand, supplementing PSMs with inorganic P may provide for the immediate nutrient need of both the microorganisms and the plants during PSM establishment in soil. On the other hand, supplementing PSMs with organic P fertilizers may alter legacy P speciation into predominantly less sparingly soluble forms. However, given that external P inputs may unavoidably add more to soil legacy P stocks, more work is needed in finding the right balance between minimizing supplemental P additions while maximizing solubilization/mineralization and consequently crop utilization of existing soil legacy P. Moreover, additional research is needed in evaluating any potential combination among PSM species capable of solubilizing P exclusively outside the roots with those capable of solubilizing P within the roots (e.g., endophytic bacteria [160]) with particular application to agricultural crops. Finally, PSM studies involving inoculations to actual test soils require molecular investigation of mechanisms of action to ascertain utilization of legacy P.

4.3. Immobilized Organic P Hydrolyzing Enzymes

Loading organic P-hydrolyzing enzymes onto clay or nanoclay supports has also been explored as a potential biofertilization strategy to mineralize organic P that could likely be applied to hydrolyze organic legacy P. Menezes-Blackburn et al. [161] loaded phytases to either montmorillonite or allophane, found that phytases produced by different microbial species differed in their activities. In their study, phytase activity derived from *Escherichia coli* showed peak activity at pH 5 in water, representing roughly 85% of free phytase activity at the same pH. However, phytase from *Aspergillus niger* showed considerably reduced phytase activity at pH below 5.5 with the tested nanoclay supports [161]. Contrary to decreased activities of immobilized enzymes compared to their freely dissolved forms, Calabi-Floody et al. [162] reported increased activities of up to 48% with allophane-immobilized compared to freely dissolved acid phosphatases. Using the Michaelis–Menten equation to describe degradation of p-nitrophenolphosphate as model organic P substrate, maximum rate of substrate degradation by the enzymes reportedly increased from 33 to 38% with the allophanic compared to montmorillonitic support. However, when composted cattle dung was used as the organic P source, none of the immobilized or the free acid phosphatase showed significant differences from the unamended dung in increasing dissolved inorganic P, which they attributed to high inorganic P concentrations contained in the dung. In contrast, Menezes-Blackburn et al. [163] treated cattle manure with phytases stabilized in allophanic nanoclays and found a significant increase in NaOH-EDTA P (commonly used to extract organic P) and Olsen-extractable inorganic P in soil. However, they did not observe significant hydrolysis of freshly added phytate with soil or manure amended with the immobilized phytases. The stability of these nanoclay-supported enzymes in soils is also of concern, as greatly reduced enzyme activity was also observed after merely 2 h of exposure to high temperature and after 1 d of exposure to protein degradation [161]. Overall, while immobilizing enzymes inside in carrier materials are promising under laboratory conditions, this approach appears to be challenged by limited accessibility of substrates and the interference of other dissolved compounds in the soil solution. Moreover, with a short lifespan, the capacity of nanoclay-immobilized enzymes to release organically bound legacy P in soils and sustain crop P needs would be highly challenged. Future research needs to focus on finding new approaches to adding these enzymes using other carrier materials [164] without negatively affecting their activities when added to soil or when combined with organic fertilizers.

5. Conclusions

Repeated P fertilizer applications over time has led to build-up of soil legacy P, which theoretically could be utilized to substitute for, if not eliminate dependence on, mineral P fertilizers. However, crop recovery of legacy P varies considerably across soils and duration. This paper reviewed (1) the occurrence and factors controlling bioaccessibility of different soil P forms, (2) transformation of different legacy P species with P fertilizer additions, and (3) currently studied strategies that could possibly be used for exploiting soil legacy P for crop use. We found that soil legacy P exists predominantly as inorganic P in either adsorbed or precipitated forms, and secondarily as organic P, mainly as monoester or diester forms. Bioaccessibility of different chemical P forms is predominantly controlled by soil pH either directly (e.g., desorption/dissolution of inorganic P species) or indirectly (e.g., enzymatic degradation of organic P). We also found that inorganic legacy P could transform into more bioaccessible forms with fertilizer applications, but this effect is highly specific to soil and fertilizer type and may diminish away from the point and time of fertilizer application. In contrast, transformation of native soil organic P forms does not appear to be impacted by fertilizer applications and cessations, suggesting lesser accumulation of organic legacy P in soils, as well as lower potential for crop utilization of native organic P in soil particularly when bioaccessible inorganic legacy P is sufficient.

Most currently pursued approaches for possibly increasing crop utility of legacy P are biological in nature. These range from natural to genetically engineered plant adaptations for enhanced soil P uptake and use efficiencies to the use of symbiotic or beneficial P-solubilizing microorganisms and immobilized organic P-hydrolyzing enzymes. However, plant- or microorganism-based strategies are limited by the high specificity of acquired adaptation mechanisms capable of solubilizing soil P, whereas immobilized enzyme efficacy is mainly limited by physical and chemical constraints. Moreover, studies investigating these approaches typically do not elucidate the mechanism of actions for utilization of soil legacy P largely due to lack of molecular soil P speciation component.

As scarcity of P resources becomes a more pressing global issue, mining legacy P from soils may emerge as a more pressing societal need. Harnessing this resource will require not only development of new and existing technologies for improved legacy P utilization but also holistic management strategies that incorporate multiple approaches tailored to specific soil environments. Overall, enhancing our ability to access soil legacy P for croplands will require a multidisciplinary approach to establish soil and crop management systems that enhance complementary use of the different chemical forms, spatial variability, and temporal changes of legacy P. A systems approach should integrate many of the strategies identified above, as well as future technologies, for accessing native soil P. As multidisciplinary advances, for example in genetic engineering and material sciences, are combined with molecular soil P speciation tools in addressing each approach's limitations, it is foreseeable that prospects for widespread utilization of legacy P as a partial or complete substitute for mineral P fertilizer may increase in the future.

Author Contributions: Conceptualization, O.W.D. and S.D.; writing—original draft preparation, Sections 1–3 and Section 5, S.D., O.W.D., L.G., E.S.M., K.G., and J.L.J.; Section 4, S.D., R.S., R.P., and L.V.d.B.; writing—review and editing, O.W.D., K.G., D.H., and E.S.M. All authors have read and agreed to the published version of the manuscript.

Funding: This work was supported by the USDA National Institute of Food and Agriculture, Hatch project NC02713. We thank the Game-Changing Research Initiative Program (GRIP) through the NC State Office of Research and Innovation (ORI), RTI International, and the Kenan Institute for Engineering, Science and Technology.

Soil Syst. **2020**, *4*, 74

Acknowledgments: We thank Sarah Powers (NC State University) for assistance with graphics.

Conflicts of Interest: The authors declare no conflict of interest.

References

1. Fraústo da Silva, J.J.R.; Williams, R.J.P. *The Biological Chemistry of the Elements: The Inorganic Chemistry of Life*, 2nd ed.; Oxford University Press: Oxford, UK, 2001.
2. Vance, C.P.; Uhde-Stone, C.; Allan, D.L. Phosphorus acquisition and use: Critical adaptations by plants for securing a nonrenewable resource. *New Phytol.* **2003**, *157*, 423–447. [CrossRef]
3. Stewart, W.M.; Hammond, L.L.; Van Kauwenbergh, S.J. Phosphorus as a Natural Resource. In *Phosphorus: Agriculture and the Environment*; Pierzynski, G.M., McDowell, R.W., Sims, J.T., Eds.; Soil Science Society of America: Madison, WI, USA, 2005; pp. 1–22.
4. Cordell, D.; White, S. Life's Bottleneck: Sustaining the World's Phosphorus for a Food Secure Future. *Ann. Rev. Environ. Resour.* **2014**, *39*, 161–188. [CrossRef]
5. Pierzynski, G.M.; McDowell, R.W.; Sims, J.T. Chemistry, Cycling, and Potential Movement of Inorganic Phosphorus in Soils. In *Phosphorus: Agriculture and the Environment*; Sims, J.T., Sharpley, A.N., Westermann, T.D., Eds.; Soil Science Society of America: Madison, WI, USA, 2005; pp. 51–86.
6. Fox, R.L.; Kamprath, E.J. Phosphate Sorption Isotherms for Evaluating the Phosphate Requirements of Soils. *Soil Sci. Soc. Am. J.* **1970**, *34*, 902–907. [CrossRef]
7. Syers, J.K.; Johnston, A.E.; Curtin, D. *Efficiency of Soil Fertilizer Phosphorus Use—Reconciling Changing Concepts of Soil Phosphorus Behavior with Agronomic Information*; Food And Agriculture Organization of The United Nations (FAO): Rome, Italy, 2008.
8. Dhillon, J.; Torres, G.; Driver, E.; Figueiredo, B.; Raun, W.R. World Phosphorus Use Efficiency in Cereal Crops. *Agron. J.* **2017**, *109*, 1670–1677. [CrossRef]
9. Bennett, E.M.; Carpenter, S.R.; Caraco, N.F. Human Impact on Erodable Phosphorus and Eutrophication: A Global Perspective: Increasing accumulation of phosphorus in soil threatens rivers, lakes, and coastal oceans with eutrophication. *BioScience* **2001**, *51*, 227–234. [CrossRef]
10. Menezes-Blackburn, D.; Giles, C.; Darch, T.; George, T.S.; Blackwell, M.; Stutter, M.; Shand, C.; Lumsdon, D.; Cooper, P.; Wendler, R.; et al. Opportunities for mobilizing recalcitrant phosphorus from agricultural soils: A review. *Plant Soil* **2018**, *427*, 5–16. [CrossRef]
11. Yang, Y.-Y.; Asal, S.; Toor, G.S. Residential catchments to coastal waters: Forms, fluxes, and mechanisms of phosphorus transport. *Sci. Total Environ.* **2020**, 142767. [CrossRef]
12. Jarvie, H.P.; Sharpley, A.N.; Withers, P.J.A.; Scott, J.T.; Haggard, B.E.; Neal, C. Phosphorus Mitigation to Control River Eutrophication: Murky Waters, Inconvenient Truths, and "Postnormal" Science. *J. Environ. Qual.* **2013**, *42*, 295–304. [CrossRef]
13. USEPA. National Lakes Assessment 2012 Results. Available online: https://www.epa.gov/national-aquatic-resource-surveys/national-lakes-assessment-2012-results (accessed on 15 June 2020).
14. CDC. Harmful Algal Bloom (HAB)-Associated Illness. Available online: https://www.cdc.gov/habs/general.html#:~{}:text=Fresh%20water%20%E2%80%94In%202014%2C%20a,or%20other%20fresh%20water%20bodies (accessed on 15 June 2020).
15. Schlesinger, W.H.; Bernhardt, E.S. Chapter 12—The Global Cycles of Nitrogen and Phosphorus. In *Biogeochemistry*, 3rd ed.; Schlesinger, W.H., Bernhardt, E.S., Eds.; Academic Press: Boston, MA, USA, 2013; pp. 445–467.
16. Kamprath, E.J. Changes in phosphate availability of ultisols with long-term cropping. *Commun. Soil Sci. Plant Anal.* **1999**, *30*, 909–919. [CrossRef]
17. Roberts, T.; Johnston, A. Phosphorus use efficiency and management in agriculture. *Resour. Conserv. Recycl.* **2015**, *105*. [CrossRef]
18. Johnston, A.E.; Poulton, P.R. Phosphorus in Agriculture: A Review of Results from 175 Years of Research at Rothamsted, UK. *J. Environ. Qual.* **2019**, *48*, 1133–1144. [CrossRef] [PubMed]
19. Kizewski, F.; Liu, Y.-T.; Morris, A.; Hesterberg, D. Spectroscopic Approaches for Phosphorus Speciation in Soils and Other Environmental Systems. *J. Environ. Qual.* **2011**, *40*, 751–766. [CrossRef] [PubMed]
20. Hou, E.; Tan, X.; Heenan, M.; Wen, D. A global dataset of plant available and unavailable phosphorus in natural soils derived by Hedley method. *Sci. Data* **2018**, *5*, 180166. [CrossRef] [PubMed]

21. Sims, J.T.; Pierzynski, G.M. Chemistry of Phosphorus in Soils. In *Chemical Processes in Soils*; Tabatabai, M.A., Sparks, D.L., Eds.; Soil Science Society of America: Madison, WI, USA, 2005; pp. 151–192.

22. Hinsinger, P. Bioavailability of soil inorganic P in the rhizosphere as affected by root-induced chemical changes: A review. *Plant Soil* **2001**, *237*, 173–195. [CrossRef]

23. Liu, J.; Hu, Y.; Yang, J.; Abdi, D.; Cade-Menun, B.J. Investigation of Soil Legacy Phosphorus Transformation in Long-Term Agricultural Fields Using Sequential Fractionation, P K-edge XANES and Solution P NMR Spectroscopy. *Environ. Sci. Technol.* **2015**, *49*, 168–176. [CrossRef]

24. Cade-Menun, B.J.; He, Z.; Zhang, H.; Endale, D.M.; Schomberg, H.H.; Liu, C.W. Stratification of Phosphorus Forms from Long-Term Conservation Tillage and Poultry Litter Application. *Soil Sci. Soc. Am. J.* **2015**, *79*, 504–516. [CrossRef]

25. Vincent, A.G.; Vestergren, J.; Gröbner, G.; Persson, P.; Schleucher, J.; Giesler, R. Soil organic phosphorus transformations in a boreal forest chronosequence. *Plant Soil* **2013**, *367*, 149–162. [CrossRef]

26. Turner, B.L.; Cheesman, A.W.; Godage, H.Y.; Riley, A.M.; Potter, B.V.L. Determination of neo- and d-chiro-Inositol Hexakisphosphate in Soils by Solution 31P NMR Spectroscopy. *Environ. Sci. Technol.* **2012**, *46*, 4994–5002. [CrossRef]

27. Ahlgren, J.; Djodjic, F.; Börjesson, G.; Mattsson, L. Identification and quantification of organic phosphorus forms in soils from fertility experiments. *Soil Use Manag.* **2013**, *29*, 24–35. [CrossRef]

28. Cade-Menun, B.J.; Carter, M.R.; James, D.C.; Liu, C.W. Phosphorus Forms and Chemistry in the Soil Profile under Long-Term Conservation Tillage: A Phosphorus-31 Nuclear Magnetic Resonance Study. *J. Environ. Qual.* **2010**, *39*, 1647–1656. [CrossRef]

29. Dou, Z.; Ramberg, C.F.; Toth, J.D.; Wang, Y.; Sharpley, A.N.; Boyd, S.E.; Chen, C.R.; Williams, D.; Xu, Z.H. Phosphorus Speciation and Sorption-Desorption Characteristics in Heavily Manured Soils. *Soil Sci. Soc. Am. J.* **2009**, *73*, 93–101. [CrossRef]

30. Koch, M.; Kruse, J.; Eichler-Löbermann, B.; Zimmer, D.; Willbold, S.; Leinweber, P.; Siebers, N. Phosphorus stocks and speciation in soil profiles of a long-term fertilizer experiment: Evidence from sequential fractionation, P K-edge XANES, and 31P NMR spectroscopy. *Geoderma* **2018**, *316*, 115–126. [CrossRef]

31. Weyers, E.; Strawn, D.G.; Peak, D.; Moore, A.D.; Baker, L.L.; Cade-Menun, B. Phosphorus Speciation in Calcareous Soils Following Annual Dairy Manure Amendments. *Soil Sci. Soc. Am. J.* **2016**, *80*, 1531–1542. [CrossRef]

32. Oh, Y.-M.; Hesterberg, D.L.; Nelson, P.V.; Niedziela, C.E. Desorption Characteristics of Three Mineral Oxides and a Non-crystalline Aluminosilicate for Supplying Phosphate in Soilless Root Media. *Commun. Soil Sci. Plant Anal.* **2016**, *47*, 753–760. [CrossRef]

33. Parfitt, R.L. The availability of P from phosphate-goethite bridging complexes. Desorption and uptake by ryegrass. *Plant Soil* **1979**, *53*, 55–65. [CrossRef]

34. Gypser, S.; Hirsch, F.; Schleicher, A.M.; Freese, D. Impact of crystalline and amorphous iron- and aluminum hydroxides on mechanisms of phosphate adsorption and desorption. *J. Environ. Sci.* **2018**, *70*, 175–189. [CrossRef]

35. Kruse, J.; Abraham, M.; Amelung, W.; Baum, C.; Bol, R.; Kühn, O.; Lewandowski, H.; Niederberger, J.; Oelmann, Y.; Rüger, C.; et al. Innovative methods in soil phosphorus research: A review. *J. Plant Nutr. Soil Sci.* **2015**, *178*, 43–88. [CrossRef]

36. Mehra, O.P.; Jackson, M.L. Iron oxide removal from soils and clays by a dithionite–citrate system buffered with sodium bicarbonate. In *Clays and Clay Minerals*; Ingerson, E., Ed.; Pergamon: Oxford, UK, 2013; pp. 317–327.

37. Nair, V.D. Soil phosphorus saturation ratio for risk assessment in land use systems. *Front. Environ. Sci.* **2014**, *2*. [CrossRef]

38. Beauchemin, S.; Simard, R.R. Soil phosphorus saturation degree: Review of some indices and their suitability for P management in Québec, Canada. *Can. J. Soil Sci.* **1999**, *79*, 615–625. [CrossRef]

39. Wang, X.M.; Liu, F.; Tan, W.F.; Li, W.; Feng, X.H.; Sparks, D.L. Characteristics of Phosphate Adsorption-Desorption Onto Ferrihydrite: Comparison With Well-Crystalline Fe (Hydr)Oxides. *Soil Sci.* **2013**, *178*, 1–11. [CrossRef]

40. Strauss, R.; Brümmer, G.W.; Barrow, N.J. Effects of crystallinity of goethite: II. Rates of sorption and desorption of phosphate. *Eur. J. Soil Sci.* **1997**, *48*, 101–114. [CrossRef]

41. Yan, Y.; Koopal, L.K.; Liu, F.; Huang, Q.; Feng, X. Desorption of myo-inositol hexakisphosphate and phosphate from goethite by different reagents. *J. Plant Nutr. Soil Sci.* **2015**, *178*, 878–887. [CrossRef]

42. Krumina, L.; Kenney, J.P.L.; Loring, J.S.; Persson, P. Desorption mechanisms of phosphate from ferrihydrite and goethite surfaces. *Chem. Geol.* **2016**, *427*, 54–64. [CrossRef]

43. Sø, H.U.; Postma, D.; Jakobsen, R.; Larsen, F. Sorption of phosphate onto calcite; results from batch experiments and surface complexation modeling. *Geochim. Cosmochim. Acta* **2011**, *75*, 2911–2923.

44. Hesterberg, D. Macroscale Chemical Properties and X-Ray Absorption Spectroscopy of Soil Phosphorus. In *Developments in Soil Science*; Singh, B., Gräfe, M., Eds.; Elsevier: Amsterdam, The Netherlands, 2010; Volume 34, pp. 313–356.

45. Gu, C.; Dam, T.; Hart, S.C.; Turner, B.L.; Chadwick, O.A.; Berhe, A.A.; Hu, Y.; Zhu, M. Quantifying Uncertainties in Sequential Chemical Extraction of Soil Phosphorus Using XANES Spectroscopy. *Environ. Sci. Technol.* **2020**, *54*, 2257–2267. [CrossRef]

46. Bray, R.H.; Kurtz, L.T. Determination of Total, Organic, and Available Forms of Phosphorus in Soils. *Soil Sci.* **1945**, *59*, 39–46. [CrossRef]

47. Mehlich, A. Mehlich 3 soil test extractant: A modification of Mehlich 2 extractant. *Commun. Soil Sci. Plant Anal.* **1984**, *15*, 1409–1416. [CrossRef]

48. Watanabe, F.S.; Olsen, S.R. Test of an Ascorbic Acid Method for Determining Phosphorus in Water and NaHCO3 Extracts from Soil. *Soil Sci. Soc. Am. J.* **1965**, *29*, 677–678. [CrossRef]

49. McIntosh, J.L. Bray and Morgan Soil Extractants Modified for Testing Acid Soils from Different Parent Materials1. *Agron. J.* **1969**, *61*, 259–265. [CrossRef]

50. Hochmuth, G.J.; Mylavarapu, R.S.; Hanlon, E.A. *Developing a Soil Test Extractant: The Correlation and Calibration Processes*; University of Florida: Gainesville, FL, USA, 2017.

51. Khatiwada, R.; Hettiarachchi, G.M.; Mengel, D.B.; Fei, M. Speciation of Phosphorus in a Fertilized, Reduced-Till Soil System: In-Field Treatment Incubation Study. *Soil Sci. Soc. Am. J.* **2012**, *76*, 2006–2018. [CrossRef]

52. Sims, J.T.; Edwards, A.C.; Schoumans, O.F.; Simard, R.R. Integrating Soil Phosphorus Testing into Environmentally Based Agricultural Management Practices. *J. Environ. Qual.* **2000**, *29*, 60–71. [CrossRef]

53. Mallarino, A.P.; Atia, A.M. Correlation of a Resin Membrane Soil Phosphorus Test with Corn Yield and Routine Soil Tests. *Soil Sci. Soc. Am. J.* **2005**, *69*, 266–272. [CrossRef]

54. Sims, J.T. *Soil Test Phosphorus: Principles and Methods In Methods of Phosphorus Analysis for Soils, Sediments, Residuals, and Waters*, 2nd ed.; Kovar, J., Pierzynski, G., Eds.; Southern Cooperative Series Bulletin 408; Southern Extension and Research Activity (SERA): Blacksburg, VA, USA, 2009; Volume 408.

55. Turner, B.L.; Engelbrecht, B.M.J. Soil organic phosphorus in lowland tropical rain forests. *Biogeochemistry* **2011**, *103*, 297–315. [CrossRef]

56. Murphy, P.N.C.; Bell, A.; Turner, B.L. Phosphorus speciation in temperate basaltic grassland soils by solution 31P NMR spectroscopy. *Eur. J. Soil Sci.* **2009**, *60*, 638–651. [CrossRef]

57. Hou, E.; Wen, D.; Kuang, Y.; Cong, J.; Chen, C.; He, X.; Heenan, M.; Lu, H.; Zhang, Y. Soil pH predominantly controls the forms of organic phosphorus in topsoils under natural broadleaved forests along a 2500km latitudinal gradient. *Geoderma* **2018**, *315*, 65–74. [CrossRef]

58. Turner, B.L.; Blackwell, M.S.A. Isolating the influence of pH on the amounts and forms of soil organic phosphorus. *Eur. J. Soil Sci.* **2013**, *64*, 249–259. [CrossRef]

59. Li, R.; Zhao, J.; Sun, C.; Lu, W.; Guo, C.; Xiao, K. Biochemical properties, molecular characterizations, functions, and application perspectives of phytases. *Front. Agric. China* **2010**, *4*, 195–209. [CrossRef]

60. Vohra, A.; Satyanarayana, T. Phytases: Microbial Sources, Production, Purification, and Potential Biotechnological Applications. *Crit. Rev. Biotechnol.* **2003**, *23*, 29–60. [CrossRef]

61. McKelvie, I.D.; Hart, B.T.; Cardwell, T.J.; Cattrall, R.W. Use of immobilized 3-phytase and flow injection for the determination of phosphorus species in natural waters. *Anal. Chim. Acta* **1995**, *316*, 277–289. [CrossRef]

62. Ullah, A.H.J.; Gibson, D.M. Extracellular Phytase (E. C. 3.1.3.8) from Aspergillus Ficuum NRRL 3135: Purification and Characterization. *Prep. Biochem.* **1987**, *17*, 63–91.

63. Konietzny, U.; Greiner, R. Molecular and catalytic properties of phytate-degrading enzymes (phytases). *Int. J. Food Sci. Technol.* **2002**, *37*, 791–812. [CrossRef]

64. Turner, B.L.; Haygarth, P.M. Phosphatase activity in temperate pasture soils: Potential regulation of labile organic phosphorus turnover by phosphodiesterase activity. *Sci. Total Environ.* **2005**, *344*, 27–36. [CrossRef] [PubMed]

65. Yan, Y.; Wan, B.; Liu, F.; Tan, W.; Liu, M.; Feng, X. Adsorption-Desorption of Myo-Inositol Hexakisphosphate on Hematite. *Soil Sci.* **2014**, *179*, 476–485. [CrossRef]

66. Saeki, K.; Kunito, T.; Sakai, M. Effects of pH, ionic strength, and solutes on DNA adsorption by andosols. *Biol. Fertil. Soils* **2010**, *46*, 531–535. [CrossRef]

67. Deiss, L.; de Moraes, A.; Maire, V. Environmental drivers of soil phosphorus composition in natural ecosystems. *Biogeosciences* **2018**, *15*, 4575–4592. [CrossRef]

68. Siebers, N.; Sumann, M.; Kaiser, K.; Amelung, W. Climatic Effects on Phosphorus Fractions of Native and Cultivated North American Grassland Soils. *Soil Sci. Soc. Am. J.* **2017**, *81*, 299–309. [CrossRef]

69. Dalai, R.C. Soil Organic Phosphorus. In *Advances in Agronomy*; Brady, N.C., Ed.; Academic Press: Cambridge, MA, USA, 1977; Volume 29, pp. 83–117.

70. Moata, M.R.S.; Doolette, A.L.; Smernik, R.J.; McNeill, A.M.; Macdonald, L.M. Organic phosphorus speciation in Australian Red Chromosols: Stoichiometric control. *Soil Res.* **2016**, *54*, 11–19. [CrossRef]

71. Spohn, M. Phosphorus and carbon in soil particle size fractions: A synthesis. *Biogeochemistry* **2020**, *147*, 225–242. [CrossRef]

72. Bowman, R.A.; Cole, C.V. Transformations of organic phosphorus substrates in soils as evaluated by NaHCO3 extraction. *Soil Sci.* **1978**, *125*, 49–54. [CrossRef]

73. Allison, V.J.; Condron, L.M.; Peltzer, D.A.; Richardson, S.J.; Turner, B.L. Changes in enzyme activities and soil microbial community composition along carbon and nutrient gradients at the Franz Josef chronosequence, New Zealand. *Soil Biol. Biochem.* **2007**, *39*, 1770–1781. [CrossRef]

74. Yadav, R.S.; Tarafdar, J.C. Phytase and phosphatase producing fungi in arid and semi-arid soils and their efficiency in hydrolyzing different organic P compounds. *Soil Biol. Biochem.* **2003**, *35*, 745–751. [CrossRef]

75. George, T.S.; Simpson, R.J.; Gregory, P.J.; Richardson, A.E. Differential interaction of Aspergillus niger and Peniophora lycii phytases with soil particles affects the hydrolysis of inositol phosphates. *Soil Biol. Biochem.* **2007**, *39*, 793–803. [CrossRef]

76. Renella, G.; Szukics, U.; Landi, L.; Nannipieri, P. Quantitative assessment of hydrolase production and persistence in soil. *Biol. Fertil. Soils* **2007**, *44*, 321–329. [CrossRef]

77. Skujiņš, J.; Burns, R.G. Extracellular Enzymes in Soil. *CRC Crit. Rev. Microbiol.* **1976**, *4*, 383–421.

78. Gerke, J. Phytate (Inositol Hexakisphosphate) in Soil and Phosphate Acquisition from Inositol Phosphates by Higher Plants. A Review. *Plants* **2015**, *4*, 253–266. [CrossRef]

79. Doolette, A.L.; Smernik, R.J.; Dougherty, W.J. Rapid decomposition of phytate applied to a calcareous soil demonstrated by a solution 31P NMR study. *Eur. J. Soil Sci.* **2010**, *61*, 563–575. [CrossRef]

80. Leytem, A.B.; Smith, D.R.; Applegate, T.J.; Thacker, P.A. The Influence of Manure Phytic Acid on Phosphorus Solubility in Calcareous Soils. *Soil Sci. Soc. Am. J.* **2006**, *70*, 1629–1638. [CrossRef]

81. Annaheim, K.E.; Rufener, C.B.; Frossard, E.; Bünemann, E.K. Hydrolysis of organic phosphorus in soil water suspensions after addition of phosphatase enzymes. *Biol. Fertil. Soils* **2013**, *49*, 1203–1213. [CrossRef]

82. Turner, B.L.; McKelvie, I.D.; Haygarth, P.M. Characterisation of water-extractable soil organic phosphorus by phosphatase hydrolysis. *Soil Biol. Biochem.* **2002**, *34*, 27–35. [CrossRef]

83. Toor, G.S.; Condron, L.M.; Cade-Menun, B.J.; Di, H.J.; Cameron, K.C. Preferential phosphorus leaching from an irrigated grassland soil. *Eur. J. Soil Sci.* **2005**, *56*, 155–168. [CrossRef]

84. Toor, G.S.; Condron, L.M.; Di, H.J.; Cameron, K.C.; Cade-Menun, B.J. Characterization of organic phosphorus in leachate from a grassland soil. *Soil Biol. Biochem.* **2003**, *35*, 1317–1323. [CrossRef]

85. Darch, T.; Blackwell, M.S.A.; Hawkins, J.M.B.; Haygarth, P.M.; Chadwick, D. A Meta-Analysis of Organic and Inorganic Phosphorus in Organic Fertilizers, Soils, and Water: Implications for Water Quality. *Crit. Rev. Environ. Sci. Technol.* **2014**, *44*, 2172–2202. [CrossRef]

86. Jarosch, K.A.; Doolette, A.L.; Smernik, R.J.; Tamburini, F.; Frossard, E.; Bünemann, E.K. Characterisation of soil organic phosphorus in NaOH-EDTA extracts: A comparison of 31P NMR spectroscopy and enzyme addition assays. *Soil Biol. Biochem.* **2015**, *91*, 298–309. [CrossRef]

87. McLaren, T.I.; Smernik, R.J.; McLaughlin, M.J.; McBeath, T.M.; Kirby, J.K.; Simpson, R.J.; Guppy, C.N.; Doolette, A.L.; Richardson, A.E. Complex Forms of Soil Organic Phosphorus–A Major Component of Soil Phosphorus. *Environ. Sci. Technol.* **2015**, *49*, 13238–13245. [CrossRef] [PubMed]

88. Pierzynski, J.; Hettiarachchi, G.M. Reactions of Phosphorus Fertilizers with and without a Fertilizer Enhancer in Three Acidic Soils with High Phosphorus-Fixing Capacity. *Soil Sci. Soc. Am. J.* **2018**, *82*, 1124–1139. [CrossRef]

89. Luo, L.; Ma, Y.; Sanders, R.L.; Xu, C.; Li, J.; Myneni, S.C.B. Phosphorus speciation and transformation in long-term fertilized soil: Evidence from chemical fractionation and P K-edge XANES spectroscopy. *Nutr. Cycl. Agroecosystems* **2017**, *107*, 215–226. [CrossRef]

90. Kar, G.; Peak, D.; Schoenau, J.J. Spatial Distribution and Chemical Speciation of Soil Phosphorus in a Band Application. *Soil Sci. Soc. Am. J.* **2012**, *76*, 2297–2306. [CrossRef]

91. Liu, J.; Yang, J.; Cade-Menun, B.J.; Hu, Y.; Li, J.; Peng, C.; Ma, Y. Molecular speciation and transformation of soil legacy phosphorus with and without long-term phosphorus fertilization: Insights from bulk and microprobe spectroscopy. *Sci. Rep.* **2017**, *7*, 15354. [CrossRef] [PubMed]

92. Ajiboye, B.; Akinremi, O.O.; Hu, Y.; Jürgensen, A. XANES Speciation of Phosphorus in Organically Amended and Fertilized Vertisol and Mollisol. *Soil Sci. Soc. Am. J.* **2008**, *72*, 1256–1262. [CrossRef]

93. Gamble, A.V.; Northrup, P.A.; Sparks, D.L. Elucidation of soil phosphorus speciation in mid-Atlantic soils using synchrotron-based microspectroscopic techniques. *J. Environ. Qual.* **2020**, *49*, 184–193. [CrossRef]

94. Abdala, D.B.; Moore, P.A.; Rodrigues, M.; Herrera, W.F.; Pavinato, P.S. Long-term effects of alum-treated litter, untreated litter and NH4NO3 application on phosphorus speciation, distribution and reactivity in soils using K-edge XANES and chemical fractionation. *J. Environ. Manag.* **2018**, *213*, 206–216. [CrossRef] [PubMed]

95. Cade-Menun, B.J. Characterizing phosphorus forms in cropland soils with solution 31P-NMR: Past studies and future research needs. *Chem. Biol. Technol. Agric.* **2017**, *4*, 19. [CrossRef]

96. Annaheim, K.E.; Doolette, A.L.; Smernik, R.J.; Mayer, J.; Oberson, A.; Frossard, E.; Bünemann, E.K. Long-term addition of organic fertilizers has little effect on soil organic phosphorus as characterized by 31P NMR spectroscopy and enzyme additions. *Geoderma* **2015**, *257-258*, 67–77. [CrossRef]

97. Cade-Menun, B.J.; Doody, D.G.; Liu, C.W.; Watson, C.J. Long-term Changes in Grassland Soil Phosphorus with Fertilizer Application and Withdrawal. *J. Environ. Qual.* **2017**, *46*, 537–545. [CrossRef]

98. Recena, R.; Torrent, J.; del Campillo, M.C.; Delgado, A. Accuracy of Olsen P to assess plant P uptake in relation to soil properties and P forms. *Agron. Sustain. Dev.* **2015**, *35*, 1571–1579. [CrossRef]

99. Mander, C.; Wakelin, S.; Young, S.; Condron, L.; O'Callaghan, M. Incidence and diversity of phosphate-solubilising bacteria are linked to phosphorus status in grassland soils. *Soil Biol. Biochem.* **2012**, *44*, 93–101. [CrossRef]

100. Richardson, A.E.; Simpson, R.J. Soil Microorganisms Mediating Phosphorus Availability Update on Microbial Phosphorus. *Plant Physiol.* **2011**, *156*, 989–996. [CrossRef]

101. Lynch, J. Root Architecture and Plant Productivity. *Plant Physiol.* **1995**, *109*, 7–13. [CrossRef] [PubMed]

102. Richardson, A.E.; Lynch, J.P.; Ryan, P.R.; Delhaize, E.; Smith, F.A.; Smith, S.E.; Harvey, P.R.; Ryan, M.H.; Veneklaas, E.J.; Lambers, H.; et al. Plant and microbial strategies to improve the phosphorus efficiency of agriculture. *Plant Soil* **2011**, *349*, 121–156. [CrossRef]

103. Niu, Y.F.; Chai, R.S.; Jin, G.L.; Wang, H.; Tang, C.X.; Zhang, Y.S. Responses of root architecture development to low phosphorus availability: A review. *Ann. Bot.* **2012**, *112*, 391–408. [CrossRef] [PubMed]

104. Lambers, H.; Shane, M.W.; Cramer, M.D.; Pearse, S.J.; Veneklaas, E.J. Root Structure and Functioning for Efficient Acquisition of Phosphorus: Matching Morphological and Physiological Traits. *Ann. Bot.* **2006**, *98*, 693–713. [CrossRef] [PubMed]

105. Lambers, H.; Finnegan, P.M.; Laliberté, E.; Pearse, S.J.; Ryan, M.H.; Shane, M.W.; Veneklaas, E.J. Phosphorus Nutrition of Proteaceae in Severely Phosphorus-Impoverished Soils: Are There Lessons To Be Learned for Future Crops? *Plant Physiol.* **2011**, *156*, 1058–1066. [CrossRef] [PubMed]

106. Williamson, L.C.; Ribrioux, S.P.C.P.; Fitter, A.H.; Leyser, H.M.O. Phosphate Availability Regulates Root System Architecture in Arabidopsis. *Plant Physiol.* **2001**, *126*, 875–882. [CrossRef] [PubMed]

107. Jin, J.; Tang, C.; Armstrong, R.; Sale, P. Phosphorus supply enhances the response of legumes to elevated CO2 (FACE) in a phosphorus-deficient vertisol. *Plant Soil* **2012**, *358*, 91–104. [CrossRef]

108. Rose, T.J.; Impa, S.M.; Rose, M.T.; Pariasca-Tanaka, J.; Mori, A.; Heuer, S.; Johnson-Beebout, S.E.; Wissuwa, M. Enhancing phosphorus and zinc acquisition efficiency in rice: A critical review of root traits and their potential utility in rice breeding. *Ann. Bot.* **2013**, *112*, 331–345. [CrossRef]

109. Lynch, J.; van Beem, J.J. Growth and Architecture of Seedling Roots of Common Bean Genotypes. *Crop Sci.* **1993**, *33*, cropsci1993.0011183X003300060028x. [CrossRef]

110. Wang, Y.L.; Almvik, M.; Clarke, N.; Eich-Greatorex, S.; Øgaard, A.F.; Krogstad, T.; Lambers, H.; Clarke, J.L. Contrasting responses of root morphology and root-exuded organic acids to low phosphorus availability in three important food crops with divergent root traits. *AoB Plants* **2015**, *7*. [CrossRef]

111. Henry, A.; Chaves, N.F.; Kleinman, P.J.A.; Lynch, J.P. Will nutrient-efficient genotypes mine the soil? Effects of genetic differences in root architecture in common bean (*Phaseolus vulgaris* L.) on soil phosphorus depletion in a low-input agro-ecosystem in Central America. *Field Crops Res.* **2010**, *115*, 67–78. [CrossRef]

112. da Silva, A.; Bruno, I.P.; Franzini, V.I.; Marcante, N.C.; Benitiz, L.; Muraoka, T. Phosphorus uptake efficiency, root morphology and architecture in Brazilian wheat cultivars. *J. Radioanal. Nucl. Chem.* **2016**, *307*, 1055–1063. [CrossRef]

113. Salvi, S. An evo-devo perspective on root genetic variation in cereals. *J. Exp. Bot.* **2017**, *68*, 351–354. [CrossRef]
114. Rich, S.M.; Watt, M. Soil conditions and cereal root system architecture: Review and considerations for linking Darwin and Weaver. *J. Exp. Bot.* **2013**, *64*, 1193–1208. [CrossRef] [PubMed]
115. Watt, M.; Moosavi, S.; Cunningham, S.C.; Kirkegaard, J.A.; Rebetzke, G.J.; Richards, R.A. A rapid, controlled-environment seedling root screen for wheat correlates well with rooting depths at vegetative, but not reproductive, stages at two field sites. *Ann. Bot.* **2013**, *112*, 447–455. [CrossRef] [PubMed]
116. White, P.J.; George, T.S.; Gregory, P.J.; Bengough, A.G.; Hallett, P.D.; McKenzie, B.M. Matching roots to their environment. *Ann. Bot.* **2013**, *112*, 207–222. [CrossRef] [PubMed]
117. Erel, R.; Bérard, A.; Capowiez, L.; Doussan, C.; Arnal, D.; Souche, G.; Gavaland, A.; Fritz, C.; Visser, E.J.W.; Salvi, S.; et al. Soil type determines how root and rhizosphere traits relate to phosphorus acquisition in field-grown maize genotypes. *Plant Soil* **2017**, *412*, 115–132. [CrossRef]
118. Li, H.; Zhang, D.; Wang, X.; Li, H.; Rengel, Z.; Shen, J. Competition between Zea mays genotypes with different root morphological and physiological traits is dependent on phosphorus forms and supply patterns. *Plant Soil* **2019**, *434*, 125–137. [CrossRef]
119. Guppy, C.N.; Menzies, N.W.; Moody, P.W.; Blamey, F.P.C. Competitive sorption reactions between phosphorus and organic matter in soil: A review. *Soil Res.* **2005**, *43*, 189–202. [CrossRef]
120. Wang, Y.; Lambers, H. Root-released organic anions in response to low phosphorus availability: Recent progress, challenges and future perspectives. *Plant Soil* **2020**, *447*, 135–156. [CrossRef]
121. Jones, D.L. Organic acids in the rhizosphere—A critical review. *Plant Soil* **1998**, *205*, 25–44. [CrossRef]
122. Oburger, E.; Leitner, D.; Jones, D.L.; Zygalakis, K.C.; Schnepf, A.; Roose, T. Adsorption and desorption dynamics of citric acid anions in soil. *Eur. J. Soil Sci.* **2011**, *62*, 733–742. [CrossRef]
123. Oburger, E.; Jones, D.L.; Wenzel, W.W. Phosphorus saturation and pH differentially regulate the efficiency of organic acid anion-mediated P solubilization mechanisms in soil. *Plant Soil* **2011**, *341*, 363–382. [CrossRef]
124. Touhami, D.; McDowell, R.W.; Condron, L.M. Role of Organic Anions and Phosphatase Enzymes in Phosphorus Acquisition in the Rhizospheres of Legumes and Grasses Grown in a Low Phosphorus Pasture Soil. *Plants* **2020**, *9*, 1185. [CrossRef]
125. Pearse, S.J.; Veneklaas, E.J.; Cawthray, G.R.; Bolland, M.D.A.; Lambers, H. Carboxylate release of wheat, canola and 11 grain legume species as affected by phosphorus status. *Plant Soil* **2006**, *288*, 127–139. [CrossRef]
126. Wang, Y.; Krogstad, T.; Clarke, J.L.; Hallama, M.; Øgaard, A.F.; Eich-Greatorex, S.; Kandeler, E.; Clarke, N. Rhizosphere Organic Anions Play a Minor Role in Improving Crop Species' Ability to Take Up Residual Phosphorus (P) in Agricultural Soils Low in P Availability. *Front. Plant Sci.* **2016**, *7*. [CrossRef] [PubMed]
127. Sun, B.; Gao, Y.; Wu, X.; Ma, H.; Zheng, C.; Wang, X.; Zhang, H.; Li, Z.; Yang, H. The relative contributions of pH, organic anions, and phosphatase to rhizosphere soil phosphorus mobilization and crop phosphorus uptake in maize/alfalfa polyculture. *Plant Soil* **2020**, *447*, 117–133. [CrossRef]
128. Lynch, J.P. Root Phenes for Enhanced Soil Exploration and Phosphorus Acquisition: Tools for Future Crops. *Plant Physiol.* **2011**, *156*, 1041–1049. [CrossRef] [PubMed]
129. Hasan, M.M.; Hasan, M.M.; Teixeira da Silva, J.A.; Li, X. Regulation of phosphorus uptake and utilization: Transitioning from current knowledge to practical strategies. *Cell. Mol. Biol. Lett.* **2016**, *21*, 7. [CrossRef] [PubMed]
130. Liu, X.; Zhao, X.; Zhang, L.; Lu, W.; Li, X.; Xiao, K. TaPht1;4, a high-affinity phosphate transporter gene in wheat (*Triticum aestivum*), plays an important role in plant phosphate acquisition under phosphorus deprivation. *Funct. Plant Biol.* **2013**, *40*, 329–341. [CrossRef] [PubMed]
131. Sun, S.; Gu, M.; Cao, Y.; Huang, X.; Zhang, X.; Ai, P.; Zhao, J.; Fan, X.; Xu, G. A Constitutive Expressed Phosphate Transporter, OsPht1;1, Modulates Phosphate Uptake and Translocation in Phosphate-Replete Rice. *Plant Physiol.* **2012**, *159*, 1571–1581. [CrossRef]
132. Liu, T.-Y.; Huang, T.-K.; Tseng, C.-Y.; Lai, Y.-S.; Lin, S.-I.; Lin, W.-Y.; Chen, J.-W.; Chiou, T.-J. PHO2-Dependent Degradation of *PHO1* Modulates Phosphate Homeostasis in *Arabidopsis*. *Plant Cell* **2012**, *24*, 2168–2183. [CrossRef]
133. Ye, Y.; Yuan, J.; Chang, X.; Yang, M.; Zhang, L.; Lu, K.; Lian, X. The Phosphate Transporter Gene OsPht1;4 Is Involved in Phosphate Homeostasis in Rice. *PLoS ONE* **2015**, *10*, e0126186. [CrossRef]
134. Zhang, F.; Wu, X.-N.; Zhou, H.-M.; Wang, D.-F.; Jiang, T.-T.; Sun, Y.-F.; Cao, Y.; Pei, W.-X.; Sun, S.-B.; Xu, G.-H. Overexpression of rice phosphate transporter gene OsPT6 enhances phosphate uptake and accumulation in transgenic rice plants. *Plant Soil* **2014**, *384*, 259–270. [CrossRef]

135. Wang, Z.; Ruan, W.; Shi, J.; Zhang, L.; Xiang, D.; Yang, C.; Li, C.; Wu, Z.; Liu, Y.; Yu, Y.; et al. Rice SPX1 and SPX2 inhibit phosphate starvation responses through interacting with PHR2 in a phosphate-dependent manner. *Proc. Natl. Acad. Sci. USA* **2014**, *111*, 14953–14958. [CrossRef] [PubMed]

136. Jia, H.; Ren, H.; Gu, M.; Zhao, J.; Sun, S.; Zhang, X.; Chen, J.; Wu, P.; Xu, G. The phosphate transporter gene OsPht1;8 is involved in phosphate homeostasis in rice. *Plant Physiol.* **2011**, *156*, 1164–1175. [CrossRef] [PubMed]

137. Heuer, S.; Gaxiola, R.; Schilling, R.; Herrera-Estrella, L.; López-Arredondo, D.; Wissuwa, M.; Delhaize, E.; Rouached, H. Improving phosphorus use efficiency: A complex trait with emerging opportunities. *Plant J.* **2017**, *90*, 868–885. [CrossRef]

138. Shimizu, A.; Kato, K.; Komatsu, A.; Motomura, K.; Ikehashi, H. Genetic analysis of root elongation induced by phosphorus deficiency in rice (*Oryza sativa* L.): Fine QTL mapping and multivariate analysis of related traits. *Theor. Appl. Genet.* **2008**, *117*, 987–996. [CrossRef]

139. Zhu, J.; Kaeppler, S.M.; Lynch, J.P. Mapping of QTL controlling root hair length in maize (Zea mays L.) under phosphorus deficiency. *Plant Soil* **2005**, *270*, 299–310. [CrossRef]

140. Liang, Q.; Cheng, X.; Mei, M.; Yan, X.; Liao, H. QTL analysis of root traits as related to phosphorus efficiency in soybean. *Ann. Bot.* **2010**, *106*, 223–234. [CrossRef]

141. Wissuwa, M.; Ae, N. Genotypic variation for tolerance to phosphorus deficiency in rice and the potential for its exploitation in rice improvement. *Plant Breed.* **2001**, *120*, 43–48. [CrossRef]

142. Li, Q.; Sapkota, M.; van der Knaap, E. Perspectives of CRISPR/Cas-mediated cis-engineering in horticulture: Unlocking the neglected potential for crop improvement. *Hortic. Res.* **2020**, *7*, 36. [CrossRef]

143. Doudna, J.A.; Charpentier, E. Genome editing. The new frontier of genome engineering with CRISPR-Cas9. *Science* **2014**, *346*, 1258096. [CrossRef]

144. Gaxiola, R.A.; Regmi, K.; Hirschi, K.D. Moving On Up: H(+)-PPase Mediated Crop Improvement. *Trends Biotechnol.* **2016**, *34*, 347–349. [CrossRef] [PubMed]

145. Rouached, H.; Stefanovic, A.; Secco, D.; Bulak Arpat, A.; Gout, E.; Bligny, R.; Poirier, Y. Uncoupling phosphate deficiency from its major effects on growth and transcriptome via PHO1 expression in Arabidopsis. *Plant J.* **2011**, *65*, 557–570. [CrossRef] [PubMed]

146. Ai, P.; Sun, S.; Zhao, J.; Fan, X.; Xin, W.; Guo, Q.; Yu, L.; Shen, Q.; Wu, P.; Miller, A.J.; et al. Two rice phosphate transporters, OsPht1;2 and OsPht1;6, have different functions and kinetic properties in uptake and translocation. *Plant J.* **2009**, *57*, 798–809. [CrossRef] [PubMed]

147. Alori, E.T.; Glick, B.R.; Babalola, O.O. Microbial Phosphorus Solubilization and Its Potential for Use in Sustainable Agriculture. *Front. Microbiol.* **2017**, *8*, 971. [CrossRef] [PubMed]

148. Calvo, P.; Nelson, L.; Kloepper, J.W. Agricultural uses of plant biostimulants. *Plant Soil* **2014**, *383*, 3–41. [CrossRef]

149. Pii, Y.; Mimmo, T.; Tomasi, N.; Terzano, R.; Cesco, S.; Crecchio, C. Microbial interactions in the rhizosphere: Beneficial influences of plant growth-promoting rhizobacteria on nutrient acquisition process. A review. *Biol. Fertil. Soils* **2015**, *51*, 403–415. [CrossRef]

150. Bargaz, A.; Lyamlouli, K.; Chtouki, M.; Zeroual, Y.; Dhiba, D. Soil Microbial Resources for Improving Fertilizers Efficiency in an Integrated Plant Nutrient Management System. *Front. Microbiol.* **2018**, *9*, 1606. [CrossRef]

151. Kafle, A.; Cope, K.R.; Raths, R.; Krishna Yakha, J.; Subramanian, S.; Bücking, H.; Garcia, K. Harnessing Soil Microbes to Improve Plant Phosphate Efficiency in Cropping Systems. *Agronomy* **2019**, *9*, 127. [CrossRef]

152. Brundrett, M.C.; Tedersoo, L. Evolutionary history of mycorrhizal symbioses and global host plant diversity. *New Phytol.* **2018**, *220*, 1108–1115. [CrossRef] [PubMed]

153. Smith, S.E.; Smith, F.A. Roles of Arbuscular Mycorrhizas in Plant Nutrition and Growth: New Paradigms from Cellular to Ecosystem Scales. *Ann. Rev. Plant Biol.* **2011**, *62*, 227–250. [CrossRef]

154. Smith, S.E.; Jakobsen, I.; Grønlund, M.; Smith, F.A. Roles of Arbuscular Mycorrhizas in Plant Phosphorus Nutrition: Interactions between Pathways of Phosphorus Uptake in Arbuscular Mycorrhizal Roots Have Important Implications for Understanding and Manipulating Plant Phosphorus Acquisition. *Plant Physiol.* **2011**, *156*, 1050–1057. [CrossRef] [PubMed]

155. Peng, S.; Eissenstat, D.M.; Graham, J.H.; Williams, K.; Hodge, N.C. Growth Depression in Mycorrhizal Citrus at High-Phosphorus Supply (Analysis of Carbon Costs). *Plant Physiol.* **1993**, *101*, 1063–1071. [CrossRef] [PubMed]

156. Krey, T.; Baum, C.; Ruppel, S.; Seydel, M.; Eichler-Löbermann, B. Organic and Inorganic P Sources Interacting with Applied Rhizosphere Bacteria and Their Effects on Growth and P Supply of Maize. *Commun. Soil Sci. Plant Anal.* **2013**, *44*, 3205–3215. [CrossRef]

157. Mercl, F.; Tejnecký, V.; Ságová-Marečková, M.; Dietel, K.; Kopecký, J.; Břendová, K.; Kulhánek, M.; Košnář, Z.; Száková, J.; Tlustoš, P. Co-application of wood ash and *Paenibacillus mucilaginosus* to soil: The effect on maize nutritional status, root exudation and composition of soil solution. *Plant Soil* **2018**, *428*, 105–122. [CrossRef]
158. Güneş, A.; Ataoğlu, N.; Turan, M.; Eşitken, A.; Ketterings, Q.M. Effects of phosphate-solubilizing microorganisms on strawberry yield and nutrient concentrations. *J. Plant Nutr. Soil Sci.* **2009**, *172*, 385–392.
159. Mosimann, C.; Oberhänsli, T.; Ziegler, D.; Nassal, D.; Kandeler, E.; Boller, T.; Mäder, P.; Thonar, C. Tracing of Two *Pseudomonas* Strains in the Root and Rhizoplane of Maize, as Related to Their Plant Growth-Promoting Effect in Contrasting Soils. *Front. Microbiol.* **2017**, *7*, 2150. [CrossRef]
160. Varga, T.; Hixson, K.K.; Ahkami, A.H.; Sher, A.W.; Barnes, M.E.; Chu, R.K.; Battu, A.K.; Nicora, C.D.; Winkler, T.E.; Reno, L.R.; et al. Endophyte-Promoted Phosphorus Solubilization in Populus. *Front. Plant Sci.* **2020**, *11*. [CrossRef]
161. Menezes-Blackburn, D.; Jorquera, M.; Gianfreda, L.; Rao, M.; Greiner, R.; Garrido, E.; de la Luz Mora, M. Activity stabilization of Aspergillus niger and Escherichia coli phytases immobilized on allophanic synthetic compounds and montmorillonite nanoclays. *Bioresour. Technol.* **2011**, *102*, 9360–9367. [CrossRef] [PubMed]
162. Calabi-Floody, M.; Velásquez, G.; Gianfreda, L.; Saggar, S.; Bolan, N.; Rumpel, C.; Mora, M.L. Improving bioavailability of phosphorous from cattle dung by using phosphatase immobilized on natural clay and nanoclay. *Chemosphere* **2012**, *89*, 648–655. [CrossRef]
163. Menezes-Blackburn, D.; Jorquera, M.A.; Gianfreda, L.; Greiner, R.; de la Luz Mora, M. A novel phosphorus biofertilization strategy using cattle manure treated with phytase–nanoclay complexes. *Biol. Fertil. Soils* **2014**, *50*, 583–592. [CrossRef]
164. Jones, J.L.; Yingling, Y.G.; Reaney, I.M.; Westerhoff, P. Materials matter in phosphorus sustainability. *MRS Bull.* **2020**, *45*, 7–10. [CrossRef]

Publisher's Note: MDPI stays neutral with regard to jurisdictional claims in published maps and institutional affiliations.

soil systems

MDPI

Article

Influence of Soil and Manure Management Practices on Surface Runoff Phosphorus and Nitrogen Loss in a Corn Silage Production System: A Paired Watershed Approach

Jessica F. Sherman [1], Eric O. Young [1,*], William E. Jokela [1], Michael D. Casler [2], Wayne K. Coblentz [1] and Jason Cavadini [3]

[1] USDA-ARS, Institute for Environmentally Integrated Dairy Management, 2615 Yellowstone Dr., Marshfield, WI 54449, USA; jessica.sherman@usda.gov (J.F.S.); jokela@wisc.edu (W.E.J.); wayne.coblentz@usda.gov (W.K.C.)
[2] USDA-ARS, Dairy Forage Research Center, 1925 Linden Dr., Madison, WI 53706, USA; michael.casler@usda.gov
[3] Marshfield Agricultural Research Station, University of Wisconsin, M605 Drake Ave., Stratford, WI 54484, USA; jason.cavadini@wisc.edu
*   Correspondence: eric.young@usda.gov

Citation: Sherman, J.F.; Young, E.O.; Jokela, W.E.; Casler, M.D.; Coblentz, W.K.; Cavadini, J. Influence of Soil and Manure Management Practices on Surface Runoff Phosphorus and Nitrogen Loss in a Corn Silage Production System: A Paired Watershed Approach. *Soil Syst.* **2021**, *5*, 1. https://doi.org/10.3390/soilsystems5010001

Received: 14 November 2020
Accepted: 24 December 2020
Published: 29 December 2020

Publisher's Note: MDPI stays neutral with regard to jurisdictional claims in published maps and institutional affiliations.

Abstract: Best management practices (BMPs) can mitigate erosion and nutrient runoff. We evaluated runoff losses for silage corn management systems using paired watershed fields in central Wisconsin. A two-year calibration period of fall-applied liquid dairy manure incorporated with chisel plow tillage (FMT) was followed by a three and a half-year treatment period. During the treatment period FMT was continued on one field, and three different systems on the others: (a) fall-applied manure and chisel tillage plus a vegetative buffer strip (BFMT); (b) a fall rye cover crop with spring manure application and chisel tillage (RSMT), both BMPs; a common system (c) fall manure application with spring chisel tillage (FMST). Year-round runoff monitoring included flow, suspended sediment (SS), total phosphorus (TP), dissolved reactive phosphorus (DRP), ammonium (NH_4^+-N), nitrate, and total nitrogen (TN). Results showed BFMT reduced runoff SS, TP, and TN concentration and load compared to FMT. The RSMT system reduced concentrations of SS, TP, and TN, but not load because of increased runoff. The FMST practice increased TP, DRP, and NH_4^+-N loads by 39, 376, and 197%, respectively. While BMPs showed mitigation potential for SS, TN, and TP, none controlled DRP, suggesting additional practices may be needed in manured corn silage fields with high runoff potential.

Keywords: best management practices; corn silage; erosion; nutrient management; liquid manure; surface runoff; water quality

1. Introduction

Corn silage is an important crop to dairy producers in the US and globally. A lack of crop residue after silage harvest increases erosion potential and phosphorus (P) and nitrogen (N) transport to surface waters compared to less erosive crops [1–7]. The US dairy industry faces increasing pressure from regulatory agencies and the public to improve on-farm nutrient use efficiency while decreasing environmental impacts [8]. While dairy manure application on silage fields provides needed nutrients and is a source of organic matter for soil quality, it also contributes to runoff P and N losses and greenhouse gas emissions [8].

Nutrient management practices such as manure application methods, tillage practices, cover crops, and grass buffers can reduce erosion and P transport in corn production systems [1–3,5,9–17], however, it is important to recognize that they can also have differential effects on sediment and P transport. For example, cover crops and reduced tillage can decrease erosion and particulate P loss but increase bioavailable P loss in the form of dissolved reactive P (DRP) transport in surface runoff [18–20]. With long-term reduced and/or no-tillage, DRP can be increased relative to conventional tillage from a combination of lower erosion, nutrient stratification, and/or dissolved nutrient release from crop residues [21–24]. Snowmelt runoff in cold climates often accounts for a substantial fraction of total annual runoff and N and P loss. Several studies conducted in cold climates report a dominance of DRP in snowmelt runoff compared to particulate P, which tends to be higher with more erosive rainfall events compared to snowmelt [19,25,26]. Establishing a winter rye cover crop after corn silage harvest in the fall can help mitigate dissolved and particulate-bound nutrients in surface runoff, including ammonium-N (NH_4^+-N) and nitrate-N (NO_3-N) [5,11,27,28]. Grass buffer strips are another commonly recommended BMP to reduce erosion and edge-of-field nutrient loss in agricultural systems [15,29]. Mayer et al. [30] compared data from 88 studies evaluating vegetative buffer N removal effectiveness and concluded adequate vegetation was a critical factor for mitigating erosion and enhancing N attenuation. While plant buffer species did not affect N removal, buffers > 50 m removed N more consistently than 0 to 25 m wide strips.

Research indicates that individual and combinations of best practices can mitigate erosion and surface runoff nutrient loss potential. Much of this is from field plots, however, which largely ignores the dynamic nature and complex hydrology of larger fields/watersheds [31–33]. Evaluating the effectiveness of practices at the landscape and watershed scales is also challenging due to the heterogeneity of runoff processes [34–36]. Paired watershed designs account for inherent physiographic differences between watersheds and can help to isolate management effects on runoff water quality [9,32,37,38]. The objective of this study was to use a paired-watershed approach to quantify the effectiveness of targeted BMPs to mitigate surface runoff nutrient transport from corn silage fields in central Wisconsin (WI) with high runoff potential. Specifically, the following management systems were compared to the typical practice of fall manure application with chisel tillage incorporation (FMT) (see Table S1 for abbreviation guide): (i) fall-applied manure and chisel tillage with the inclusion of a grass-legume buffer strip along the lower field edge (BFMT); (ii) a fall rye cover crop with spring manure application and chisel tillage (RSMT); (iii) fall manure application with spring chisel tillage (FMST).

2. Materials and Methods

2.1. Research Site

This study was conducted at the University of Wisconsin/USDA-ARS Marshfield Agricultural Research Station near Stratford, WI. The fields are mapped as somewhat poorly drained Withee silt loam (fine-loamy, mixed, superactive, frigid Aquic Glossudalfs; 1–3% slope). These soils have a dense, compacted B-horizon at approximately 50 cm resulting in high runoff potential. The 30-year average annual temperature and precipitation at the site are 6.9 °C and 831 mm, respectively. The site is comprised of four forage crop production fields separated by earthen berms directing runoff to individual flumes, with each field functioning as a small watershed. Each field is approximately 1.6 ha and are referred to individually as M1, M2, M3, and M4. The average field slope is approximately 2% for M1, M2, and M4. The average slope for field M3 is 0.25% for the lower third of the field area and 3% for the upper 2/3 of the field. The lower area of M3 is more imperfectly drained with visible standing water after heavy rainfall and snowmelt events. More details on site establishment and calibration period (2006–2008) results are presented in Jokela and Casler [39].

2.2. Field Treatments and Agronomic Considerations

During the calibration period (2006–2008), all fields were managed identically with fall liquid dairy manure application (at a rate to meet 80% of annual corn N need) and chisel plow tillage to a depth of 15 cm (2.7 m wide seven-shank plow Landoll Farm Equipment, Brillion, WI, USA) the same day. Field cultivation to a depth of 7.5 cm (6.2 m wide John Deere, Moline, IL, USA) was performed in the spring before planting corn.

The same management was continued on the designated control field (M1) during the treatment period, but management was changed on the others. Field M4 (BFMT) was treated identically to M1 but with the addition of a mixed grass-legume buffer strip (9.8 m wide) along the lower side of the field above the drainageway. The buffer strip was planted on 3 October 2008 using a no-till drill (John Deere 1590, Moline, IL, USA) with 112 kg ha^{-1} winter rye (*Secale cereale* L.), 2.2 kg ha^{-1} alsike clover (*Trifolium hybridum* L.), 9.0 kg ha^{-1} tall fescue (*Festuca arundinaceum*), and 3.4 kg ha^{-1} smooth bromegrass (*Bromus inermis* L.). While the vegetative buffer was implemented as a conservation practice, it can also provide an economic return as harvestable hay forage crop for dairy animals and removes soil N and P. In our study the buffer vegetation was harvested approximately twice per year with field-scale equipment, but yield and nutrient data are incomplete. In M2, a winter rye cover crop was planted in early Oct after the silage corn harvest at 112 kg ha^{-1} with a no-till drill (John Deere 1590, Moline, IL, USA) and manure application and chisel tillage was delayed until spring (RSMT). On field M3, liquid dairy manure was broadcast on the surface in the fall (without incorporation) and chisel tilled the following spring (FMST). While not considered a BMP because of the lack of manure incorporation, it is still a relatively common practice, so we chose to include it to assess potential effects on surface runoff water quality.

Liquid dairy manure was sampled at each application and analyzed for N, P, potassium (K), NH$_4^+$-N, and dry matter content by the University of WI Soil and Forage Lab [40] (Table 1). Manure application rates averaged 45,880 L ha^{-1} and ranged from 3.2% to 18% dry matter content. Nutrient application rates are presented in Table 1. Average total N and P application rates from manure were approximately 155 and 24 kg ha^{-1} year^{-1}, respectively. This N application met approximately 60% of crop needs (based on N availability of 50% in the first year and 10% in the second) slightly less than during calibration due to lower manure N contents in 2010 and 2011 than expected. Each year, corn was planted on all fields on the same day. Corn (*Zea Mays* L.) (2905RB; 89-day RM; YGCB RR, in 2009 and 2010; RK212GT; 81-day RM; RR, in 2011) was planted in May or late April at 87,500 seeds ha^{-1} with 112 kg ha^{-1} of 9-11-30-6S-1Zn starter fertilizer applied as a band (50 mm to the side of the seed row and 50 mm deep) via the planter. Additional fertilizer N was applied in June or July as needed (Table 1) based on a pre-sidedress nitrate soil test [41]. Corn was harvested for silage on the same day for all fields between mid-Sept and early Oct (Table 1). Yields were estimated by hand-harvesting above-ground corn biomass samples (all but the bottom 25 cm of stalk) from nine randomly selected, 3 m row-length sub-plots from each field when whole-plant DM content had reached approximately 350 g kg^{-1}. Subsamples were taken and a composite plot sample was dried at 55 °C, ground to pass 1 mm, and analyzed for total N (Elementar Variomax CN analyzer, Ronkonkoma, NY, USA) and P content after nitric acid digestion by ICP-OES (University of WI Soil and Forage Lab) to estimate corn N and P removal. Soil samples were collected: nine 2.5-cm diameter, 20-cm deep samples were taken in each watershed with a hand sampler, every fall throughout the treatment phase, and in the spring of 2012. Plant-available P was extracted using the Bray 1 solution (0.03 N ammonium fluoride + 0.025 N hydrochloric acid; [42]). Phosphorus in extracts was determined colorimetrically (ammonium molybdate solution) using standard techniques (abbreviated as B1P).

Table 1. Field activities and manure composition during the 2008–2011 treatment period.

Event	Manure Nutrient [†]	2008	2009	2010	2011	2012
Fall Rye Planting (M2/RSMT [‡])		3 October	8 October	7 October	4 October	
Fall Manure Application (M1/FMT; M3/FMST; M4/BFMT)		3 November	10 November	4 November	9 November	
				$kg\,ha^{-1}$		
	N	200	164	59	130	
	P	28	29	10	24	
	K	86	152	81	143	
	NH_4-N	96	85	35	65	
	DM	8715	6303	1085	5570	
Spring Manure Application (M2/RSMT)			8 May	27 April	27 May	1 May
				$kg\,ha^{-1}$		
	N		204	157	147	136
	P		25	27	26	22
	K		114	148	134	19
	NH_4-N		94	67	70	64
	DM		9718	10531	5350	3992
Corn Planting (all watersheds)			11 May	28 April	31 May	
Sidedress N (kg ha^{-1})			39	71	42; 92 [†‡]	
Corn Silage Harvest (all watersheds)			30 September	14 September	5 October	

[†] N = nitrogen, P = phosphorus, K = potassium, NH_4-N = ammonium nitrogen; [‡] RSMT = fall rye (cover crop) with spring applied manure and chisel tillage; FMST = fall applied manure with spring tillage; BFMT = fall applied manure/chisel tillage with grass buffer; [†‡] 42 kg ha^{-1} on M1 and M4; 92 kg ha^{-1} on M2 and M3.

2.3. Hydrology and Runoff Measurements

Details on runoff instrumentation and monitoring are found in Jokela and Casler [39] and are briefly described here. Runoff was sampled and monitored at gauging stations located at the low elevation point of each field. Original flume design and monitoring procedures were based on those of the US Geological Survey with slight modifications [43]. A 60 cm fiberglass H-flume (Tracom, Inc., Alpharetta, GA, USA) was attached to pressure-treated plywood wingwalls (driven to approximately 60 cm deep and extending approximately 3 m on each side). In November 2007, 1.8 m long channels were installed between the wingwall and the flume to provide more uniform flow entering the flume and a greater distance for deposition of sediment ahead of the flume [44]. Plywood wingwalls were replaced with steel sheet pilings placed 1.2 m deep at M2 on 11 May 2012 and M3 on 24 Mar 2010 because of failure due to frost heaving. Flumes were mounted with threaded rods for leveling as needed. Shallow earthen berms directed surface runoff to each flume. During late fall through early spring, plywood enclosures were attached to the approach channel/flume and a quartz heater was used as needed to prevent freezing of sample lines. Instrumentation was housed inside a $1.8 \times 2.1 \times 2.0\,m^3$ high shed (Niagara model, Yardmate Series, Royal Outdoor Products, Inc., Middleburg Heights, OH, USA) equipped with AC power for data loggers, sampling equipment, heaters, and heat tape with battery backup power.

Runoff volume was determined by the measuring stage in the H-flumes with an air bubbler/pressure transducer flow meter (ISCO Model 4230, Teledyne Isco, Inc., Lincoln,

NE, USA). A bubbler PVC tube (3.175 mm i.d) was attached to the floor of the flume 40 mm back from the outlet. Staff gages were also installed in the H-flumes to allow simultaneous comparison of the stage with that from the flow meter. Time-based runoff samples were collected at intervals based on estimated event runoff quantity by an automated 24-bottle (1 L) refrigerated sampler (ISCO 6712SR, Teledyne Isco, Inc., Lincoln, NE, USA). A sampling tube (9.3 mm i.d.) was attached to the flume floor near the flume outlet and extended approximately 2 m to the automated sampler inside the enclosure (protected from freezing by heat tape and foam insulation). A CR10X datalogger (Campbell Scientific, Inc., Logan, UT, USA) was used to read and store data and control the runoff sampling collection scheme. A weather station (Campbell Scientific, Inc., Logan, UT, USA) was located 1000 m from the site and measured precipitation (tipping bucket), air temperature, humidity, wind, and solar irradiance. Real-time, two-way radio telemetry allowed remote communication with each station and the weather station. A Campbell scientific software program (PC208W) was used for real-time communication to modify sampling intervals as needed.

2.4. Nutrient and Runoff Water Quality Measures

Samples from individual autosampler bottles were combined into a flow weighted composite for each runoff event. Samples were analyzed for suspended sediment (SS; gravimetric method 3977-97B) [45], total P (TP; block digestion, method 4500 P F; [46]), and total Kjeldhal N (TKN; block digester automated colorimetric, 4500 NH_3 G; [46]), a filtered (0.45 μm) subsample was analyzed for DRP (automated colorimetric method 4500 P F; [46]), nitrate + nitrite-N; abbreviated as NO_3-N (automated Cd reduction; 4500 NO_3 F5; [46]), and NH_4^+-N (automated phenolate, 4500 NH_3 G; [46]). Values for TKN and NO_3-N were added to provide an estimate of total N (TN) since NO_3-N is not measured in the TKN procedure.

2.5. Statistical Analysis

Paired watershed analysis involves edge-of-field runoff water quality monitoring of two or more fields similar in soil characteristics with hydrologic isolation to enable fields to function as individual small watersheds. Watersheds are treated identically during the calibration/baseline period to account for runoff variation among watersheds, and event-based runoff data is collected. During the treatment phase, the control watershed continues with same treatment while treatment designated watersheds receive different management treatments. Treatment and calibration regression equations are then tested statistically for differences to determine treatment effects. The direction and magnitude of the effect are determined by comparing values predicted by calibration regression to those observed during the treatment period.

In the present study, the treatment period runoff water quality data are presented in relation to previously reported calibration period results [39]. The number of paired runoff events for statistical analysis ranged from n = 34 to 39 during the calibration period and n = 55 to 61 events during the treatment period. All dependent runoff water quality variables measured (and model residuals) were non-normally distributed. While log transformation is commonly done for non-normality and heteroscedasticity (e.g., [10,19]), log transformations rarely normalized variables and did not adequately resolve variance heterogeneity for this study. A generalized linear mixed modeling approach (log-link function) was also conducted [47] with several distribution types, but heteroscedasticity remained an issue. Given these limitations, permutation tests [48,49] were conducted (rather than analysis of covariance/regression) to test two hypotheses for each dependent variable: (1) calibration period mean equals treatment period mean and (2) calibration period slope equals treatment period slope. A total of 1000 permutations were generated by randomly reassigning calibration and treatment data to the two time periods followed by one-way analysis of variance (Hypothesis 1) or randomly sorting x-axis values followed by estimation of permuted slopes (Hypothesis 2). Hypothesis tests for means and slopes were conducted as two-tailed tests with p-values set at 0.25, 0.10, 0.05, and 0.01 to define significance levels. Predicted constituent concentrations/loads were based on calibration regression equations of Jokela and Casler [39]. The relative impact of BMPs on water quality was expressed as the percent difference between predicted (i.e., from calibration phase equations) and measured values for the treatment. Corn silage yield and soil nutrient data were analyzed by mixed models analysis of variance [50].

3. Results and Discussion

3.1. Weather and Corn Silage Yield

Precipitation during the treatment period (2008–2011) was slightly greater (10%) compared to the calibration period (2006–2008) and more pronounced in the growing season (36% greater than the calibration period; Table 2). Treatment phase growing season rainfall was similar to the 30-year average, whereas the calibration period was 32% drier. During the calibration period, measured runoff averaged 14% of total precipitation. For the treatment phase, the mean runoff fraction was considerably higher (26% of precipitation), most likely due to the overall wetter conditions increasing runoff potential in these somewhat poorly drained soils. Average temperatures for calibration and treatment periods were similar to the 30-year mean.

Corn yield averaged across fields was 16.2 Mg ha^{-1} for the treatment period and in the range of expected yields for central Wisconsin. There were some yield differences among fields with notably lower yield for FMST, which could have been related to greater NH_3 volatilization due to the lack of fall incorporation. However, the lower third of field M3 becomes more imperfectly drained as the main field slope levels out and runoff water collects. It is therefore likely that greater N loss from not incorporating manure and the more imperfectly drained section of M3 contributed to the lower average yield. Averaged across the treatment period, yields ranged from 15 to 17 Mg ha^{-1}; N and P uptake ranged from 177 to 192 and 21.9 to 25.4 kg ha^{-1}, respectively. Yield, N, and P removal differences among fields were also noted for the calibration phase [39]. For the calibration period average yield (17.3 Mg ha^{-1}), N and P removal (197 and 34.7 kg ha^{-1}, respectively) were highest for field M2 (RSMT for the present study), while average yields for the other fields were similar (16 to 16.6 Mg ha^{-1}).

Table 2. Monthly average temperatures (°C) and precipitation totals (cm) at the Marshfield Agricultural Research Station, Stratford, Wisconsin.

Month	2006	2007	2008	2009	2010	2011	2012	Average 1981–2010	2006	2007	2008	2009	2010	2011	2012	Average 1981–2010
					°C								cm			
January		−6.8	−10.7	−14.4	−7.9	−11.1	−6.9	−9.1		2.3	2.9	1.1	2.2	1.9	3.0	2.29
February		−11.8	−11.6	−6.3	−5.2	−7.9	−4.2	−6.7		2.5	2.9	1.9	0.7	1.6	0.4	2.13
March		1.0	−4.8	−1.3	3.6	−2.5	7.3	−0.6		4.2	1.4	3.2	1.7	4.8	3.3	4.45
April		6.1	5.6	6.7	10.6	5.2	7.4	7.2		2.7	10.7	7.3	2.0	9.0	5.8	7.11
May		14.9	11.5	13.7	14.6	12.7		13.7		10.3	8.4	9.4	5.8	8.1		9.32
June		19.4	17.9	18.2	18.6	18.4		18.9		8.8	9.2	6.0	13.2	10.5		11.3
July		20.7	20.6	18.2	22.2	23.1		21.2		7.9	10.7	1.2	25.1	20.7		10.1
August		20.2	19.3	18.9	21.8	20.8		20.1		11.0	4.6	16.6	8.8	6.9		10.9
September	13.4	15.9	15.8	16.8	14.0	14.1		15.4	2.2	10.3	3.8	1.0	19.5	9.2		10.0
October	5.6	11.6	8.3	5.4	10.4	10.1		8.6	4.7	10.7	3.2	12.8	5.6	5.9		6.63
November	2.2	0.1	0.9	4.6	−9.2	2.2		0.8	2.2	0	3.5	1.0	2.6	2.2		5.64
December	−3.6	−9.1	−11.3	−7.5	−16.8	−4.4		−6.8	4.8	7.6	5.7	5.4	0	3.4		3.18
Average Temperature or Total Precipitation		6.9	5.1	6.1	6.4	6.7		6.9	13.9	78.3	67	66.9	87.2	84.2	12.5	83.1

3.2. Runoff Water Quantity

Cumulative runoff over the calibration and treatment period ranged from 660 to 980 mm among the four fields (Figure 1). In the two treatments with spring chisel tillage (RSMT and FMST), runoff increased by 131 and 34%, respectively, compared to the fall-tilled control (FMT), whereas runoff for BFMT was similar to the control (FMT), both of which were fall chisel plowed (Table 3). These results are consistent with those of Stock et al. [14], who reported that chisel tillage significantly reduced runoff and nutrient loading in surface runoff from silt loam soils in southern WI. They attributed the lower runoff to greater surface roughness and increased water storage/infiltration compared to no-tillage. The lower increase in runoff for FMST may be due to physical protection of soil structure, essentially a mulching effect aiding infiltration [51,52]. While several studies show cover crops can contribute to lower surface runoff from greater evapotranspiration and soil structural effects [53–55], delaying tillage for RSMT resulted in more runoff in our study. As stated, the decreased surface roughness from delayed plowing likely contributed to increased runoff. However, while delaying tillage until spring could help reduce erosion potential during the non-growing season when runoff risk is elevated, any such effect would probably be minimal with the limited crop residue remaining after silage harvest. Furthermore, any potential beneficial effect of the rye cover crop on runoff in our study was minimized because of its limited fall growth (seeded in October). Percentage cover of rye and other live plant material was 68, 29, and 32 in November 2008, 2010, and 2011, respectively, (2009 data missing), and rye biomass was 500, 338, and 399 kg ha^{-1} in late Apr or early May 2010, 2011, and 2012, respectively, before manure application, tillage, and corn planting.

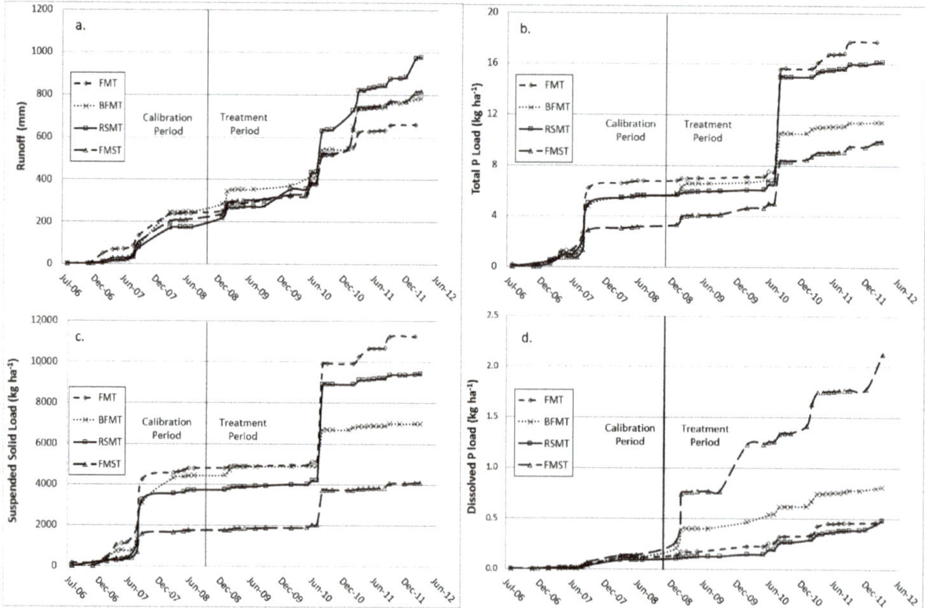

Figure 1. (**a**) Cumulative runoff (mm), (**b**) total P load (kg ha^{-1}), (**c**) suspended sediment load (kg ha^{-1}), and (**d**) dissolved P load (mg kg^{-1}) over the calibration (2006–2008) and treatment periods (2008–2012) for each watershed; FMT = fall applied manure with chisel tillage, RSMT = fall rye (cover crop) with spring-applied manure and chisel tillage; FMST = fall applied manure with spring tillage; BFMT = fall applied manure/chisel tillage with grass buffer.

Table 3. Event-based means and regression slopes for calibration and treatment periods, predicted treatment periods means, and hypothesis test results for runoff volume, suspended sediment, and nutrient concentrations.

Measure	Treatment [†]	Calibration Period Mean	Treatment Period Mean	Predicted Treatment Mean	Change [‡]	Mean p-Value [†‡]	Calibration Period Slope	Treatment Period Slope	Slope p-Value
			mm		%				
Mean Surface Runoff	FMT	4.5	5.0						
	BFMT	5.5	6.3	6.44	−2	NS	1.32	0.91	+
	RSMT	3.99	8.9	3.86	131	*	0.92	1.41	+
	FMST	4.67	7.0	5.19	34	#	0.88	0.82	NS
			mg L^{-1}						
Suspended Sediment	FMT	3560	953						
	BFMT	2375	405	828	−48	**	0.68	0.28	+
	RSMT	2975	390	766	−44	**	0.87	0.39	#
	FMST	1210	316	431	−26	**	0.29	0.27	NS
Total P	FMT	3.87	1.48						
	BFMT	3.33	0.93	1.62	−39	**	0.87	0.30	*
	RSMT	3.56	0.89	1.44	−40	**	0.85	0.43	#
	FMST	1.67	1.18	0.69	46	+	0.46	0.20	+
Dissolved Reactive P	FMT	0.05	0.1						
	BFMT	0.05	0.17	0.08	88	**	0.61	0.33	NS
	RSMT	0.05	0.09	0.08	−30	NS	0.71	0.48	NS
	FMST	0.06	0.54	0.09	312	**	0.57	1.15	NS
Total N	FMT	23.8	10.3						
	BFMT	17.7	7.00	9.89	−27	**	0.73	0.58	NS
	RSMT	23.0	6.22	14.0	−54	**	0.60	0.24	*
	FMST	12.1	7.30	8.57	−23	**	0.28	0.23	NS

Table 3. *Cont.*

Measure	Treatment [†]	Calibration Period Mean	Treatment Period Mean	Predicted Treatment Mean	Change [‡]	Mean p-Value [†‡]	Calibration Period Slope	Treatment Period Slope	Slope p-Value
Nitrate-N	FMT	9.9	4.40						
	BFMT	6.39	2.98	3.75	−21	**	0.63	0.80	NS
	RSMT	10.4	2.34	4.99	−48	**	0.86	0.07	**
	FMST	5.61	2.43	4.58	−55	**	0.25	0.13	NS
Ammonium-N	FMT	1.03	0.71						
	BFMT	0.85	0.79	0.73	8	NS	0.84	0.46	+
	RSMT	0.82	0.40	0.54	−24	**	0.73	0.13	*
	FMST	0.58	1.42	0.45	184	**	0.57	0.73	NS

[†] FMT = fall applied manure with chisel tillage; BFMT = fall applied manure/chisel tillage with grass buffer; RSMT = fall rye (cover crop) with spring applied manure and chisel tillage; FMST = fall applied manure with spring tillage; [‡] change is the percent difference between predicted means and treatment period arithmetic means; [†‡] p-values for testing calibration vs. treatment period: ** $p \leq 0.01$; * $p \leq 0.05$; + $p \leq 0.10$; # $p \leq 0.25$, NS ≥ 0.25.

3.3. Runoff Water Quality

Similar to other studies [22,36,56,57], we found significant relationships between runoff volume and SS, TP, and TN loads (calibration and treatment periods combined). A stronger relationship between TN load and runoff (R^2 = 0.75) may reflect greater mobility compared to SS and TP (R^2 = 0.26; 0.50, respectively); a large portion (20 to >50%) of TN load was NO_3-N further supporting this hypothesis (Table 4). Event runoff TP and TN loads were closely associated with SS loads (R^2 = 0.83; 0.74, respectively), indicating sediment was an important transporter of P and N in runoff. Jokela and Casler [39] also reported a strong relationship between runoff SS and TP loads (R^2 = 0.82), with the exception of two high flow events in early spring 2007 with very high SS.

The influence of management systems on P transport and overall runoff water quality during the treatment period differed. Establishment of a vegetative buffer along with fall manure and tillage (BFMT) significantly decreased SS, TP, TN, and NO_3-N concentrations (Table 3). Loads of SS, TP, TN, NO_3-N and NH_4^+-N were reduced by 55, 28, 34, 41, and 22%, respectively (Table 4), however, mean DRP concentration and load increased by 88 and 15%, respectively, compared to predicted means (Tables 3 and 4). In addition to using means to compare treatment effects, regression slopes (treatments vs. control watersheds) for calibration and treatment periods were also compared for constituent loads and concentrations (Tables 3 and 4). As an example, effects of BFMT on runoff and TP load were based on slope comparisons, as illustrated in Figure 2. Plots of the deviation between predicted and observed values for SS and TP load show minimal effects until well into the treatment period (September 2010; Figure 3), perhaps related to time for development of buffer vegetation. While vegetated buffers are generally effective at mitigating surface runoff sediment and particulate P, DRP removal efficiency is generally lower and in some cases positive [58–60]. Reports of episodic DRP release from biomass after freeze-thaw events, presumably due to cell lysis and runoff mobilization [20,21,61–63], may have also contributed to elevated DRP loads for BFMT.

The combination of a post-harvest rye cover crop and delay of manure application/chisel tillage until spring (RSMT) significantly decreased SS, TP, TN, NO_3-N, and NH_4^+-N concentrations by 24 to 54% (Table 3), suggesting potential benefit. However, because of the increased runoff volume, discussed above, there were no decreases in runoff load of any of these parameters. The limited rye growth because of late planting, noted earlier, likely contributed to this. The lack of load reduction of DRP, NO_3-N, and NH_4^+-N is not unexpected because a rye cover crop is considered more effective at reducing erosion and sediment-bound nutrient loss as compared to dissolved nutrients. Unfortunately, some confounding results may have occurred because manure applied in the spring (RSMT) had a 10 to 22% greater nutrient content than that applied in fall treatments, most pronounced in fall 2010/spring 2011 when differences in N, P, and NH_4^+-N applied were two-fold or more (Table 1).

The practice of fall-applied manure with spring chisel tillage (FMST) is not considered a BMP, however, it is a fairly common practice for central WI dairy farms and, therefore, the evaluation was deemed important. Not unexpectedly, there were large concentration increases (46, 312, and 184% increases for TP, DRP, and NH_4^+-N, respectively) and large loading increases for DRP and NH_4^+-N (376 and 197%); TP load also showed a numerical increase of 39%, but was NS (Tables 3 and 4; Figure 3). Previous studies have also demonstrated the importance of manure incorporation or injection to mitigate runoff N and P in corn and hay forage systems [51,52,64,65]. There was a significant decrease in SS, TN, and NO_3-N concentration, presumably due to the protective soil mulching effect described above regarding runoff effects, but there were no load decreases, except for NO_3-N, because of the increase in runoff.

Table 4. Event-based means and regression slopes for calibration and treatment periods, predicted treatment periods means, and hypothesis test results for suspended sediment and nutrient loads.

Measure	Treatment [†]	Calibration Period Mean	Treatment Period Mean	Predicted Treatment Mean	Change [‡]	Mean p-Value [††‡]	Calibration Period Slope	Treatment Period Slope	Slope p-Value
		kg ha^{-1}			%				
Suspended Sediment	FMT	146	111						
	BFMT	107	43.7	96.4	−55	*	0.35	0.39	NS
	RSMT	113	60	54.5	10	NS	0.72	0.95	+
	FMST	52.8	38.5	41.6	−8	NS	0.34	0.33	NS
Total P	FMT	0.15	0.14						
	BFMT	0.13	0.09	0.12	−28	NS	0.72	0.49	+
	RSMT	0.13	0.11	0.09	20	NS	0.85	1.01	NS
	FMST	0.07	0.09	0.07	39	NS	0.44	0.41	NS
Dissolved Reactive P	FMT	0.002	0.006						
	BFMT	0.003	0.008	0.007	15	**	1.06	0.29	**
	RSMT	0.002	0.004	0.004	3	+	0.85	0.28	**
	FMST	0.003	0.030	0.006	376	**	0.91	2.54	NS
Total N	FMT	1.12	0.77						
	BFMT	1.01	0.48	0.73	−34	*	0.85	0.50	+
	RSMT	0.96	0.61	0.47	28	NS	0.82	0.96	NS
	FMST	0.55	0.45	0.43	5	NS	0.36	0.33	NS
Nitrate-N	FMT	0.51	0.21						
	BFMT	0.50	0.15	0.25	−41	**	0.86	0.45	#
	RSMT	0.48	0.18	0.16	13	+	0.88	0.30	+
	FMST	0.29	0.09	0.17	−46	**	0.36	0.10	#
Ammonium-N	FMT	0.07	0.05						
	BFMT	0.09	0.05	0.06	−22	#	1.39	0.15	*
	RSMT	0.05	0.03	0.03	25	NS	0.86	0.14	+
	FMST	0.04	0.08	0.03	197	*	0.41	0.13	#

[†] FMT = fall applied manure with chisel tillage; BFMT = fall applied manure/chisel tillage with grass buffer; RSMT = fall rye (cover crop) with spring applied manure and chisel tillage; FMST = fall applied manure with spring tillage; [‡] change is the percent difference between predicted calibration means and treatment period arithmetic means; [††‡] p-values for testing calibration vs. treatment period: ** $p \leq 0.01$; * $p \leq 0.05$; + $p \leq 0.10$; # $p \leq 0.25$, NS ≥ 0.25.

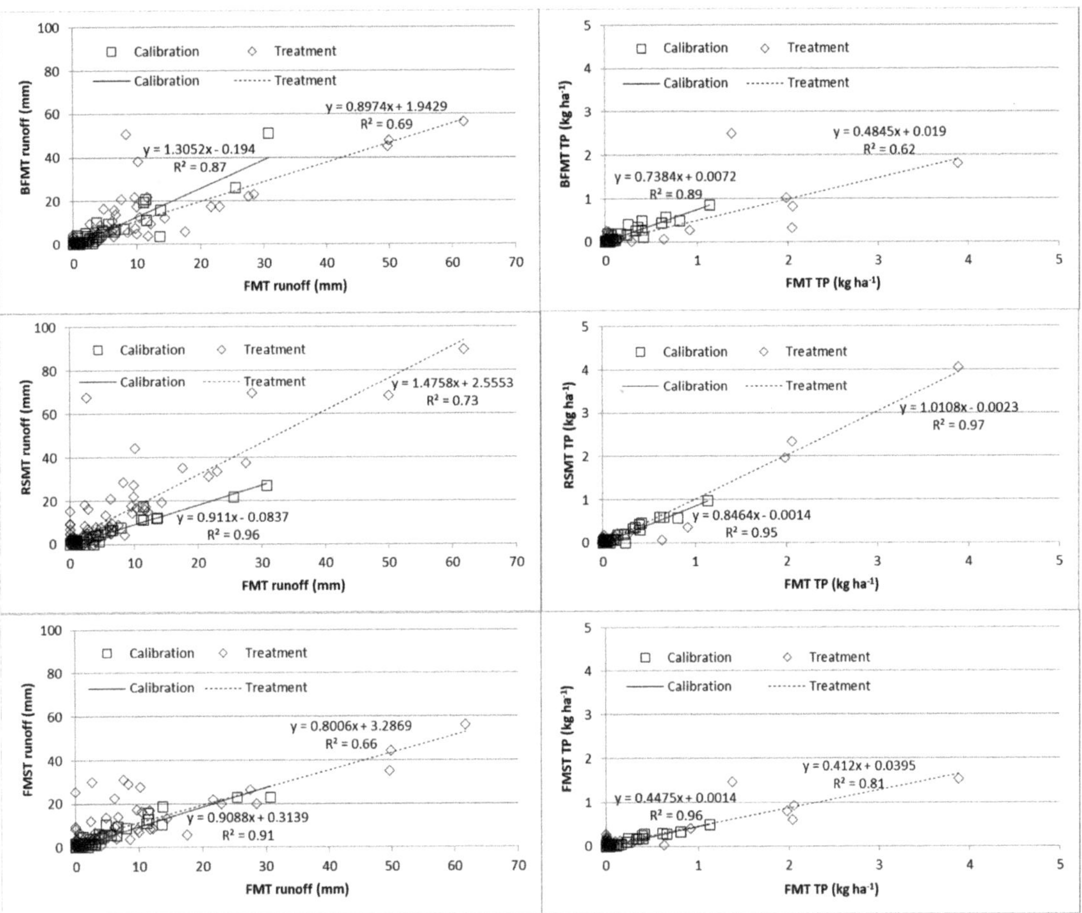

Figure 2. Relationships between each watershed (BFMT = fall applied manure/chisel tillage with grass buffer (**top**); RSMT = fall rye (cover crop) with spring-applied manure and chisel tillage (**center**); or FMST = fall applied manure with spring tillage (**bottom**) and FMT (fall applied manure with chisel tillage) during the calibration and treatment period for runoff quantity (mm) (**left**) and total P (TP) load (kg ha^{-1}) (**right**). The solid and dashed lines are linear regressions for the calibration and treatment periods, respectively.

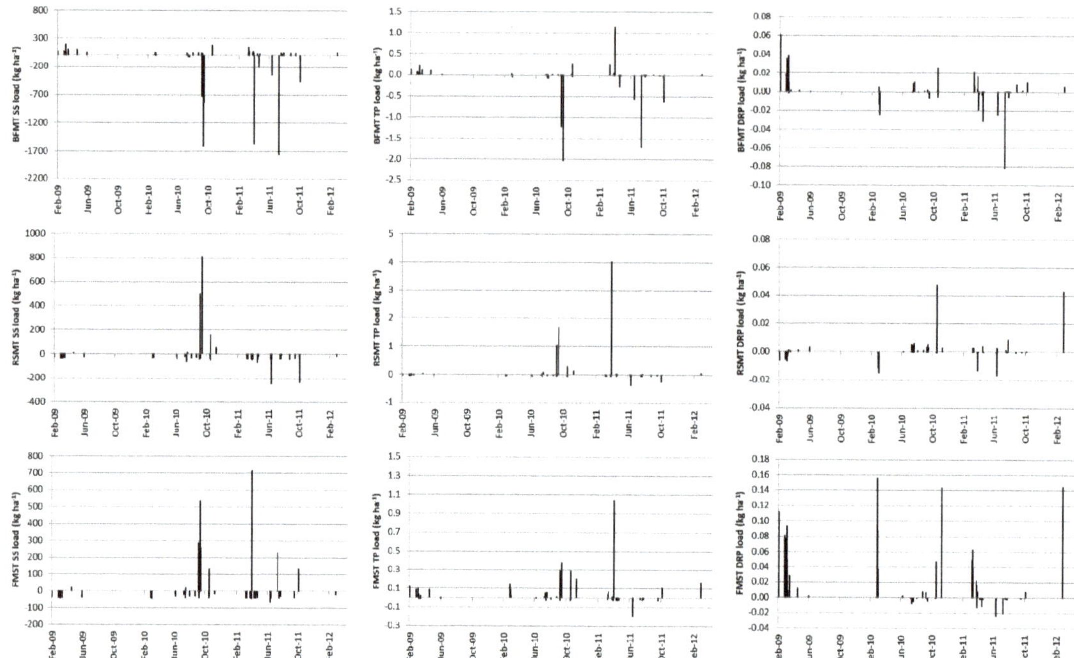

Figure 3. Observed minus predicted values for runoff events throughout the treatment period for each system BFMT = fall applied manure/chisel tillage with grass buffer (**top**); RSMT = fall rye (cover crop) with spring applied manure and chisel tillage (**center**); or FMST = fall applied manure with spring tillage (**bottom**) for suspended solid (SS) load (mg kg^{-1}) (**left**); total P (TP) load (mg kg^{-1}) (**middle**); and dissolved reactive P (DRP) load (mg kg^{-1}) (**right**).

The distribution of runoff events in a given hydrologic year is often skewed, with a few large events contributing a majority of the annual runoff, sediment, and/or nutrient loss [10,66,67]. We observed a similar phenomenon, with much of the cumulative SS and TP load over the course of the trial coming from a small number of events in early fall after silage harvest when soil is exposed (Figure 1). As a result, the effectiveness of conservation practices may be greatly reduced if they are overwhelmed by one or a few large events.

Much of the annual runoff in cold climate regions is from snowmelt [11,13,34,35,39,68,69]. An average of 53% of annual runoff (averaged across watersheds and control/treatment periods) was derived from snowmelt. We also found greater surface runoff yields during the non-growing season when soils are wetter due to lower evapotranspiration rates. In our study, snowmelt runoff had a much greater proportion of nutrients in dissolved form, for example, 70% of the TP load for FMST was DRP for snowmelt events as compared to 17% for rain events. This is consistent with other studies reporting greater DRP/TP ratios for snowmelt compared to rainfall [13,14,17,39,70,71], potentially exacerbating water quality risk due to coincident elevated DRP mobility and high runoff volumes. Runoff DRP and NH_4^+-N loads were generally larger for snowmelt (Table S2) and early spring rainfall events, a loss pattern noted elsewhere [13,19,69,72,73].

Separate analyses of snowmelt and rainfall events showed some differences in relative treatment effects, as indicated by the change between observed and predicted treatment period means (Tables S2 and S3). For rain events, BFMT showed significant decreases in mean runoff concentrations of SS, TP, and all N species (40 to 68%) and in TP, DRP, and NH_4^+-N loads (34 to 84%; Table S3). In contrast, snowmelt events had no significant decreases (some concentrations increased) with significant load decreases only for SS and NO_3-N (Table S2). While BFMT was much less effective for snowmelt than rain, the 80%

reduction in SS load indicates the potential of a vegetative buffer to mitigate particulate P and SS for snowmelt runoff events [11,13,14,17,19,26,35,67,69]. However, while a vegetated buffer may contribute to a reduction in SS and particulate P transport year-round, buffers are less effective at DRP removal, particularly under conditions of frozen or partially frozen soils [29]. In addition, studies show that snowmelt runoff tends to have a greater proportion of DRP compared to rainfall-induced surface runoff [26,67]. In our study, approximately 43% of TP was in DRP form averaged across watersheds versus 3% for rain (Tables S2 and S3).

In contrast to BFMT, RSMT had significantly lower mean concentrations of SS and most P and N species than predicted in both rain and snowmelt runoff, with somewhat greater reductions in snowmelt. However, because of increased runoff, these were translated into significant decreases in loads only for SS, TN, and NO_3-N in snowmelt. Overall, there were only modest treatment effects on P loads and the other measures for either rainfall or snowmelt events, (although some slopes indicated modest changes), indicating the greatest impact of a rye cover crop occurred outside the growing season, even with minimal rye growth. Spring manure application and tillage minimally impacted nutrient loss likely due to timing, occurring immediately prior to planting, typically during a drier window of time. Only DRP had a slightly (7%) but significantly greater mean load in the treatment period, most of the higher DRP concentrations found during events occurring post snowmelt but prior to planting (Figure 4).

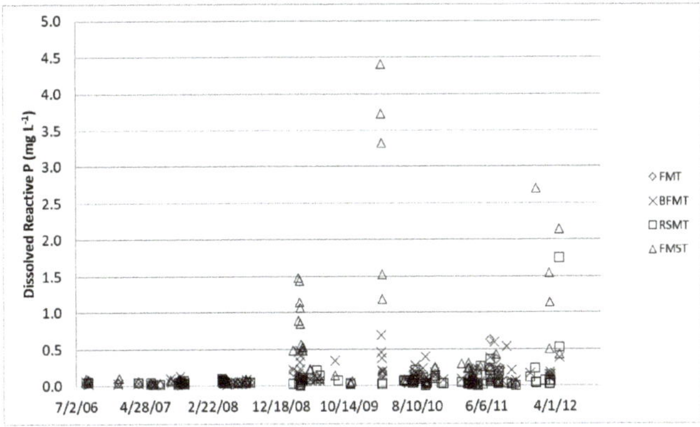

Figure 4. Concentrations (mg L^{-1}) of dissolved reactive P (DRP) over the study period.

Broadcast application of manure in the fall with tillage delayed until spring (FMST) leaves manure on the soil surface through the non-growing season, resulting in marked water quality contrasts between rain and snowmelt runoff. In rain events, the small (11%) reduction in SS concentration led to 17% and 2% decreases in TN concentration and load, respectively, but no other decreases. There were increases in DRP and NH_4^+-N concentration of about 70% and loads of 190 and 16%, respectively. In snowmelt runoff, there were significant decreases in SS, TN, and NO_3-N concentrations and SS and NO_3-N load. Perhaps the most striking effects were the very large increases in TP, DRP, and NH_4^+-N (579, 647, and 232% concentrations, respectively, and 605, 784, and 411% loads, respectively), emphasizing the runoff water quality risk posed by unincorporated manure left on the soil surface over winter.

3.4. Research Implications

Both BFMT and RSMT significantly reduced runoff SS, TP, TN, and NO_3-N concentrations while BFMT also reduced SS, TP, TN, and NO_3-N loads; however, neither significantly reduced DRP concentration or load. Similar tradeoffs with respect to practices designed to control erosion have been identified in the Lake Erie and other watersheds, where practices such as reduced tillage and tile drains can mitigate erosion and TP loss in surface runoff but have the unintended consequence of contributing to elevated runoff DRP risk [74,75]. Encouragingly, RSMT significantly reduced concentrations of SS and all P and N species and loads of SS, TN, and NO_3-N for snowmelt events (Table S2), suggesting a rye cover crop in the fall (with spring tillage) may indeed have potential in corn silage systems to mitigate sediment and P loss in surface runoff. Our results indicate that FMST is not an advisable practice, given its much higher risk of surface runoff DRP loss, as evidenced by the large total and dissolved P load increases (Table 4; Figure 3), especially in snowmelt runoff (on average 605% and 784%, respectively).

An important result from our study was the relatively large runoff DRP concentrations, especially during the treatment period. Average event runoff DRP concentrations (for all events and/or snowmelt events) were at or above the reported eutrophication threshold of 0.05 mg L^{-1} (Table 3, Figure 4), indicating runoff would present a freshwater eutrophication risk if discharged directly to streams [76,77]. Overall, DRP increased from about 2.5% of runoff TP in the calibration period to 12.2% during the treatment phase. Rye cover crop biomass and/or grass buffer strips may have contributed to increased runoff DRP from those treatments as a result of DRP release from biomass after freeze-thaw events, as previously noted [21,29,61–63]; grass buffer and rye biomass likely altered SS and particulate P transport. However, we hypothesize P from manure and labile soil P forms were more important P sources to runoff water. Manure can be a direct source of nutrients to runoff, especially when left on the surface (i.e., FMST) where labile N and P forms are vulnerable to mobilization in surface runoff. The soluble NH_4^+-N form was about half of total N in manure (Table 1) and may partially explain the large increase, especially for FMST, whereas more variable effects were noted for TN (which includes low soluble organic forms associated with sediment). It is also well established that dairy manure applications increase labile soil P concentrations as measured by agronomic soil tests [78,79]. Soil samples were taken in 2009 and 2012; mean B1P in fall 2009 was 22 ± 2 mg kg^{-1} and increased to 29.5 ± 1.9 mg kg^{-1} (37% increase) by spring 2012. This P increase represents a shift from "optimum" to "high" with respect to the University of Wisconsin soil test P fertility categories [41]. Several studies show that P release potential to surface runoff increases with greater labile soil P concentrations [80–82]. In WI, B1P is used as an agronomic indicator of plant-available P and is the basis for determining needed crop P inputs; it is also an important component in the WI P index, which is used to determine P loss risk and develop manure management plans [17,26,80,83]. For a range of soil series in WI, Laboski and Lamb [78] showed that both water-soluble P concentrations and the degree of P saturation increased linearly with B1P, while P sorption strength decreased. It is likely that annual manure applications contributed to the measured B1P increases and runoff DRP concentrations observed in our study. We consider manure P and soil-bound P to be the two main sources of runoff P in our study, further highlighting the importance of conservation practices to mitigate both particulate and dissolved P forms.

The lack of effective control of DRP in runoff was not totally unexpected because the practices evaluated are known to be more effective for erosion and particulate P reduction than dissolved P, suggesting a need for alternative approaches to control dissolved P. One option is the use of injection or other low-disturbance manure application methods, which have been shown to reduce DRP losses substantially compared to surface manure or incorporation by tillage [23,51]. Another is to limit supplementary P addition to livestock feed, which has resulted in significant reductions in runoff DRP from the application of dairy manure [18,84,85]. While our study highlights the importance of controlling erosion and SS transport in order to mitigate P loss, it also demonstrates the inadequacy of the evaluated practices to effectively reduce DRP accumulation in surface soils and the associated greater risk of P transport in surface runoff. Depending on the proximity to open water or other surface water conveyances, our results indicate that additional practices may be needed to mitigate both particulate and dissolved P transport simultaneously and to a degree necessary to protect water quality.

4. Conclusions

Fall manure application and incorporation via chisel plow or other tillage is a common practice on many dairy farms in central WI and other parts of the US. Results from our paired watershed study in central WI indicate that each of the three manure-tillage management systems tested had different potential impacts on surface runoff water quality compared to the control of fall-applied manure with chisel tillage. Both BMPs (BFMT and RSMT) reduced runoff SS, TN, and TP concentrations, but effects on loads were variable, and DRP loads increased slightly. Addition of a vegetative buffer with fall manure application/chisel tillage reduced SS, TP, and total and dissolved N losses, which made it the most effective BMP. A fall rye cover crop with spring manure application and chisel tillage also shows potential for mitigating sediment, N, and P transport, but it would be more effective with earlier seeding to achieve more fall growth and ground cover. Ineffective control of DRP losses suggests the need for additional management practices, such as low-disturbance manure application and limiting livestock dietary P. Fall-applied manure with spring tillage incorporation is not recommended due to high runoff NH_4^+-N and DRP losses.

Supplementary Materials: The following are available online at https://www.mdpi.com/2571-8789/5/1/1/s1, Table S1: Summary of Abbreviations used, Table S2: Event-based means and regression slopes for calibration and treatment periods, predicted treatment periods means, and hypothesis test results for runoff volume, suspended sediment, and nutrient concentrations and loads for snowmelt events only, Table S3: Event-based means and regression slopes for calibration and treatment periods, predicted treatment periods means, and hypothesis test results for runoff volume, suspended sediment, and nutrient concentrations and loads for rain events only.

Author Contributions: This research project and article were contributed to by the authors in the following way: conceptualization, W.E.J.; methodology, W.E.J.; formal analysis, M.D.C.; investigation, W.E.J.; resources, J.C. and W.K.C.; data curation, J.F.S.; writing—original draft preparation, J.F.S.; writing—review and editing, E.O.Y., W.E.J., and M.D.C.; supervision, W.K.C. All authors have read and agreed to the published version of the manuscript.

Funding: This research was supported in part by the U.S. Department of Agriculture, Agricultural Research Service.

Institutional Review Board Statement: Not applicable.

Informed Consent Statement: Not applicable.

Data Availability Statement: Data is contained within the article or supplementary material. The data presented in this study are available [insert article and supplementary material here].

Acknowledgments: The authors would like to thank Craig Simson, Matt Volenec, and Ashley Braun for excellent technical support in the field and lab and Mike Bertram and the UW MARS staff for equipment operation and assistance with field maintenance. The findings and conclusions in this publication are those of the authors and should not be construed to represent any official USDA or U.S. Government determination or policy.

Conflicts of Interest: The authors declare no conflict of interest.

References

1. Grande, J.D.; Karthikeyan, K.G.; Miller, P.S.; Powell, J.M. Corn residue level and manure application timing effects on phos-phorus losses in runoff. *J. Environ. Qual.* **2005**, *34*, 1620–1631. [CrossRef] [PubMed]
2. Grande, J.D.; Karthikeyan, K.; Miller, P.S.; Powell, J.M. Residue Level and Manure Application Timing Effects on Runoff and Sediment Losses. *J. Environ. Qual.* **2005**, *34*, 1337–1346. [CrossRef] [PubMed]
3. Sharpley, A.N.; Chapra, S.C.; Wedepohl, R.; Sims, J.T.; Daniel, T.C.; Reddy, K.R. Managing agricultural phosphorus for pro-tection of surface waters: Issues and options. *J. Environ. Qual.* **1994**, *23*, 437–451. [CrossRef]
4. Shipatitalo, M.J.; Edwards, W.M. Runoff and erosion control with conservation tillage and reduced-input practices on cropped watersheds. *Soil Till. Res.* **1998**, *46*, 1–12. [CrossRef]
5. Strock, J.S.; Porter, P.M.; Russelle, M.P. Cover Cropping to Reduce Nitrate Loss through Subsurface Drainage in the Northern U.S. Corn Belt. *J. Environ. Qual.* **2004**, *33*, 1010–1016. [CrossRef]
6. Thoma, D.P.; Gupta, S.; Strock, J.S.; Moncrief, J.F. Tillage and Nutrient Source Effects on Water Quality and Corn Grain Yield from a Flat Landscape. *J. Environ. Qual.* **2005**, *34*, 1102–1111. [CrossRef]

7. Udawatta, R.P.; Motavalli, P.P.; Garrett, H.E. Phosphorus loss and runoff characteristics in three adjacent agricultural watersheds with clay-pan soils. *J. Environ. Qual.* **2004**, *33*, 1709–1719. [CrossRef]
8. Holly, M.A.; Kleinman, P.J.; Bryant, R.B.; Bjorneberg, D.L.; Rotz, C.A.; Baker, J.M.; Boggess, M.V.; Brauer, D.K.; Chintala, R.; Feyereisen, G.W.; et al. Identifying challenges and opportunities for improved nutrient manage-ment through the USDA's Dairy Agroecosystem Working Group. *J. Dairy Sci.* **2008**, *101*, 6632–6641. [CrossRef]
9. Bishop, P.L.; Hively, W.D.; Stedinger, J.R.; Rafferty, M.R.; Lojpersberger, J.L.; Bloomfield, J.A. Multivariate Analysis of Paired Watershed Data to Evaluate Agricultural Best Management Practice Effects on Stream Water Phosphorus. *J. Environ. Qual.* **2005**, *34*, 1087–1101. [CrossRef]
10. Clausen, J.C.; Jokela, W.E.; Potter, F.I.; Williams, J.W. Paired Watershed Comparison of Tillage Effects on Runoff, Sediment, and Pesticide Losses. *J. Environ. Qual.* **1996**, *25*, 1000–1007. [CrossRef]
11. Griffith, K.E.; Young, E.O.; Klaiber, L.B.; Kramer, S.R. Winter Rye Cover Crop Impacts on Runoff Water Quality in a Northern New York (USA) Tile-Drained Maize Agroecosystem. *Water Air Soil Pollut.* **2020**, *231*, 84. [CrossRef]
12. Jokela, W.E.; Clausen, J.C.; Meals, D.W.; Sharpley, A.N. Effectiveness of agricultural best management practices in reducing phosphorus loading to Lake Champlain. In *Lake Champlain: Partnerships and Research in the New Millennium*; Manley, T.O., Manley, P.L., Mihuc, T.B., Eds.; Kluwer Academic Publisher: New York, NY, USA, 2004; pp. 39–52.
13. Liu, J.; Baulch, H.M.; Macrae, M.L.; Wilson, H.F.; Elliott, J.A.; Bergström, L.; Glenn, A.J.; Vadas, P.A. Agricultural Water Quality in Cold Climates: Processes, Drivers, Management Options, and Research Needs. *J. Environ. Qual.* **2019**, *48*, 792–802. [CrossRef] [PubMed]
14. Stock, M.N.; Arriaga, F.J.; Vadas, P.A.; Good, L.W.; Casler, M.D.; Karthikeyan, K.G.; Zopp, Z. Fall Tillage Reduced Nutrient Loads from Liquid Manure Application during the Freezing Season. *J. Environ. Qual.* **2019**, *48*, 889–898. [CrossRef] [PubMed]
15. Udawatta, R.P.; Garrett, H.E.; Kallenbach, R. Agroforestry Buffers for Nonpoint Source Pollution Reductions from Agricultural Watersheds. *J. Environ. Qual.* **2011**, *40*, 800–806. [CrossRef] [PubMed]
16. Uusi-Kämppä, J.; Braskerud, B.; Jansson, H.; Syversen, N.; Uusitalo, R. Buffer Zones and Constructed Wetlands as Filters for Agricultural Phosphorus. *J. Environ. Qual.* **2000**, *29*, 151–158. [CrossRef]
17. Vadas, P.A.; Stock, M.N.; Arriaga, F.J.; Good, L.W.; Karthikeyan, K.G.; Zopp, Z.P. Dynamics of Measured and Simulated Dissolved Phosphorus in Runoff from Winter-Applied Dairy Manure. *J. Environ. Qual.* **2019**, *48*, 899–906. [CrossRef]
18. Jokela, W.E.; Coblentz, W.K.; Hoffman, P.C. Dairy Heifer Manure Management, Dietary Phosphorus, and Soil Test P Effects on Runoff Phosphorus. *J. Environ. Qual.* **2012**, *41*, 1600–1611. [CrossRef]
19. Tiessen, K.H.D.; Elliott, J.A.; Yarotski, J.; Lobb, D.A.; Flaten, D.N.; Glozier, N.E. Conventional and conservation tillage: In-fluence on seasonal runoff, sediment, and nutrient losses in the canadian praries. *J. Environ. Qual.* **2010**, *39*, 964–980. [CrossRef]
20. Ulen, B. Nutrient losses by surface run-off from soils with winter cover crops and spring-ploughed soils in the south of Swe-den. *Soil Till. Res.* **1997**, *44*, 165–177. [CrossRef]
21. Bechmann, M.; Kleinman, P.J.A.; Sharpley, A.N.; Saporito, L.S. Freeze-Thaw Effects on Phosphorus Loss in Runoff from Manured and Catch-Cropped Soils. *J. Environ. Qual.* **2005**, *34*, 2301–2309. [CrossRef]
22. Bundy, L.G.; Andraski, T.W.; Powell, J.M. Management Practice Effects on Phosphorus Losses in Runoff in Corn Production Systems. *J. Environ. Qual.* **2001**, *30*, 1822–1828. [CrossRef] [PubMed]
23. Daverede, I.C.; Kravchenko, A.N.; Hoeft, R.G.; Nafziger, E.D.; Bullock, D.G.; Warren, J.J.; Gonzini, L.C. Phosphorus Runoff from Incorporated and Surface-Applied Liquid Swine Manure and Phosphorus Fertilizer. *J. Environ. Qual.* **2004**, *33*, 1535–1544. [CrossRef] [PubMed]
24. Sharpley, A.N.; Smith, S. Wheat tillage and water quality in the Southern plains. *Soil Tillage Res.* **1994**, *30*, 33–48. [CrossRef]
25. Little, J.L.; Nolan, S.C.; Casson, J.P.; Olson, B.M. Relationships between soil and runoff phosphorus in small Alberta water-sheds. *J. Environ. Qual.* **2007**, *36*, 1289–1300. [CrossRef] [PubMed]
26. Vadas, P.A.; Good, L.W.; Jokela, W.E.; Karthikeyan, K.; Arriaga, F.J.; Stock, M. Quantifying the Impact of Seasonal and Short-term Manure Application Decisions on Phosphorus Loss in Surface Runoff. *J. Environ. Qual.* **2017**, *46*, 1395–1402. [CrossRef]
27. Feyereisen, G.W.; Wilson, B.N.; Sands, G.; Strock, J.S.; Porter, P.M. Potential for a Rye Cover Crop to Reduce Nitrate Loss in Southwestern Minnesota. *Agron. J.* **2006**, *98*, 1416–1426. [CrossRef]
28. Siller, A.R.S.; Albrecht, K.A.; Jokela, W.E. Soil Erosion and Nutrient Runoff in Corn Silage Production with Kura Clover Living Mulch and Winter Rye. *Agron. J.* **2016**, *108*, 989–999. [CrossRef]
29. Kieta, K.A.; Owens, P.N.; Lobb, D.A.; Vanrobaeys, J.A.; Flaten, D.N. Phosphorus dynamics in vegetated buffer strips in cold climates: A review. *Environ. Rev.* **2018**, *26*, 255–272. [CrossRef]
30. Mayer, P.M.; Reynolds, S.K., Jr.; McCutchen, M.D.; Canfield, T.J. Meta-Analysis of Nitrogen Removal in Riparian Buffers. *J. Environ. Qual.* **2007**, *36*, 1172–1180. [CrossRef]
31. Buda, A.R.; Kleinman, P.J.A.; Srinivasan, M.; Bryant, R.B.; Feyereisen, G.W. Effects of Hydrology and Field Management on Phosphorus Transport in Surface Runoff. *J. Environ. Qual.* **2009**, *38*, 2273–2284. [CrossRef]
32. Daniels, M.; Sharpley, A.; Harmel, R.; Anderson, K. The utilization of edge-of-field monitoring of agricultural runoff in addressing nonpoint source pollution. *J. Soil Water Conserv.* **2018**, *73*, 1–8. [CrossRef]
33. Gburek, W.J.; Sharpley, A.N. Hydrologic Controls on Phosphorus Loss from Upland Agricultural Watersheds. *J. Environ. Qual.* **1998**, *27*, 267–277. [CrossRef]

34. Danz, M.E.; Corsi, S.R.; Brooks, W.R.; Bannerman, R.T. Characterizing response of total suspended solids and total phosphorus loading to weather and watershed characteristics for rainfall and snowmelt events in agricultural watersheds. *J. Hydrol.* **2013**, *507*, 249–261. [CrossRef]

35. Good, L.W.; Carvin, R.; Lamba, J.; Fitzpatrick, F.A. Seasonal Variation in Sediment and Phosphorus Yields in Four Wisconsin Agricultural Watersheds. *J. Environ. Qual.* **2019**, *48*, 950–958. [CrossRef] [PubMed]

36. Lemke, A.M.; Kirkham, K.G.; Lindenbaum, T.T.; Herbert, M.E.; Tear, T.H.; Perry, W.L.; Herkert, J.R. Evaluating Agricultural Best Management Practices in Tile-Drained Subwatersheds of the Mackinaw River, Illinois. *J. Environ. Qual.* **2011**, *40*, 1215–1228. [CrossRef]

37. Clausen, J.C.; Spooner, J. *Paired Watershed Study Design*; US Environmental Protection Agency, Office of Water: Washington, DC, USA, 1993.

38. Mulla, D.J.; Birr, A.S.; Kitchen, N.R.; David, M.B. Limitations of evaluating the effectiveness of agricultural management practices at reducing nutrient losses to surface waters. In *Final Report: Gulf Hypoxia and Local Water Quality Concerns Wrokshop*; American Society of Agricultural and Biological Engineers: St. Joseph, MI, USA, 2008; pp. 189–212.

39. Jokela, W.E.; Casler, M. Transport of phosphorus and nitrogen in surface runoff in a corn silage system: Paired watershed methodology and calibration period results. *Can. J. Soil Sci.* **2011**, *91*, 479–491. [CrossRef]

40. Peters, J. *Recommended Methods of Manure Analysis*; University of Wisconsin-Extension, Ed.; University of Wisconsin-Extension: Madison, WI, USA, 2003.

41. Laboski, C.A.M.; Peters, J.B. *Nutrient Application Guidelines for Field, Vegetable, and Fruit Crops in Wisconsin*; University of Wisconsin-Extension, Cooperative Extension: Madison, WI, USA, 2012.

42. Bray, R.H.; Kurtz, L.T. Determination of total, organic, and available forms of phosphorus in soils. *Soil Sci.* **2006**, *59*, 39–46. [CrossRef]

43. Stuntebeck, T.D.; Komiskey, M.J.; Owens, D.W.; Hall, D.W. *Methods of Data Collection, Sample Processing, and Data Analysis for Edge-of-Field, Streamgaging, Subsurface-Tile, and Meteorological Stations at Discovery Farms and Pioneer Farm in Wiscon-Sin, 2001–2007*; US Geological Survey: Reston, VA, USA, 2008.

44. Brakensiek, D.L.; Osborn, H.B.; Rawls, W.J. (Eds.) *Field Manual for Research in Agricultural Hydrology*; Agricultural Handbook 224; US Department of Agriculture: Washington, DC, USA, 1979.

45. ASTM International. *Standard Test Method for Determining Sediment Concentration in Water Samples*; ASTM International, Ed.; ASTM International: West Conshohocken, PA, USA, 2000.

46. APHA. *Standard Methods for the Examination of Water and Wastewater*, 19th ed.; American Public Health Association, American Water Works Association, and Water Environment Federation: Washington, DC, USA, 1995.

47. Stroup, W.W. *Generalized Linear Mixed Models: Modern Concepts, Methods and Applications*; CRC Press: Boca Raton, FL, USA, 2012.

48. Good, P. *Permutation Tests: A Practical Guide to Resampling Methods for Testing Hypotheses*; Springer: New York, NY, USA, 1995; p. 226.

49. Pesarin, F.; Salmaso, L. The permutation testing approach: A review. *Statistica* **2010**, *70*, 481–509. [CrossRef]

50. SAS Institute Inc. *SAS 9.4 Guide to Software Updates*; SAS Institute Inc.: Cary, NC, USA, 2013.

51. Jokela, W.; Sherman, J.; Cavadini, J. Nutrient Runoff Losses from Liquid Dairy Manure Applied with Low-Disturbance Methods. *J. Environ. Qual.* **2016**, *45*, 1672–1679. [CrossRef]

52. Sherman, J.F.; Young, E.O.; Coblentz, W.K.; Cavadini, J. Runoff water quality after low-disturbance manure application in an alfalfa–grass hay crop forage system. *J. Environ. Qual.* **2020**, *49*, 663–674. [CrossRef]

53. Jokela, W.E.; Grabber, J.H.; Karlen, D.L.; Balser, T.C.; Palmquist, D.E. Cover Crop and Liquid Manure Effects on Soil Quality Indicators in a Corn Silage System. *Agron. J.* **2009**, *101*, 727–737. [CrossRef]

54. Karlen, D.L.; Cambardella, C.A.; Kovar, J.L.; Colvin, T.S. Soil quality response to long-term tillage and crop rotation practic-es. *Soil Till. Res.* **2013**, *133*, 54–64. [CrossRef]

55. Kaspar, T.C.; Radke, J.K.; Laflen, J.M. Small grain cover crops and wheel traffic effects on infiltration, runoff, and erosion. *J. Soil Water Conserv.* **2001**, *56*, 160–164.

56. Eghball, B.; Gilley, J.E. Phosphorus and Nitrogen in Runoff following Beef Cattle Manure or Compost Application. *J. Environ. Qual.* **1999**, *28*, 1201–1210. [CrossRef]

57. Yague, M.R.; Andraski, T.W.; Laboski, C.A.M. Manure composition and incorporation effects on phosphorus in runoff fol-lowing corn biomass removal. *J. Environ. Qual.* **2011**, *40*, 1963–1971. [CrossRef]

58. Dorioz, J.M.; Wang, D.; Poulenard, J.; Trévisan, D. The effect of grass buffer strips on phosphorus dynamics: A critical re-view and synthesis as a basis for application in agricultural landscapes in France. *Agric. Ecosyst. Environ.* **2006**, *117*, 4–21. [CrossRef]

59. Roberts, W.M.; Stutter, M.I.; Haygarth, P.M. Phosphorus Retention and Remobilization in Vegetated Buffer Strips: A Review. *J. Environ. Qual.* **2012**, *41*, 389–399. [CrossRef]

60. Udawatta, R.P.; Krstansky, J.J.; Henderson, G.S.; Garrett, H.E. Agroforestry practices, runoff, and nutrient loss: A paired watershed comparison. *J. Environ. Qual.* **2002**, *31*, 1214–1225. [CrossRef]

61. Aronsson, H.; Hansen, E.M.; Thomsen, I.K.; Liu, J.; Ogaard, A.F.; Känkänen, H.; Ulén, B. The ability of cover crops to reduce nitrogen and phosphorus losses from arable land in southern Scandinavia and Finland. *J. Soil Water Conserv.* **2016**, *71*, 41–55. [CrossRef]

62. Roberson, T.; Bundy, L.G.; Andraski, T.W. Freezing and Drying Effects on Potential Plant Contributions to Phosphorus in Runoff. *J. Environ. Qual.* **2007**, *36*, 532–539. [CrossRef]
63. Wendt, R.C.; Corey, R.B. Phosphorus Variations in Surface Runoff from Agricultural Lands as a Function of Land Use. *J. Environ. Qual.* **1980**, *9*, 130–136. [CrossRef]
64. Maguire, R.O.; Kleinman, P.J.A.; Dell, C.J.; Beegle, D.B.; Brandt, R.C.; McGrath, J.M.; Ketterings, Q.M. Manure Application Technology in Reduced Tillage and Forage Systems: A Review. *J. Environ. Qual.* **2011**, *40*, 292–301. [CrossRef] [PubMed]
65. Sherman, J.F.; Young, E.O.; Jokela, W.E.; Cavadini, J. Influence of low-disturbance fall liquid dairy manure application on corn silage yield, soil nitrate, and rye cover crop growth. *J. Environ. Qual.* **2020**, *49*, 1298–1309. [CrossRef] [PubMed]
66. Coelho, B.B.; Murray, R.; Lapen, D.; Topp, E.; Bruin, A.J. Phosphorus and sediment loading to surface waters from liquid swine manure application under different drainage and tillage practices. *Agric. Water Manag.* **2012**, *104*, 51–61. [CrossRef]
67. Klaiber, L.B.; Kramer, S.R.; Young, E.O. Impacts of Tile Drainage on Phosphorus Losses from Edge-of-Field Plots in the Lake Champlain Basin of New York. *Water* **2020**, *12*, 328. [CrossRef]
68. Hoffman, A.; Polebitski, A.S.; Penn, M.R.; Busch, D.L. Long-term Variation in Agricultural Edge-of-Field Phosphorus Transport during Snowmelt, Rain, and Mixed Runoff Events. *J. Environ. Qual.* **2019**, *48*, 931–940. [CrossRef]
69. Liu, J.; Macrae, M.L.; Elliott, J.A.; Baulch, H.M.; Wilson, H.F.; Kleinman, P.J. Impacts of Cover Crops and Crop Residues on Phosphorus Losses in Cold Climates: A Review. *J. Environ. Qual.* **2019**, *48*, 850–868. [CrossRef]
70. Ginting, D.; Moncrief, J.F.; Gupta, S.C.; Evans, S.D. Interaction between Manure and Tillage System on Phosphorus Uptake and Runoff Losses. *J. Environ. Qual.* **1998**, *27*, 1403–1410. [CrossRef]
71. Ulén, B. Concentrations and transport of different forms of phosphorus during snowmelt runoff from an illite clay soil. *Hydrol. Process.* **2002**, *17*, 747–758. [CrossRef]
72. Hansen, N.C.; Daniel, T.C.; Sharpley, A.N.; Lemunyon, J.L. The fate and transport of phosphorus in agricultural Systems. *J. Soil Water Conserv.* **2002**, *57*, 408–417.
73. Glozier, N.E.; Elliott, J.A.; Holliday, B.; Yarotski, J.; Harker, B. *Water Quality Characteristics and Trends in a Small Agricultural Watershed: South Tobacco Creek, Manitoba, 1992–2001*; Environment Canada: Saskatoon, SK, Canada, 2006.
74. Jarvie, H.P.; Johnson, L.T.; Sharpley, A.N.; Smith, D.R.; Baker, D.B.; Bruulsema, T.W.; Confesor, R. Increased Soluble Phosphorus Loads to Lake Erie: Unintended Consequences of Conservation Practices? *J. Environ. Qual.* **2017**, *46*, 123–132. [CrossRef]
75. Smith, D.R.; King, K.W.; Johnson, L.R.; Francesconi, W.; Richards, P.; Baker, D.; Sharpley, A.N. Surface Runoff and Tile Drainage Transport of Phosphorus in the Midwestern United States. *J. Environ. Qual.* **2015**, *44*, 495–502. [CrossRef] [PubMed]
76. Carpenter, S.R.; Caraco, N.F.; Correll, D.L.; Howarth, R.W.; Sharpley, A.N.; Smith, V.H. Nonpoint Pollution of surface waters with phosphorus and nitrogen. *Ecol. Appl.* **1998**, *8*, 559–568. [CrossRef]
77. Sharpley, A.N.; Troeger, W.W.; Smith, S.J. The Measurement of Bioavailable Phosphorus in Agricultural Runoff. *J. Environ. Qual.* **1991**, *20*, 235–238. [CrossRef]
78. Laboski, C.A.M.; Lamb, J.A. Changes in soil test phosphorus concentration after application of manure or fertilizer. *Soil Sci. Soc. Am. J.* **2003**, *67*, 544–554. [CrossRef]
79. Sharpley, A.; McDowell, R.W.; Kleinman, P.J.A. Amounts, Forms, and Solubility of Phosphorus in Soils Receiving Manure. *Soil Sci. Soc. Am. J.* **2004**, *68*, 2048–2057. [CrossRef]
80. Good, L.W.; Vadas, P.; Panuska, J.C.; Bonilla, C.A.; Jokela, W.E. Testing the Wisconsin Phosphorus Index with Year-Round, Field-Scale Runoff Monitoring. *J. Environ. Qual.* **2012**, *41*, 1730–1740. [CrossRef]
81. Magdoff, F.; Hryshko, C.; Jokela, W.E.; Durieux, R.P.; Bu, Y. Comparison of Phosphorus Soil Test Extractants for Plant Availability and Environmental Assessment. *Soil Sci. Soc. Am. J.* **1999**, *63*, 999–1006. [CrossRef]
82. Vadas, P.A.; Kleinman, P.J.A.; Sharpley, A.N.; Turner, B.L. Relating Soil Phosphorus to Dissolved Phosphorus in Runoff: A Single Extraction Coefficient for Water Quality Modeling. *J. Environ. Qual.* **2005**, *34*, 572–580. [CrossRef]
83. Andraski, T.W.; Bundy, L.G. Relationship between phosphorus levels in soil and in runoff from corn production systems. *J. Environ. Qual.* **2003**, *32*, 310–316. [CrossRef]
84. Ebeling, A.M.; Bundy, L.G.; Powell, J.M.; Andraski, T.W. Dairy Diet Phosphorus Effects on Phosphorus Losses in Runoff from Land-Applied Manure. *Soil Sci. Soc. Am. J.* **2002**, *66*, 284–291. [CrossRef]
85. Hanrahan, L.P.; Jokela, W.E.; Knapp, J.R. Dairy Diet Phosphorus and Rainfall Timing Effects on Runoff Phosphorus from Land-Applied Manure. *J. Environ. Qual.* **2009**, *38*, 212–217. [CrossRef] [PubMed]

soil systems

MDPI

Review

Phosphorus Transport along the Cropland–Riparian–Stream Continuum in Cold Climate Agroecosystems: A Review

Eric O. Young [1,*], Donald S. Ross [2], Deb P. Jaisi [3] and Philippe G. Vidon [4]

1 USDA-ARS, Institute for Environmentally Integrated Dairy Management, Marshfield, WI 54449, USA
2 Department of Plant and Soil Science, University of Vermont, Burlington, VT 05405, USA; dross@uvm.edu
3 Department of Plant and Soil Sciences, University of Delaware, Newark, DE 19716, USA; jaisi@udel.edu
4 Department of Sustainable Resources Management, SUNY College of Environmental Science and Forestry, Syracuse, NY 13201, USA; pgvidon@esf.edu
* Correspondence: eric.young@usda.gov

Abstract: Phosphorus (P) loss from cropland to ground and surface waters is a global concern. In cold climates (CCs), freeze–thaw cycles, snowmelt runoff events, and seasonally wet soils increase P loss potential while limiting P removal effectiveness of riparian buffer zones (RBZs) and other practices. While RBZs can help reduce particulate P transfer to streams, attenuation of dissolved P forms is more challenging. Moreover, P transport studies often focus on either cropland or RBZs exclusively rather than spanning the natural cropland–RBZ–stream gradient, defined here as the cropland–RBZ–stream continuum. Watershed P transport models and agronomic P site indices are commonly used to identify critical source areas; however, RBZ effects on P transport are usually not included. In addition, the coarse resolution of watershed P models may not capture finer-scale soil factors affecting P mobilization. It is clear that site microtopography and hydrology are closely linked and important drivers of P release and transport in overland flow. Combining light detection and ranging (LiDAR) based digital elevation models with P site indices and process-based models show promise for mapping and modeling P transport risk in cropland-RBZ areas; however, a better mechanistic understanding of processes controlling mobile P species across regions is needed. Broader predictive approaches integrating soil hydro-biogeochemical processes with real-time hydroclimatic data and risk assessment tools also hold promise for improving P transport risk assessment in CCs.

Keywords: phosphorus; agriculture; biogeochemistry; riparian buffers; critical source areas; nutrient management; overland flow; hydropedology; snowmelt; streamflow; tile drainage; water quality

Citation: Young, E.O.; Ross, D.S.; Jaisi, D.P.; Vidon, P.G. Phosphorus Transport along the Cropland–Riparian–Stream Continuum in Cold Climate Agroecosystems: A Review. *Soil Syst.* **2021**, *5*, 15. https://doi.org/10.3390/soilsystems5010015

Academic Editors: Rick Wilkin and Peter Leinweber

Received: 1 January 2021
Accepted: 5 March 2021
Published: 9 March 2021

Publisher's Note: MDPI stays neutral with regard to jurisdictional claims in published maps and institutional affiliations.

1. Introduction

Phosphorus (P) is an essential biosphere component and integral to cellular energy currency in the form of adenosine triphosphate. Phosphate molecules also form the backbone of deoxyribiose nucleic acid and other important biological molecules. In addition to imposing important limits on both terrestrial plant and crop productivity, P availability is also the main factor affecting freshwater eutrophication risk [1]. Unlike carbon (C) and nitrogen (N), P does not undergo substantial atmospheric loss. Phosphine (PH_3) is the only known gaseous P form on Earth and its formation is not considered a substantial P loss mechanism from most soils or aquatic sediments [2]. In soil–water systems, pentavalent P forms appear to be most common (P^{5+}); however, water-soluble reduced organic and inorganic P species have also been reported [3].

Once in solution, P acts as a weak Lewis acid with strong affinity for positively charged surface metal ligands, most notably aluminum (Al), iron (Fe), and manganese (Mn) hydroxides, often as organic matter-metal-P complexes [4,5]. Orthophosphate is bioavailable once in solution with maximum availability to (micro)organisms in soils and aquatic sediments near pH 7.0. Variably charged Al and Fe hydroxides are protonated at lower pH (and thus are highly soluble at lower pH), sorbing P from solution more efficiently [5]. As pH

173

increases above 7.0, Ca and Mg phosphate formation is thermodynamically favorable; however, a range of metal-P species occur over a wide pH range in soils and sediments [4–7]. The term legacy P refers to accumulation of P in soils/sediments over time accelerated by anthropogenic activities including P inputs from agriculture. Part of the challenge in sustainable water quality improvement is that legacy P stocks can function as a variable but continual source of P release, hampering the efficacy of remediation efforts.

Agricultural P sources are a leading cause of water quality impairment in US rivers and lakes [8]. Managing P for the dual purpose of profitable agriculture and water quality is a major challenge and is pivotal in the water–energy–food security nexus [8–11]. Once viewed as relatively immobile and subject to mainly erosional transport, carrier-facilitated P transport as particulate or colloidal P in addition to dissolved P forms are all vulnerable to transport in Dunne and Hortonian overland flow (a.k.a., overland flow or surface runoff), interflow, subsurface tile drainage, and shallow groundwater flow [4–6,10–20]. Soil physical properties impose important physical transport constraints on P fluxes from upland agricultural and forested landscapes to riparian buffer zones (RBZs) and streams [15,17–19]. While overland flow is an important P transport mechanism in many settings, P is also mobilized in shallow subsurface flows where it has the potential to contribute P to open waters including ditches, streams, rivers, lakes, wetlands, and RBZs.

Cold climates characterize a large number of agriculturally productive regions globally and can be qualitatively defined by areas where a snowpack and frozen soils substantially influence hydrology [19]. Managing P transport in CCs is uniquely challenged by the combination of short growing seasons, high snowmelt runoff, and seasonally wet and/or partially frozen soils [20]. Recent literature highlights gaps in our current understanding of P transport in CCs, suggesting new approaches are needed to more effectively mitigate P transport from cropland to streams and better understand RBZs effects on P speciation and fluxes [21–23].

Water quality is intimately connected to the landscapes through which streams flow. RBZs are widely recognized for their stream water quality benefits, however, their impacts on P transport are variable and site-specific. Traditionally, P transport research has tended to focus on cropland or RBZs exclusively, with relatively few studies evaluating P dynamics in both cropland and RBZs and/or along their natural hydrologic gradients. Since RBZs and cropland often have a close hydrologic connection with similar processes regulating P transport, in this review we focus on factors influencing P transport in surface and subsurface runoff flows along the continuum from cropland through RBZs to streamflow, defined here as the cropland–RBZ–stream continuum. We primarily draw on studies from the USA and Canada over the last two decades.

Sections 2 and 3 focus on the relationship among agronomic nutrient management, assessing agronomic P transport potential, and an overview of hydroclimatic and agricultural management factors influencing P transport. Sections 4 and 5 discuss the critical source area concept and the importance of soil properties for P transport modeling, mapping and risk assessment. The cropland–RBZ–stream hydrologic continuum concept is introduced in Section 6, followed by a review of RBZ impacts on P transport in overland and subsurface flow (interflow and shallow groundwater), including a Section 7 describing stream bank erosion effects on P loading to streams. Section 8 concludes with future research suggestions and some examples from the literature illustrating new approaches combining hydrologic modeling with geographic information system tools for mapping runoff flow pathways in cropland–RBZ–stream systems.

2. Agricultural Nutrient Management

2.1. Agronomic Phosphorus Site Indices

Agricultural nutrient management plans (NMPs) specify the form, method, rate, and timing of crop nutrient applications with the goal of increasing crop nutrient use efficiency while minimizing environmental losses and crop production risk. In the US, regulated livestock farms must follow nutrient management guidelines developed by state Land Grant Universities and the USDA—Natural Resources Conservation Service (NRCS) (Figure 1). The amount of plant-available soil P (i.e., soil test P concentration) is a main driver of agronomic P recommendations. Unlike P, NMPs estimate plant-available N release from mineralization of soil organic matter, manure, and previous crops (using static rate estimates independent of in-season weather conditions). While NMPs account for total P inputs from manure applications, plant-available P release from mineralization of soil organic P is not considered. Similarly, while potentially ecologically important in some regions, atmospheric depositions of P (and N) are not considered.

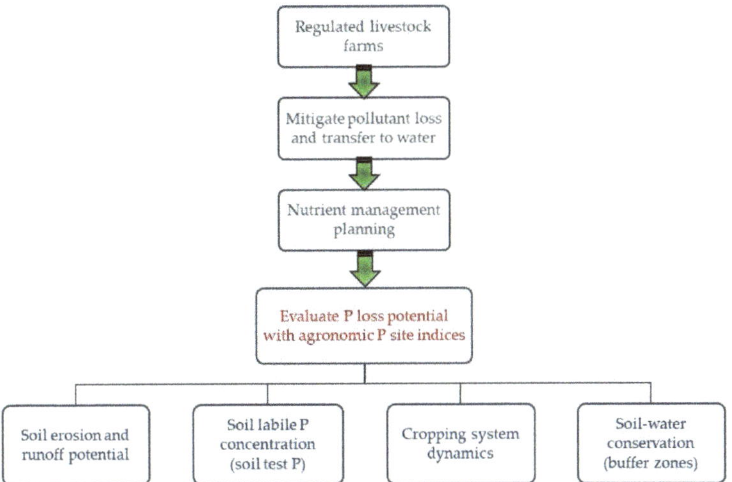

Figure 1. US livestock farms subject to federal Clean Water Act regulations or receiving grant monies must implement cropland nutrient management plans (NMPs) to reduce nonpoint source pollutant loss to open waters. Agronomic P site indices (PSIs) capture soil and management factors affecting annual P loss potential in overland flows and are used to rank P loss potential by fields.

Agronomic NMPs specify field-by-field crop nutrient needs and must include delineation of field characteristics related to erosion and nutrient loss potential, including modeled erosion estimates, presence of concentrated overland flow areas, and proximity to streams/ditches and other landscape features that affect water and nutrient movement (tile drains, karst topography, springs, swales, surface drain inlets). In general, these are also areas where manure and fertilizer P are not recommended during times of high runoff potential and, in some cases, are not to receive any further P applications. Watershed agencies may place further restrictions on land application of manure and fertilizer if farms are in priority watersheds with public drinking water supplies (i.e., New York City watershed, US Great Lakes, Lake Champlain).

Most NMPs in the US require a formal field site assessment of P loss potential using a research based, Land Grant University and NRCS-approved agronomic P site index (PSI). Agronomic PSIs include various rubrics for quantifying P source and transport factors to assign a P loss potential for individual fields based on soil and management factors [24] (Figure 1). Whereas some PSIs include more detailed runoff processes with calibration from edge-of-field runoff P data, many remain qualitative.

Recent US national guidance indicates that agronomic PSIs must establish threshold water quality risks to identify fields not to receive further P inputs. There is also a general consensus that, despite best efforts, P management practices are underperforming with respect to necessary water quality improvement and that there is a need to better account for site-specific hydrology, farm management, and biogeochemical processes influencing P fate and transport [10,11,19,22].

2.2. Precision Agriculture and Phosphorus Management

The ability to manage the timing and placement of crop nutrients in accordance with variable soil and weather conditions can help increase crop P uptake while minimizing losses in runoff. Precision agriculture takes advantage of known field spatial variability (from sampling) by using geographic information systems (GIS) to facilitate autonomous equipment navigation, real-time crop yield monitoring, and variable rate nutrient application. These tools also offer economic advantages for larger farms and are now fairly common [25]. Variable-rate fertilizer application technologies differentially apply P and other nutrients as soil and crop conditions vary across fields [26]. With variable rate application, auxiliary data important for P transport are also routinely collected including soil type boundaries, drainage features, erosion/runoff potential, and other spatially varying soil properties (soil test P, pH, organic matter content). These data can be used to refine P fertility for individual fields and used as inputs for PSIs and other P transport decision support tools aimed at better quantifying P transport potential.

3. Evaluating Cropland Phosphorus Transport Potential

3.1. Agricultural and Hydroclimatic Factors

Managing P inputs from manure and fertilizers for optimal crop production while protecting water quality is a challenge in CC agroecosystems. Livestock manure is an important source of C, N, and P for crops and has beneficial physicochemical effects on soil quality, however, P from manure can contribute to excessive soil P concentrations over time and can be readily transported by overland flow, particularly if not incorporated via tillage or injected beneath the soil surface [27–29]. Dairy manure contains relatively high P content with speciation and total P content dependent on animal species, age, diet, and other farm-specific factors [29]. However, once applied to soils, research indicates that much of the organic P transforms fairly rapidly to inorganic P [30,31] and subject to transport in runoff [10,18,20,22]. Recent research suggests that dairy manure application can be associated with larger and more variable overland flow P losses compared to fields receiving similar rates of fertilizer P [32].

It is clear that a range of P forms can be transported in both overland and shallow sub-surface flow in a variety of crop production systems receiving a mix of fertilizer and organic P mainly in the form of livestock manure [5,10,12–22,27–49]. While agricultural operations often account for a major nonpoint P source in the watershed via the combination of land disturbance and P applications, it is also important to recognize that streambank erosion and runoff from forested lands can contribute to loading to streams [50–53]. Irrespective of original source, landscape position, or form, P transfer risk to streams is greater during the non-growing season, when much of the annual runoff occurs in CC regions [15,18–22,34,35,45–47,52,54–60]. Biogeochemical reactions removing P from solution (sorption and plant and microbial assimilation) also diminish during the non-growing season, contributing to greater overall P mobility and the non-growing season is also a period of elevated overland flow potential. Frozen surface soil layers all but eliminate surface water infiltration and exacerbate overland flows during snow melting or mixed precipitation events. Additionally, decreased soil–water interaction in frozen or partially frozen soils contributes to lower P sorption and greater P mobility in overland flow compared to unfrozen soils. On the other hand, when soils are not frozen and infiltration is possible, greater soil–water interaction increases P removal from solution via sorption reactions and metabolic uptake prior to overland flow reaching streamflow.

Climate and the amount, form, and intensity of precipitation are important factors affecting overland flow, erosion, and P transport potential, and varies regionally in CCs. Hoffman et al. [35] monitored overland flow from five small agricultural watersheds (4 to 30 ha) over a 12-yr period in southwestern Wisconsin (WI) and showed that mixed precipitation events had greater mean dissolved reactive P (DRP; assumed to be mainly orthophosphate and bioavailable) concentrations (2.2 mg L^{-1}) than snow (1.9 mg L^{-1}) or rainfall events (1.2 mg L^{-1}). They also reported that snow (74%) and mixed (84%) events had nearly two-fold greater proportions of DRP in overland flow compared to rainfall (39%), stressing the importance of field-specific interactions among precipitation types and soil physical conditions, temperature, and depth of frozen layers.

Vadas et al. [54] used 108 site years of edge-of-field overland flow data from WI and a calibrated P transport model (SurPhos) to evaluate P loss potential with differing soil hydrologic and P management. Unlike many current P transport models, SurPhos attempts to simulate snowmelt runoff dynamics and processes regulating DRP transfer from soil, fertilizer, and manure P sources using daily weather data. Their simulations indicated site hydrology was the overriding factor influencing P loss with winter application increasing P loss potential by 2.5 to 3.6 times relative to unfrozen soils. They reported that P loss potential was greatest in late January and early February (from melting events) and that P loss potential was reduced by a factor of 3.4 to 7.5-fold by applying manure to fields with a lower overland flow potential.

In a similar geographic region, Zopp et al. [60] used regression tree analysis to determine factors affecting flow-weighted mean total P (TP) and dissolved P concentrations/loads in the upper Midwest using a large regional edge-of-field overland flow and P export data set from WI and Minnesota with 26 fields, 123 site-yr of data, and >20 additional hydroclimatic and management variables. They reported that, when soils were frozen, the majority of overland flow TP was dissolved. Overall, labile soil P concentration at 0–5 cm was the most important predictor of flow weighted mean TP and DRP concentrations in frozen conditions. Soil labile P content is often highly correlated with overland runoff flow DRP concentrations [61] and a critical input for P transport models and PSIs. Additionally, recent edge-of-field runoff research suggests that surface soil P concentration is a main factor affecting DRP transfer to overland flows [62,63], emphasizing the need for NMP strategies to consider practices that slow down the rate of P accumulation in surface soils in addition to focusing on applying manure/fertilizer to fields under low P loss risk conditions (i.e., when soils are unfrozen).

3.2. Cropping System Impacts on Phosphorus Loss Potential

Soil erosion and total P loss in overland runoff flows are both generally greater under annually tilled crops compared to perennial forage crops or pasture due to mechanical disturbance of tillage operations and lack of continuous vegetative cover [55]. Despite this effect, dissolved P loss can still be substantial in overland flow from perennial forage and no-till systems due to P accumulation in surface soils [55,56]. On the other hand, in annually tilled systems, there is a wide range of impacts on erosion, overland flow, and P loss potential. Besides greater aeration and other potential agronomic benefits, tillage can decrease overland runoff flows compared to no-till by increasing surface roughness in finer-textured soils [57–59]. While there are well-known tradeoffs between greater erosion/particulate P loss with tillage versus lower erosion/particulate P loss with no-till, it is important to note that in some soils, tillage can decrease overland runoff potential, however, this effect is site-specific and depends on several other variables including the consistency and duration of no-till practices. Pasture land often comprises a substantial fraction of agricultural land and generally results in less erosion and particulate P transport compared to row crops; dissolved P forms can still be vulnerable to transport in overland flow (see Sections 4, 5 and 7 for more discussion). While beyond the scope of this review, it is important to recognize that pastured livestock with direct stream access can pose serious water quality challenges [55].

4. Critical Source Areas of Phosphorus

Source and Transport Factors

The critical source area concept assumes P transport potential is a function of hydrologic loss mechanisms interacting with P sources on the landscape at any given time [22,32]. Agricultural P sources subject to transfer in runoff pathways and streams along the cropland–RBZ–stream continuum include soil, manure, and fertilizer. From a watershed biogeochemical perspective, RBZ sources must also be considered as potential P sources to streams in the form of overland and subsurface flows or via stream bank erosion [50–53]. Determining where and when P sources interact with hydrologic flow paths to physically transfer P to RBZs and streamflow is integral to critical source area and watershed "hotspot/hot moment" approaches and derives from distributed hydrologic modeling theory, now more commonly known as variable source area hydrology [64–67]. Variable source area hydrology posits that the amount and timing of overland flows are driven by topographic and soil moisture gradients [66–70]. Studies indicate that incorporating variable source area hydrology routines into watershed P transport models show promise for improving overland runoff flow P fluxes [15,22,32,66–70]. Overland flow sources to streamflow include cropland areas but also near-stream areas subject to variable soil moisture regimes and overland flow generation (i.e., RBZs, swales, springs/seeps, and other wetlands) [68,70,71]. Since topographic features are an important control on both overland flow generation and groundwater hydrology, accurate characterization of cropland–RBZ–stream topographic complexity is critical for developing realistic models and indices of P transport that can better account for RBZs impacts on P transport.

Both spring snowmelt and storm events are important times for P transfer from cropland to surface waters and from variable sources areas to streams [10,14,18,32,35]. Part of the difficulty of controlling CC cropland P transport resides in the seasonal asymmetry between greater non-growing season runoff potential and concomitant decreased P sorption potential and biological assimilation driven by lower soil temperatures, effectively increasing dissolved P availability to overland flow. Recognizing this asymmetry between elevated runoff potential and diminished P removal capacity is a critical aspect for NMPs to consider in CC regions to better manage cropland P loss risk and more effectively target P-specific best practices for mitigating P transport to streams.

5. Importance of Soil Properties for Evaluating Phosphorus Transport Potential
Modeling and Mapping

GIS tools and digital soil survey data are routinely used in agriculture to develop NMPs and to support other agronomic and environmental objectives. These tools can help identify and manage soil-related factors affecting crop yields while providing important input data for P transport risk assessment tools [72–74]. For example, digital elevation models (DEMs) are routinely used in P transport models and PSIs for estimating field slopes for erosion assessment. Agronomic PSIs and several P transport models [Agricultural Policy/Environmental eXtender Model (APEX); Environmental Policy Integrated Climate (EPIC) model; Soil and Water Assessment Tool (SWAT); Surface Runoff Phosphorus Model (SurPhos)] use soil survey data or measured properties as model inputs [22,32,54,59,69,70,75,76]. Riparian biogeochemical models including the Riparian Ecosystem Management Model (REMM) and RZ-TRADEOFF also use soil survey data [77–80].

While soil survey maps are useful for many applications, it is important to note that a number of them were performed in the 1960 and 1970's. The maps were also done using different methods and mapping scales. Early mapping focused on agricultural areas with less emphasis on forested and stream areas in general. A soil survey conducted at 1:20,000 scale (a fairly representative soil survey scale) generally relates to a minimum mapping area (termed 'mapping unit area' by USDA-NRCS) of approximately 1.2 ha, implying variation <1.2 ha cannot be included. Thus, soil variation may or may not be reflected at a given map scale depending on a given objective. Moreover, US soil surveys operate on the general assumption that up to 15% of any soil series may be comprised of a taxonomically distinct series.

Modern soil survey mapping techniques can integrate traditional soil survey field data with a range of computational tools to predict soil associations (using classification, fuzzy logic, generalized linear modeling, geostatistics, neural networks, regression trees, machine learning/artificial intelligence algorithms and hybrid models) [81,82]. These approaches are collectively aimed at improving digital soil mapping precision with the ability to integrate expert knowledge and prediction uncertainty [81,82]. Light detection and ranging (LiDAR) derived DEMs can provide detailed microtopographic information to help map soils and overland and subsurface hydrologic flow pathways. A few recent studies have used LiDAR-derived DEMs with hydrologic modeling to design variable-width RBZs based on landscape attributes to optimize agricultural land efficiency and stream water quality protection [71,82–85]. Given the strong association between soils, water movement and P biogeochemistry, characterization of soil hydrology and properties affecting P desorption potential (pH, redox, labile P status) is paramount for better understanding and predicting P fate and transport in CC cropland–RBZ–stream environments.

6. Cropland–Riparian–Stream Hydrologic Continuum

Hydrologic processes are important for understanding the relative contribution of different P sources to watersheds and the relative effectiveness of P transport mitigation practices such as RBZs [15–22,38–49,69,75,86–91]. Site hydrology is also critical for P transport from cropland to RBZs and from RBZs to streamflow. We define the cropland–RBZ–stream continuum as the contiguous land area between active cropland (including pastures) and the nearest perennial or intermittent stream capable of transporting P to downstream systems. Runoff pathways along the cropland–RBZ–stream continuum contributing to streamflow (Q_{sf}) include Hortonian and Dunne overland flow (Q_{of}), groundwater flow (Q_{gw}), and interflow (Q_{if}) (Figure 2). Subsurface tile drainage is a fairly common practice for farms in CCs with poorly drained soils to improve agronomic performance. Tile drain flows are a mix of shallow groundwater and vadose zone water fluxes. Since tile drainage represents a form of subsurface lateral flow, it is included with Q_{if} for simplicity. Stream flow (Q_{sf}) at any given time is thus the sum of individual flow components:

$$Q_{sf} = Q_{of} + Q_{if} + Q_{gw} \tag{1}$$

Figure 2. Hydrologic pathways contributing to streamflow along the cropland–riparian–stream continuum.

Note that groundwater flow components (Q_{gw}) are lumped for simplicity and likely include a mix of both shallow/younger and deeper/older flow paths. Stream baseflow is defined as those times when Q_{gw} is the main flow source contributing to Q_{sf}. Interflow (Q_{if}) includes infiltrated water subject to gravitationally driven lateral movement in the unsaturated zone often induced by the presence of a flow boundary. To reiterate, the uncultivated area between cropland edge-of-field areas and stream bank edges is defined as the RBZ (Figure 2).

RBZs include both semi-natural and unmanaged systems in addition to designed and well managed RBZs. The critical assumption is that RBZs must have permanent vegetation maintained with no agricultural operations (i.e., no tillage or agrichemical applications) occurring. Riparian areas are largely owned and managed by farms in the US and exist in a wide range of field conditions. The grass buffer in Figure 2 is approximately 2 to 3 m wide, which is narrow relative to NRCS riparian forest buffer specifications (minimum width = 10.7 m). Maintaining minimum width RBZs is mandatory in some US states. For example, in Vermont, State Required Agricultural Practices mandate a 3 m permanent RBZ along drainage ditches and 7.6 m wide RBZ along perennial streams and lakes.

7. Riparian Buffer Zone Impacts on Phosphorus Transport

7.1. Phosphorus Transport in Surface Runoff (Q_{of})

Overland flow is an important P transport pathway in cropland as previously highlighted and also critical for P transport in RBZs, along with subsurface components (Figure 3). Properly maintained RBZs can contribute to improved stream water quality and other ecosystem benefits including fish habitat and biodiversity [73,92,93]. Research also indicates RBZs of varying width and composition can attenuate sediment and P fluxes in Q_{of} from upland agricultural areas [38,64,76,78,79,82,83,85,86,92]. In general, a curvilinear relationship is found between RBZ width and TP removal in Q_{of}; however, RBZ width impacts on dissolved P fluxes are less clear. Research also indicates RBZ effects on dissolved P are more variable, with several studies noting dissolved P increases in RBZs [37,78,86,92,94–97]. Fixed width RBZs may not be the most efficient for mitigating P since landscape heterogeneity plays an important role in both cropland P loss and RBZ-P attenuation potential [82,86–90,94,97]. While RBZ width is an important consideration, other factors can have equal or greater importance on P transport from cropland to RBZs [75–79,86–92,94–100].

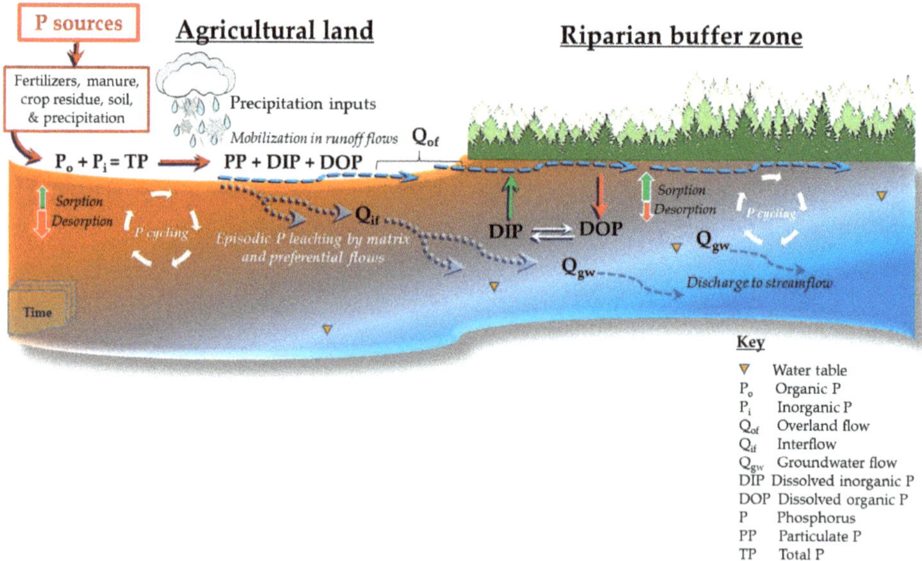

Figure 3. Conceptual diagram depicting hydro-biogeochemical and management factors driving phosphorus (P) fate and transport in cold climates (CCs). Dotted lines represent hydrologic flow pathways that transport P and other solutes. Red arrows represent P inputs into the system or P release via desorption reactions. Green arrows represent P removal via sorption reactions or metabolic uptake of dissolved inorganic P from solution. White arrows indicate biogeochemical processes affecting P bioavailability including pH fluctuations, redox reactions, organic matter cycling (mineralization), and hydrolysis of organic P that affect net P release and fluxes. Note presence of tile drains, stream bank erosion, and other aspects discussed in the text are omitted due to space limitations.

Adequately controlling dissolved and particle-bound P species in Q_{of} is a challenge in both agricultural fields and RBZs. Kieta et al. [94] reported wide variation in P removal efficiencies (from −36% to +89%) for vegetated buffer strips and concluded that both soil P accumulation and freeze–thaw cycle effects on P release from vegetation were important variables related to P removal effectiveness in Q_{of}. They emphasized the difficulty in using vegetated buffers to control P transport in CC agroecosystems where frozen soils and snowmelt-runoff processes limit soluble P removal in Q_{of}, compared to warmer climates where plants and soils remain more biologically active in the non-growing season.

In a review of 41 field studies of crop biomass residue effects on P transport in cropland Q_{of} conducted in CC regions, Liu et al. [98] reported wide ranging biomass P concentrations with substantial P inputs in some cases (0.03 to 51.7 kg P ha^{-1}); however, 45 to >99% of P was retained by soil. Fields with lower erosion potential and biomass residue tended to increase DRP concentrations in Q_{of} compared to fields without residue, suggesting that biomass itself or the interaction of biomass residues with soils increased net P flux to Q_{of}. A similar process may operate in RBZ soils dominated by grass species, whereby a portion of organic P from vegetation and roots is recycled and contributes to the labile inorganic P pool. Labile soil P concentration modified crop residue effects on P transport; fields with lower soil test P and presumably greater sorption capacity tended to retain a greater fraction of P released compared to fields with higher soil P [98]. Beyond highlighting the importance of crop residue effects on P mobilization, these results support the idea that labile soil P concentration is a critical factor affecting P release to Q_{of} in both cropland and RBZs [100–105].

The combination of permanent vegetation and little disturbance in RBZs tends to result in net organic C and P accumulation [86–90,94,97]. Likewise, forest and long-term grassland soils often display organic C and P stratification with enriched surface layers. Dissolved inorganic P in O_{of} is important since it is immediately bioavailable; however, a substantial fraction of P in overland and subsurface flows can be organic in all of these systems [23,45,86,103–105]. Bol et al. [23] reviewed P fluxes in temperate forested ecosystems and reported total soil solution P concentrations of 1 to 400 µg P L^{-1}. Dissolved organic P was the main form, mainly composed of orthophosphate monoesters (phytic acid and its degradation productions). Both labile and more strongly sorbed organic P forms can also be important in RBZ soils. Young et al. [104] reported that 78% of the mean water-extractable total P in surface RBZ soils was organic, nearly half of which was hydrolyzed to DRP after phosphatase enzyme addition, suggesting a substantial fraction of water-soluble organic P in Q_{of} could be bioavailable [104].

Strong linkages between soil C and P biogeochemical cycling have long been recognized by pedology and forest soils literature [106–110], however, as highlighted by Bol [23], little progress has been made on developing a quantitative framework to move static P measures. Dissolved and particle-bound organic P are covalently bound to C and partially account for correlations between soil C and P, however, organic C and other factors like pH and redox potential alters inorganic P solubility and orthophosphate sorption/desorption dynamics [104,111]. Several studies report significant correlations between labile soil inorganic P availability and soil organic C attributing the effect to dissolved organic C competing for P sorption sites [4,5,86–88,103,104,110]. It is well known that carboxylic acid (R-COOH; where R = an alkyl group) and other organic acids compete for P binding sites on soil surfaces after oxidation to carboxylate (COO$^-$), which is difficult to disentangle from inherent correlations between C and P. A certain fraction of organic P is also dynamically hydrolyzed to inorganic P, further confounding relations between C and P.

7.2. Streambank Erosion and P Loading to Streamflow

Streambank erosion is another potentially important P contributor to Q_{sf} with implications for legacy P transport in fluvial systems, aquatic P biogeochemistry and water quality [50–53,89,112]. Ishee et al. [51] combined GIS imagery and field sampling to track streambank erosion rates with field soil P analyses (n = 76 sites) to estimate P inputs from streambank erosion over a 4-yr period. Approximately 6 to 30 % of the total P loading to Q_{sf} among sites was due to streambank erosion. The authors hypothesized that eroding streambanks could act as a sink for P since labile P concentrations were low compared to agricultural land uses. In the Mad River basin of Vermont (a subwatershed of Lake Champlain), Ross et al. [53] used aerial imagery and post-storm sampling to estimate P loading from Tropical Storm Irene in 2011. An area from six sites (0.87-km length of stream bank) contributed an estimated 17.6×10^3 Mg of sediment and 15.8 Mg of total P, similar to average annual watershed P export. Substantial streambank erosion and P loading has also been documented in the Midwestern US. Zaimes et al. [112] measured streambank erosion and associated P loads along forest RBZs, grass dominated buffers, pasture (stratified by continuous, rotational, and intensive rotational) and row-cropped fields for three distinct physiographic regions in Iowa where grazing is common. Forested RBZs had the lowest streambank erosion and P loss rates (2 to 6 kg P km^{-1} $year^{-1}$), followed by grass RBZs (9 to 15 kg km^{-1} $year^{-1}$). The greatest P loading rates were associated with pasture (range: 37 to 123 km^{-1} $year^{-1}$) and row-cropped fields (108 kg km^{-1} $year^{-1}$).

Collectively, results indicate that high P loading rates from streambank erosion can overwhelm TP loss inputs to Q_{sf} compared to other sources. Increased Q_{sf} from greater precipitation extremes related to climate change along with land use/cover effects (ie., tile drainage/ditching cultivation of native prairie and wetlands) have also contributed to greater runoff flows to Q_{sf} and exacerbated nutrient loss [113–115]. For example, river-bank sediments were reported to be the major P source for the Lake Pepin sediment P pool before 1850, which then switched to both a source and carrier of anthropogenic P after European settlements in 1850 [50]. Similarly, sediment–bound P from streambank erosion and river sediment fluxes to coastal estuaries can be a net P source under steady state conditions with the extent of P desorption related to changes in pH and redox potential [48,49,52,116]. These and other studies indicate that streambank erosion itself can be an important P source to Q_{sf} compared to other sources, particularly if widespread throughout the watershed. However, whether or not these sediments ultimately act as a DRP source or sink is inherently dynamic and difficult to predict given the array of watershed scale land use management and variables influencing watershed P speciation and fluxes [10,11,19,22,23,38,40,42–44,66,69,75,86,94,115].

Using high frequency monitoring of Q_{sf} in two predominately forested watersheds of the Piedmont physiographic region in Maryland, USA, Inamdar et al. [52] showed winter storms after freeze–thaw cycles exported high loads of suspended sediment and particulate C and N, with peak suspended sediment and particulate N concentrations >5000 mg L^{-1} and >15 mg L^{-1}. Based on their data and observations from other USGS monitoring stations, the authors speculated that much of the Q_{sf} sediment was derived from streambank erosion and fluvial sources. Inamdar et al. [116] sampled streambank legacy sediments in the Chesapeake Bay watershed, USA, along with upland soils, and evaluated P release potential using laboratory based measures with reducing and oxidizing conditions. Streambank legacy sediments had low average labile P concentrations and equilibrium P concentrations and might therefore act as a net P sink; however, sediments incubated under reducing conditions had nearly 5-fold greater DRP concentrations, suggesting legacy sediments could readily desorb P to Q_{sf} under conditions of low redox potential due to dissolution of Fe-P compounds. The authors highlighted the need for P transport models and indices to better account for spatially variable P legacy sediment impacts on aquatic ecosystems. In summary, while it is apparent that streambank erosion and fluvial transport of legacy sediments can contribute P to Q_{sf}, the relative water quality risk for downstream open waters depends on the amount, speciation and timing of P fluxes relative to other P sources, in addition to sediment characteristics (i.e, labile P content/speciation, P sorption capacity, pH, organic C) and biogeochemical changes in differing RBZ soil and Q_{sf} environments.

7.3. Riparian Zone Impacts on Subsurface Phosphorus Transport (Q_{if} and Q_{gw})

While there is considerable RBZ-P attenuation uncertainty surrounding Q_{of}, there is wider variation for shallow groundwater and vadose zone P attenuation [86,99]. Despite the fact that RBZs are commonly recommended for reducing P transport, information on RBZ effects on P speciation and fluxes is lacking, particularly in CC regions where hydro-biogeochemical processes and agricultural/riparian management interactions largely control nutrient fluxes [94,99]. Moreover, the hydrologic and soil processes driving P transport from upland agricultural areas to RBZs and Q_{sf} are themselves highly spatially and temporally variable, particularly during freeze–thaw cycles with diurnally fluctuating air temperatures and soil physical conditions (frozen/partially frozen) that complicate water infiltration, subsurface water movement, and therefore P transport [45–47,52,91,94,99,105,116]. An improved understanding of coupled soil hydro-biogeochemical processes driving P transport from RBZs to Q_{sf} in CC regions is needed, in addition to developing a broader set of predictive tools that can accommodate the multivariate and dynamic nature of subsurface P transport and subsequent movement and potential transfer to Q_{sf}.

Hydrology, soils and vegetation are intimately linked and their interactions largely control localized physicochemical environments and biogeochemical mechanisms regulating P availability to both matrix and macropore soil water flows [117,118] (Figure 3). This observation helps partially explain studies reporting mixed efficacy for RBZ subsurface P attenuation [13,37,78,86,95,96,117–124]. In shallow Q_{gw} of RBZs from eastern Canada, Carlyle and Hill [95] reported that RBZ shallow Q_{gw} with lower dissolved oxygen concentrations had higher ferrous iron (Fe^{2+}) and DRP concentrations, and suggested that Q_{gw} redox potential was a main factor affecting the likelihood of P release to Q_{gw} and discharging Q_{sf}. Young and Briggs [13] monitored P concentrations in soil solution (sampled via tension lysimeters, representing Q_{if}) and shallow Q_{gw} for 16 paired cropland-RBZ plots for >2-yr in Central New York. Mean DRP concentrations in Q_{gw} and Q_{if} were lower for RBZs compared to corn and hayfields; however, poorly drained RBZs had greater particulate reactive and dissolved unreactive P concentrations in Q_{gw}, suggesting poorly drained RBZs with elevated water tables and low to moderate labile P status were vulnerable to P release and transport in Q_{gw} compared to more oxidizing Q_{gw} zones. The importance of soil hydrology on P biogeochemistry was also supported by ammonium and nitrate-N patterns. Shallow Q_{gw} zones with lower dissolved oxygen concentrations had lower nitrate-N, higher ammonium-N, and significantly greater DRP concentrations, which suggests denitrification zones could be episodic P flux hotspots [13,125].

Gu et al. [126] combined Q_{gw}, Q_{if}, and Q_{of} measures in the Kervidy-Naizin catchment of Northwestern France over 4-yr with P concentrations and speciation to elucidate transport mechanisms in shallow subsurface flows (Q_{if} + Q_{gw}). The authors hypothesized that the main P transport mechanisms were related to soil hydrology via: (i) reductive dissolution of ferric (Fe^{3+}) phosphates during episodic saturation events (hot moment) and, (ii) P mobilization in soil water flows associated with rainfall events following dry periods (hot moment). The degree and duration of soil saturation is a critical factor affecting P release from RBZ soils since prolonged saturation can elicit both reductive dissolution of Fe-P and Mn-P compounds and encourages dissolution of Al-P, Ca-P, and other P complexes [86,87,111,127,128]. Changes in pH during saturation also affect release of dissolved organic P and other C-P complexes that may be more vulnerable to movement in Q_{if} and/or Q_{gw} due to lower affinity for P sorption sites compared to free orthophosphate [4,5,13,37,78,86,103,104,111,116–118,127,128]. Shallow Q_{gw} residence time is also an important factor influencing thermodynamic conditions and P release and retention in RBZs, particularly via the Fe-P redox cycle [86,111].

7.4. Artificial Subsurface Tile Drainage and Phosphorus Loss Potential

Installation of subsurface agricultural tile drainage systems (a.k.a., tile drains) is relatively common in CC agricultural regions with poorly drained soils [129]. Modern tile drainage systems consist of perforated plastic drainpipe (typically 10 cm ID for lateral field lines) typically installed at a depth of 1.0 to 1.5 m deep with variable lateral spacing and designs. Hydrologically, the main objective is lowering the seasonally high ground water table elevation, which facilitates more rapid gravitational (macropore) soil water drainage compared to an undrained condition in a similar setting. Tile drains have long been recognized for their multiple agronomic benefits (e.g., greater yields, earlier planting/harvesting) and erosion mitigation potential [36,129]. Typically, tile-drained fields outlet to some type of surface ditch or directly to streams or open waters.

While accelerated nitrate-N loss via tile drainage flows has long been recognized, P leaching and transport in tile systems has gained more attention over the last two decades [12,16,36,45,63,105]. In a recent review, King et al. [36] discuss P transport dynamics in tile drained systems and the role of soil and nutrient management factors in controlling P concentrations and fluxes to tile drained soils (mainly the US and Canada). In addition to preventing soil P accumulation to high or excessive levels, the authors stressed the importance of soil type and the propensity for macropore flow in regulating P movement to tile drain flow. Unlike matrix soil water flow characterized by advection and dispersion mechanisms, macropore flow is much more rapid, decreasing the opportunity for P sorption reactions that might otherwise bind P and reduce transfer potential to tile flows [12,16,36,41,130].

While P leaching and transport to tile flow is a concern in some settings, it is also important to recognize that tile drains in general significantly reduce Q_{of} and as such can mitigate particulate and/or DRP transport in Q_{of} compared to undrained conditions in some settings [12,36,45,105]. From this standpoint, tile drains are part of the set of solutions to help mitigate P transport in Q_{of} using combinations of practices, while also potentially contributing to less P transfer to down-slope RBZs [131]. Therefore, while not considered environmentally beneficial with respect to N, tile drains may offer site-specific benefits for reducing erosion, Q_{of}, and P transport in Q_{of}. Early RBZ research with N suggested tile drains could lower the water table sufficiently to reduce interaction of cropland Q_{gw} with upper RBZ soil horizons, thus contributing to lower nitrate-N attenuation in the RBZ. However, the full scope of tile drain impacts in RBZ hydrology and P transport is far from clear since few studies have explicitly investigated the impacts of tile drainage designs on P loss compared to undrained conditions.

8. Future Research Considerations

Phosphorus Transport Modeling and Site Indices

Calibrated field and watershed-scale P transport models help in allocating P load estimates to different land uses and broad scale targeting of P transport mitigation practices. Incorporating variable source area hydrology algorithms into P models and agronomic PSIs show promise for improving P transport risk predictions [10,32,69,132]. However, watershed scale models are often designed to predict P transport over long time periods and over relatively large areas using historical weather and management data, potentially limiting their effectiveness as a dynamic P loss risk tool at the field scale without substantial modification. Additionally, model routines that can better capture snowmelt runoff processes and soil freeze–thaw dynamics in relation to water flow and P mobility are needed [35,59,99]. Given these potential limitations and the fact that large runoff events tend to dominate P losses from cropland to streams, developing tools that can better predict event based and real-time P fluxes and include RBZ hydro-biogeochemical impacts on P transport will be important, especially in high priority watersheds with chronic P pollution.

Combining LiDAR-based DEMs with hydrologic models and GIS tools show promise for enhancing agroecosystem services by creating opportunities to optimize agricultural land while maintaining RBZ water quality functions. For example, Shrivastav et al. [84] and Thomas et al. [71] combined LiDAR-DEMs and GIS tools to map and ground-truth Q_{of} pathways in cropland–riparian–stream settings (Figure 4a,b). These and other hydrologic studies have clearly demonstrated the tendency for Q_{of} heterogeneity in agricultural areas, highlighting the critical importance of targeting RBZs at known "delivery points" to intercept dissolved and entrained P in Q_{of} prior to reaching Q_{sf}. Kuglerová et al. [82] used a high resolution LiDAR-DEM and a hydrologic model to establish variable width forest RBZs based on soil and landscape characteristics, whereby recharge areas more vulnerable to solute leaching had wider RBZs (Figure 4c,d).

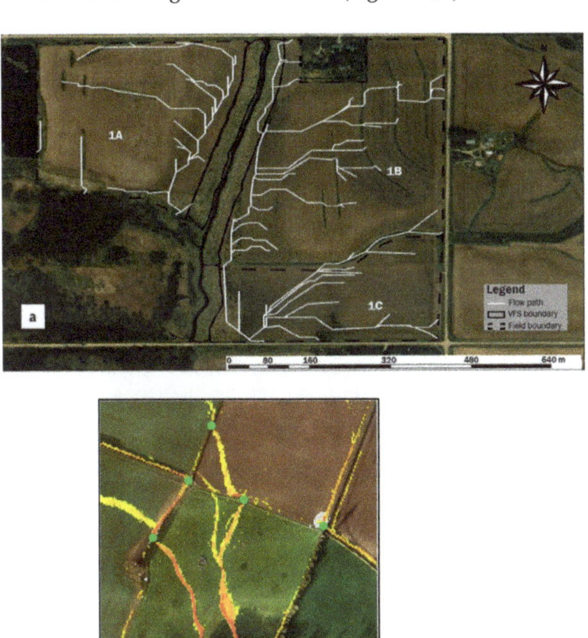

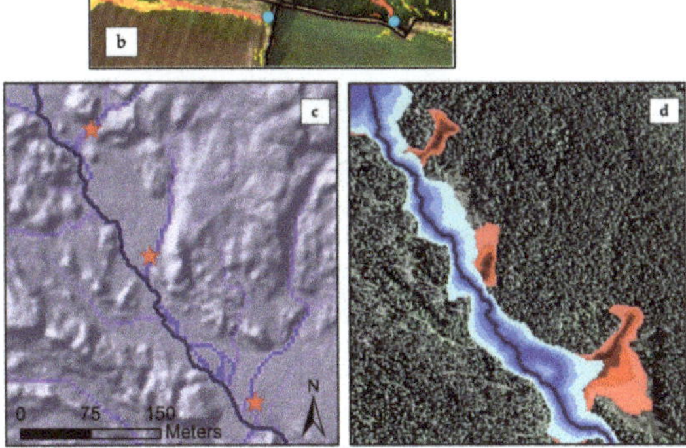

Figure 4. Model estimates of overland runoff flow (Q_{of}) pathways using light detection and ranging (LiDAR) based digital elevation models from Shrivastav et al. [84] (**a**) and Thomas et al. [71] (**b**) illustrating the tendency for non-uniform Q_{of}, and highlights the critical importance of targeting riparian buffers at known O_{of} delivery points to intercept sediment and phosphorus prior to reaching streams. Yellow and red in 4b indicate Q_{of} areas and blue circles indicate Q_{of} delivery points. Variable width riparian forest buffer zones predicted by a LiDAR based groundwater hydrology model by Kuglerová et al. [82] (**c,d**). Panel 4c shows a sunlit LiDAR image of a stream section; blue lines are small streams and red stars are runoff collection points. In panel 4d, blue layers indicate model predicted groundwater discharge zones (darker blues indicate greater fluxes) and red areas are intermittent streams. All figures reproduced with permission.

While existing watershed P transport models and PSIs will remain important tools, simplified process-based models that can readily integrate LiDAR and other digital data will be important for simulating site-specific hydrologic and P transport processes. In a review of nutrient dynamics in CC agricultural catchments, Costa et al. [99] suggested that more parsimonious P transport models that simulate major soil and hydroclimatic processes governing runoff generation and P transport may be more advantageous than larger, more complex models. Several investigators have combined process-based model outputs with Bayesian networks, machine learning, and other artificial intelligence algorithms to develop predictive hydrologic and nutrient flux models along with uncertainty estimates [133–137].

In addition to innovative predictive tools that can account for more of the weather driven and seasonal dynamics of P transport, longer term management practices aimed at reducing P imbalances and soil P accumulation are needed. As previously indicated, current efforts are not sufficiently attenuating P transport or eutrophication risk in the US and other countries, and that new tools and practices are needed to further curb P transport from cropland to streams. While RBZs will remain an important practice, modifications may be needed to improve soluble P removal efficiencies. Not unlike cropland, RBZs must also be managed for optimal performance if P removal is a desired ecosystem service [138,139]. To this end, more widespread and routine soil P testing of RBZs is suggested (similar to P testing for NMPs). Routinely testing RBZs for soil P status as part of agronomic NMPs could be a simple and cost-effective way to provide a baseline indicator of labile P status. Additionally, soil P data could be combined with hydrologic data to further characterize P transport potential.

Given the strong relationship between pH and P availability in soils and legacy sediments [140,141], lowering or raising pH to decrease P availability (something commonly done on cropland to increase soil P availability) in RBZ soils offers a way to further decrease DRP transport from RBZs to Q_{sf}. However, altering soil pH has implications for plant communities, organic C cycling, and other ecological considerations. Careful research is necessary to evaluate potential tradeoffs between enhancing P sorption in RBZs via pH alterations and maintaining overall ecological integrity.

Riparian vegetation plays an important but poorly understood role in P transfer to Q_{sf}. More research to better understand RBZ soil-vegetation interactions and their impacts on P transport is another area of need [115]. Several studies suggest the periodic harvesting of RBZ vegetation to reduce labile soil P concentrations, remove P, and presumably reduce DRP release and transport potential [86,94,97,101,115,123,138,139]. While it is clear that RBZ vegetation can affect P biogeochemistry and physical transport, it is far from clear what the optimum soil-vegetation combinations are for maximizing P attenuation. More research dedicated to soil-vegetation interactions with the goal of maximizing DRP attenuation is needed (particularly for P sensitive watersheds) to enable prescriptive management combinations to mitigate P transport. Lastly, where appropriate, a broader set of predictive approaches should be considered (i.e., Bayesian neural networks, artificial intelligence, machine learning algorithms and various hybrid models) to P loss prediction and develop real-time, dynamic P transport prediction tools that can simultaneously quantify risk and uncertainty.

Author Contributions: This article was contributed to by the authors in the following way: conceptualization, E.O.Y.; methodology, E.O.Y., D.S.R., D.P.J., P.G.V.; formal analysis, E.O.Y., D.S.R., D.P.J., P.G.V.; investigation, E.O.Y.; writing—original draft preparation, E.O.Y.; review and editing, D.S.R., D.P.J., P.G.V. All authors have read and agreed to the published version of the manuscript.

Funding: This research was supported in part by the U.S. Department of Agriculture, Agricultural Research Service.

Institutional Review Board Statement: Not applicable.

Informed Consent Statement: Not applicable

Data Availability Statement: Data are contained within the article or cited articles in the review.

Acknowledgments: The authors would like to thank Barbara C. Storandt for her assistance with formatting references and proofreading the manuscript.

Conflicts of Interest: The authors declare no conflict of interest.

References

1. Wetzel, R.G. *Limnology*; WB Sounders and Company: Philadelphia, PA, USA, 1983.
2. Pasek, M.A.; Sampson, J.M. Redox Chemistry in the Phosphorus Biogeochemical Cycle. *Proc. Natl. Acad. Sci. USA* **2014**, *111*, 15468. [CrossRef]
3. Greaves, J.S.; Richards, A.M.S.; Bains, W.; Rimmer, P.B.; Sagawa, H.; Clements, D.L.; Seager, S.; Petkowski, J.J.; Sousa-Silva, C.; Ranjan, S.; et al. Phosphine Gas in the Cloud Decks of Venus. *Nat. Astron.* **2020**. [CrossRef]
4. Sims, J.T.; Pierzynski, G.M. Chemistry of Phosphorus in Soils. In *Chemical Processes in Soils*; Tabatabai., M.A., Sparks, D.L., Eds.; SSSA: Madison, WI, USA, 2015; pp. 151–192.
5. Pierzynski, G.M.; McDowell, R.W.; Sims, J.T. Chemistry, Cycling, and Potential Movement of Inorganic Phosphorus in Soils. In *Phosphorus: Agriculture and the Environment*; Sims, J.T., Sharpley, A.N., Eds.; ASA; CSSA; SSSA: Madison, WI, USA, 2005; pp. 53–86.
6. Smith, L.; Watzin, M.C.; Druschel, G. Relating Sediment Phosphorus Mobility to Seasonal and Diel Redox Fluctuations at the Sediment-Water Interface in a Eutrophic Freshwater Lake. *Limnol. Oceanogr.* **2011**, *56*, 2251–2264. [CrossRef]
7. Kruse, J.; Abraham, M.; Amelung, W.; Baum, C.; Bol, R.; Kühn, O.; Lewandowski, H.; Niederberger, J.; Oelmann, Y.; Rüger, C.; et al. Innovative Methods in Soil Phosphorus Research-A Review. *J. Plant Nutr. Soil Sci.* **2015**, *1*, 43–88. [CrossRef] [PubMed]
8. United States Environmental Protection Agency (USEPA). *National Water Quality Inventory Report*; USEPA: Washington, DC, USA, 2000. Available online: http://www.epa.gov/305b/2000report/ (accessed on 8 October 2007).
9. Carpenter, S.; Caraco, N.F.; Correll, D.L.; Horwath, R.W.; Sharpley, A.N.; Smith, V.H. Nonpoint Pollution of Surface Waters With Phosphorus and Nitrogen. *Ecol. Appl.* **1998**, *8*, 559–568. [CrossRef]
10. Kleinman, P.J.A.; Fanelli, R.M.; Hirsch, R.M.; Buda, A.R.; Easton, Z.M.; Wainger, L.A.; Shenk, G.W. Phosphorus and the Chesapeake Bay: Lingering Issues and Emerging Concerns for Agriculture. *J. Environ. Qual. 2019 48*, 1191–1203. [CrossRef]
11. Jarvie, H.P.; Sharpley, A.N.; Flaten, D.; Kleinman, P.J.A.; Jenkins, A.; Simmons, T. The Pivotal Role of Phosphorus in a Resilient Water–Energy–Food Security Nexus. *J. Environ. Qual.* **2015**, *44*, 1049–1062. [CrossRef]
12. Sims, J.T.; Simard, R.R.; Joern, B.S. Phosphorus Loss in Agricultural Drainage: Historical Perspective and Current Research. *J. Environ. Qual.* **1998**, *27*, 277–293. [CrossRef]
13. Young, E.O.; Briggs, R.D. Phosphorus Concentrations in Soil and Subsurface Water: A Field Study among Cropland and Riparian Buffers. *J. Environ. Qual.* **2008**, *37*, 69–78. [CrossRef]
14. Vidon, P.; Hubbard, H.; Cuadra, P.; Hennessy, M. Storm Phosphorus Concentrations and Fluxes in Artificially Drained Landscapes of the US Midwest. *Agric. Sci.* **2012**, *3*, 474–485. [CrossRef]
15. Gburek, W.J.; Sharpley, A.N. Hydrologic Controls on Phosphorus Loss from Upland Agricultural Watersheds. *J. Environ. Qual.* **1998**, *27*, 267–277. [CrossRef]
16. Simard, R.R.; Beauchemin, S.; Haygarth, P.M. Potential for Preferential Pathways of Phosphorus Transport. *J. Environ. Qual.* **2000**, *29*, 97–104. [CrossRef]
17. Bryant, R.D.; Gburek, W.J.; Veith, T.L.; Hively, W.D. Perspectives on the Potential for Hydropedology to Improve Watershed Modeling of Phosphorus Loss. *Geoderma* **2006**, *131*, 299–307. [CrossRef]
18. Buda, A.R.; Kleinman, P.J.A.; Srinivasan, M.S.; Bryant, R.B.; Feyereisen, G.W. Effects of Hydrology and Field Management on Phosphorus Transport in Surface Runoff. *J. Environ. Qual.* **2009**, *38*, 2273–2284. [CrossRef] [PubMed]
19. Kleinman, P.J.A.; Sharpley, A.N.; Buda, A.R.; McDowell, R.W.; Allen, A.L. Soil Controls of Phosphorus in Runoff: Management Barriers and Opportunities. *Can. J. Soil Sci.* **2011**, *91*, 329–338. [CrossRef]
20. Liu, J.; Baulch, H.M.; Macrae, M.L.; Wilson, H.F.; Elliott, J.A.; Bergström, L.; Glenn, A.J.; Vadas, P.A. Agricultural Water Quality in Cold Climates: Processes, Drivers, Management Options, and Research Needs. *J. Environ. Qual.* **2019**, *48*, 792–802. [CrossRef]
21. Outram, F.N.; Cooper, R.J.; Suennenberg, G.; Hiscock, K.M.; Lovett, A.A. Antecedent Conditions, Hydrological Connectivity and Anthropogenic Inputs: Factors Affecting Nitrate and Phosphorus Transfers to Agricultural Headwater Streams. *Sci. Total Environ.* **2016**, *545–546*, 184–199. [CrossRef]
22. Sharpley, A.N.; Kleinman, P.J.A.; Flaten, D.N.; Buda, A.R. Critical Source Area Management of Agricultural Phosphorus: Experiences, Challenges, and Opportunities. *Water Sci. Technol.* **2011**, *64*, 945–952. [CrossRef]
23. Bol, R.; Julich, D.; Brödlin, D.; Siemens, J.; Kaiser, K.; Dippold, M.A.; Spielvogel, S.; Zilla, T.; Mewes, D.; von Blanckenburg, F.; et al. Dissolved and Colloidal Phosphorus Fluxes in Forest Ecosystems—An Almost Blind Spot in Ecosystem Research. *J. Plant Nutr. Soil Sci.* **2016**, *179*, 425–438. [CrossRef]
24. Lemunyon, J.L.; Gilbert, R.G. The Concept and Need for a Phosphorus Assessment Tool. *J. Prod. Agric.* **1993**, *6*, 483–486. [CrossRef]
25. Gebbers, R.; Adamchuk, V.I. Precision Agriculture and Food Security. *Science* **2010**, *327*, 828–831. [CrossRef]
26. Cassman, K.G. Ecological Intensification of Cereal Production: Yield Potential, Soil Quality, and Precision Agriculture. *Proc. Natl. Acad. Sci. USA* **1999**, *96*, 5952. [CrossRef]

27. Kleinman, P.J.A. Effect of Mineral and Manure Phosphorus Sources on Runoff Phosphorus. *J. Environ. Qual.* **2002**, *31*, 2026–2033. [CrossRef]

28. Jokela, W.E.; Sherman, J.; Cavadini, J. Nutrient Runoff Losses from Liquid Dairy Manure Applied with Low-Disturbance Methods. *J. Environ. Qual.* **2016**, *45*, 1672–1679. [CrossRef] [PubMed]

29. Hanrahan, L.R.; Jokela, W.E.; Knapp, J.R. Dairy Diet Phosphorus and Rainfall Timing Effects on Runoff Phosphorus from Land-Applied Manure. *J. Environ. Qual.* **2009**, *38*, 212–217. [CrossRef]

30. Hill, J.E.; Cade-Menun, B.J. Phosphorus-31 Nuclear Magnetic Resonance Spectroscopy Transect Study of Poultry Operations on the Delmarva Peninsula. *J. Environ. Qual.* **2009**, *37*, 1–9. [CrossRef] [PubMed]

31. Stout, L.M.; Nguyen, T.; Jaisi, D.P. Relationship of Phytate, Phytate Mineralizing Bacteria, and Beta-Propeller Genes along a Coastal Tributary to the Chesapeake Bay. *Soil Sci. Soc. Am. J.* **2016**, *80*, 84–96. [CrossRef]

32. Collick, A.S.; Veith, T.L.; Fuka, D.R.; Kleinman, P.J.A.; Buda, A.R.; Weld, J.L.; Bryant, R.B.; Vadas, P.A.; White, M.J.; Harmel, R.D.; et al. Improved Simulation of Edaphic and Manure Phosphorus Loss in SWAT. *J. Environ. Qual.* **2014**, *45*, 1215–1225. [CrossRef] [PubMed]

33. Good, L.W.; Vadas, P.; Panuska, J.C.; Bonilla, C.A.; Jokela, W.E. Testing the Wisconsin P Index with Year-Round, Field-Scale Runoff Monitoring. *J. Environ. Qual.* **2012**, *41*, 1730–1740. [CrossRef]

34. Good, L.W.; Carvin, R.; Lamba, J.; Fitzpatrick, F.A. Seasonal Variation in Sediment and Phosphorus Yields in Four Wisconsin Agricultural Watersheds. *J. Environ. Qual.* **2019**, *48*, 950–958. [CrossRef]

35. Hoffman, A.R.; Polebitski, A.S.; Penn, M.R.; Busch, D.L. Long-Term Variation in Agricultural Edge-of-Field Phosphorus Transport During Snowmelt, Rain, and Mixed Runoff events. *J. Environ. Qual.* **2019**, *48*, 931–940. [CrossRef] [PubMed]

36. King, K.W.; Williams, M.R.; Macrae, M.L.; Fausey, N.R.; Frankenberger, J.; Smith, D.R.; Kleinman, P.J.A.; Brown, L.C. Phosphorus Transport in Agricultural. Subsurface Drainage: A Review. *J. Environ. Qual.* **2015**, *44*, 467–485. [CrossRef] [PubMed]

37. Dunne, E.J.; Reddy, K.R.; Clark, M.W. Biogeochemical Indices of Phosphorus Retention and Release by Wetland Soils and Adjacent Stream Sediments. *Wetlands* **2006**, *26*, 1026–1041. [CrossRef]

38. Kronvang, B.; Bechman, M.; Lundekvam, H.; Behrendt, H.; Rubæk, G.H.; Schoumans, O.F.; Syversen, N.; Andersen, H.E.; Hoffmann, C.C. Phosphorus Losses From Agricultural Areas in River Basins: Effects and Uncertainties of Targeted Mitigation Measures. *J. Environ. Qual.* **2005**, *34*, 2129–2144. [CrossRef] [PubMed]

39. Withers, P.J.A.; Haygarth, P.M. Agriculture, Phosphorus and Eutrophication: A European Perspective. *Soil Use Manage* **2007**, *23*, 1–4. [CrossRef]

40. Jarvie, H.P.; Sharpley, A.N.; Spears, B.; Buda, A.R.; May, L.; Kleinman, P.J.A. Water Quality Remediation Faces Unprecedented Challenges from Legacy Phosphorus. *Environ. Sci. Technol.* **2013**, *47*, 8997–8998. [CrossRef]

41. Michaud, A.R.; Poirier, S.C.; Whalen, J.K. Tile Drainage as a Hydrologic Pathway for Phosphorus Export From an Agricultural Subwatershed. *J. Environ. Qual.* **2018**, *48*, 64–72. [CrossRef]

42. Mingus, K.A.; Liang, X.; Massoudieh, A.; Jaisi, D.P. Stable Isotopes and Bayesian Modeling Methods of Tracking Sources and Differentiating Bioavailable and Recalcitrant Phosphorus Pools in Suspended Particulate Matter. *Environ. Sci. Technol.* **2018**, *53*, 69–76. [CrossRef]

43. Li, Q.; Yuan, H.; Li, H.; Wang, D.; Jin, Y.; Jaisi, D.P. Loading and Bioavailability of Colloidal Phosphorus in the Estuarine Gradient of the Deer Creek-Susquehanna River Transect in the Chesapeake Bay. *J. Geophys. Res. Biogeosci.* **2019**, *124*, 3717–3726. [CrossRef]

44. Stackpoole, S.M.; Stets, E.G.; & Sprague, L.A. Variable Impacts of Contemporary Versus Legacy Agricultural Phosphorus on US River Water Quality. *Proc. Natl. Acad. Sci. USA* **2019**, *116*, 20562–20567. [CrossRef]

45. Klaiber, L.B.; Kramer, S.R.; Young, E.O. Impacts of Tile Drainage on Phosphorus Losses from Edge-of-Field Plots in the Lake Champlain Basin of New York. *Water* **2020**, *12*, 328. [CrossRef]

46. Su, J.J.; van Bochove, E.; Thériault, G.; Novotna, B.; Khaldoune, J.; Denault, J.T.; Zhou, J.; Nolin, M.C.; Hu, C.X.; Bernier, M.; et al. Effects of Snowmelt on Phosphorus and Sediment Losses from Agricultural Watersheds in Eastern Canada. *Agric. Water Manag.* **2011**, *98*, 867–876. [CrossRef]

47. Danz, M.E.; Corsi, S.R.; Brooks, W.R.; Bannerman, R.T. Characterizing Response of Total Suspended Solids and Total Phosphorus Loading to Weather and Watershed Characteristics for Rainfall and Snowmelt Events in Agricultural Watersheds. *J. Hydrol.* **2013**, *507*, 249–261. [CrossRef]

48. Jaisi, D.P.; Mingus, K.A.; Joshi, S.R.; Upreti, K.; Sun, M.; McGrath, J. Massudieh Linking Sources, Transformation, and Loss of Phosphorus in the Soil-Water Continuum in a Coastal Environment. In *Multi-Scale Biogeochemical Processes in Soil Ecosystems: Critical Reactions and Resilience to Climate Changes*; Yang, Y., Keiluweit, M., Senesi, N., Xing, B., Eds.; Wiley: Hoboken, NJ, USA, 2021.

49. Upreti, K.; Joshi, S.R.; McGrath, J.; Jaisi, D.P. Factors Controlling Phosphorus Mobilization in a Coastal Plain Tributary to the Chesapeake Bay. *Soil Sci. Soc. Am. J.* **2015**, *79*, 815–825. [CrossRef]

50. Grundtner, A.; Gupta, S.; Bloom, P. River Bank Materials as a Source and as Carriers of Phosphorus to Lake Pepin. *J. Environ. Qual.* **2014**, *43*, 1991–2001. [CrossRef]

51. Ishee, E.R.; Ross, D.S.; Garvey, K.M.; Bourgault, R.R.; Ford, C.R. Phosphorus Characterization and Contribution From Eroding Streambank Soils of Vermont's Lake Champlain Basin. *J. Environ. Qual.* **2015**, *44*, 1745–1753. [CrossRef]

52. Inamdar, S.; Johnson, E.; Rowland, R.; Warner, D.; Walter, R.; Merritts, D. Freeze-Thaw Processes and Intense Rainfall: The One-Two Punch for High Sediment and Nutrient Loads from Mid-Atlantic Watersheds. *Biogeochemistry* **2018**, *141*, 333–349. [CrossRef]

53. Ross, D.S.; Wemple, B.C.; Willson, L.J.; Balling, C.M.; Underwood, K.L.; Hamshaw, S.D. Impact of an Extreme Storm Event on River Corridor Bank Erosion and Phosphorus Mobilization in a Mountainous Watershed in the Northeastern United States. *JGR Biogeosci.* **2019**, *124*, 18–32.

54. Vadas, P.A.; Good, L.W.; Jokela, W.E.; Karthikeyan, K.G.; Arriaga, F.J.; Stock, M. Quantifying the Impact of Seasonal and Short-Term Manure Application Decisions on Phosphorus Loss in Surface Runoff. *J. Environ. Qual.* **2017**, *46*, 1395–1402. [CrossRef]

55. Bilotta, G.S.; Brazier, R.E.; Haygarth, P.M. The Impacts of Grazing Animals on the Quality of Soils, Vegetation, and Surface Waters in Intensively Managed Grasslands. *Adv. Agron.* **2007**, *94*, 237–280.

56. Vadas, P.A.; Busch, D.L.; Powell, J.M.; Brink, G.E. Monitoring Runoff from Cattle-Grazed Pastures for a Phosphorus Loss Quantification Tool. *Agric. Ecosyst. Environ.* **2015**, *199*, 124–131. [CrossRef]

57. Hansen, N.C.; Gupta, S.C.; Moncrief, J.F. Snowmelt Runoff, Sediment, and Phosphorus Losses under Three Different Tillage Systems. *Soil Tillage Res.* **2000**, *57*, 93–100. [CrossRef]

58. Stock, M.N.; Arriaga, F.J.; Vadas, P.A.; Good, L.W.; Casler, M.D.; Karthikeyan, K.G.; Zopp, Z. Fall Tillage Reduced Nutrient Loads from Liquid Manure Application During the Freezing Season. *J. Environ. Qual.* **2019**, *48*, 889–898. [CrossRef]

59. Vadas, P.A.; Stock, M.N.; Arriaga, F.J.; Good, L.W.; Karthikeyan, K.G.; Zopp, Z.P. Dynamics of Measured and Simulated Dissolved Phosphorus in Runoff from Winter-Applied Dairy Manure. *J. Environ. Qual.* **2019**, *48*, 899–906. [CrossRef] [PubMed]

60. Zopp, Z.P.; Ruark, M.D.; Thompson, A.M.; Stuntebeck, T.D.; Cooley, E.; Radatz, A.; Radatz, T. Effects of Manure and Tillage on Edge-of-Field Phosphorus Loss in Seasonally Frozen Landscapes. *J. Environ. Qual.* **2019**, *48*, 966–977. [CrossRef] [PubMed]

61. Vadas, P.A.; Kleinman, P.J.A.; Sharpley, A.N.; Turner, B.L. Relating Soil Phosphorus to Dissolved Phosphorus in Runoff: A Single Extraction Coefficient for Water Quality Modeling. *J. Environ. Qual.* **2005**, *34*, 572–580. [CrossRef] [PubMed]

62. Wilson, H.; Elliott, J.; Macrae, M.; Glenn, A. Near-Surface Soils as a Source of Phosphorus in Snowmelt Runoff from Cropland. *J. Environ. Qual.* **2019**, *48*, 921–930. [CrossRef]

63. Osterholz, W.; King, K.; Williams, M.; Hanrahan, B.; Duncan, E. Stratified Soil Sampling Improves Predictions of P Concentration in Surface Runoff and Tile Discharge. *Soil Syst.* **2020**, *4*, 67. [CrossRef]

64. Vidon, P.; Allan, C.; Burns, D.; Duval, T.P.; Gurwick, N.; Inamdar, S.; Lowrance, R.; Okay, J.; Scott, D.; Sebestyen, S. Hot Spots and Hot Moments in Riparian Zones: Potential for Improved Water Quality Management. *J. Am. Water Resour. Assoc.* **2010**, *46*, 278–298. [CrossRef]

65. Beven, K.J.; Kirkby, M.J. A Physically-Based Variable Contributing Area Model of Basin Hydrology. *Hydrol. Sci. Bull.* **1979**, *24*, 43–69. [CrossRef]

66. Gburek, W.J.; Drungil, C.C.; Srinivasan, M.S.; Needelman, B.A.; Woodward, D.E. Variable-Source-Area Controls on Phosphorus Transport: Bridging the Gap between Research and Design. *J. Soil Water Conserv.* **2020**, *57*, 534–543.

67. Walter, M.T.; Steenhuis, T.S.; Mehta, V.K.; Thongs, D.; Zion, M.; Schneiderman, E. Refined Conceptualization of TOPMODEL for Shallow Subsurface Flows. *Hydrol. Process.* **2002**, *16*, 2014–2046. [CrossRef]

68. Walter, M.T.; Walter, M.F.; Brooks, E.S.; Steenhuis, T.S.; Boll, J.; Weiler, K.R. Hydrologically Sensitive Areas: Variable Source Area Hydrology Implications for Water Quality Risk Assessment. *J. Soil Water Conserv.* **2000**, *55*, 277–284.

69. Easton, Z.M.; Fuka, D.R.; Walter, M.T.; Cowan, D.M.; Schneiderman, E.M.; Steenhuis, T.S. Re-conceptualizing the Soil and Water Assessment Tool (SWAT) Model to Predict Runoff from Variable Source Areas. *J. Hydrol.* **2008**, *348*, 279–291. [CrossRef]

70. Agnew, L.J.; Lyon, S.; Gérard-Marchant, P.; Collins, V.B.; Lembo, A.J.; Steenhuis, T.S.; Walter, M.T. Identifying Hydrologically Sensitive Areas: Bridging Science and Application. *J. Environ. Mgt.* **2006**, *78*, 64–76.

71. Thomas, I.A.; Jordan, P.; Mellander, P.E.; Fenton, O.; Shine, O.; ÓhUalQdUalQdUalláchaín, D.; Creamer, R.; McDonald, N.T.; Dunlop, P.; Murphy, P.N.C. Improving the Identification of Hydrologically Sensitive Areas Using LiDAR DEMs for the Delineation and Mitigation of Critical Source Areas of Diffuse Pollution. *Sci. Total Environ.* **2016**, *556*, 276–290. [CrossRef] [PubMed]

72. Tilman, D.; Balzer, C.; Hill, J.; Befort, B.L. Global Food Demand and the Sustainable Intensification of Agriculture. *Proc. Natl. Acad. Sci. USA* **2011**, *108*, 20260–20264. [CrossRef]

73. Tomer, M.D.; Porter, S.A.; James, D.E.; Boomer, K.M.B.; Kostel, J.A.; McLellan, E. Combining Precision Conservation Technologies into a Flexible Framework to Facilitate Agricultural Watershed Planning. *J. Soil Water Conserv.* **2013**, *68*, 113A–120A. [CrossRef]

74. Young, E.O. Soil Nutrient Management: Fueling Agroecosystem Sustainability. *Int. J. Agr. Sustain.* **2020**, *18*, 444–448. [CrossRef]

75. Nelson, N.O.; Parsons, J.E. Basic Approaches to Modeling Phosphorus Leaching. In *Modeling Phosphorus in the Environment*; Radcliffe, D.E., Cabrera, M.L., Eds.; CRC Press: Boca Raton, FL, USA, 2007; pp. 81–103.

76. Lowrance, R.; Altier, L.S.; Williams, R.G.; Inamdar, S.P.; Sheridan, J.M.; Bosch, D.D.; Hubbard, R.K.; Thomas, D.L. REMM: The Riparian Ecosystem Management Model. *J. Soil Water Conserv.* **2000**, *55*, 27–34.

77. Wang, Z.; Zhang, T.Q.; Tan, C.S.; Wang, X.; Taylor, R.A.J.; Qi, Z.M.; Yang, J.W. Modeling the Impacts of Manure on Phosphorus Loss in Surface Runoff and Subsurface Drainage. *J. Environ. Qual.* **2019**, *48*, 39–46. [CrossRef]

78. Vidon, P.G.; Welsh, M.K.; Hassanzadeh, Y.T. Twenty Years of Riparian Zone Research (1997–2017): Where to Next? *J. Environ. Qual.* **2019**, *48*, 248–260. [CrossRef] [PubMed]

79. Hassanzadeh, Y.T.; Vidon, P.G.; Gold, A.J.; Pradhanang, S.M.; Addy Lowder, K. RZ-TRADEOFF: A New Model to Estimate Riparian Water and Air Quality Functions. *Water* **2019**, *11*, 769. [CrossRef]
80. Zhu, A.X.; Hudson, B.; Burt, J.; Lubich, K.; Simonson, D. Soil Mapping Using GIS, Expert Knowledge, and Fuzzy Logic. *Soil Sci. Soc. Amer. J.* **2001**, *65*, 1463–1472. [CrossRef]
81. McBratney, A.B.; Mendonca Santos, M.L.; Minasny, B. On Digital Soil Mapping. *Geoderma* **2003**, *117*, 3–52. [CrossRef]
82. Kuglerová, L.; Agren, A.; Jansson, R.; Laudon, H. Towards Optimizing Riparian Buffer Zones: Ecological and Biogeochemical Implications for Forest Management. *For. Ecol. Manag.* **2014**, *334*, 74–84.
83. Wallace, C.W.; McCarty, G.; Lee, S.; Brooks, R.P.; Veith, T.L.; Kleinman, P.J.A.; Sadeghi, A.M. Evaluating Concentrated Flowpaths in Riparian Forest Buffer Contributing Areas Using LiDAR Imagery and Topographic Metrics. *Remote Sens.* **2018**, *10*, 614. [CrossRef]
84. Shrivastav, M.; Mickelson, S.K.; Webber, D. Using ArcGIS Hydrologic Modeling and LiDAR Digital Elevation Data to Evaluate Surface Runoff Interception Performance of Riparian Vegetative Filter Strip Buffers in Central Iowa. *J. Soil Water Conserv.* **2020**, *75*, 123–129. [CrossRef]
85. Webber, D.F.; Bansal, M.; Mickelson, S.K.; Helmers, M.J.; Arora, K.; Gelder, B.K.; Shrivastav, M.; Judge, C.J. Assessing Surface Flowpath Interception by Vegetative Buffers Using ArcGIS Hydrologic Modeling and Geospatial Analysis for Rock Creek Watershed, Central Iowa. *Trans. Am. Soc. Agric. Biol. Eng.* **2018**, *61*, 273–283. [CrossRef]
86. Hoffman, C.C.; Kjaergaard, C.; Uusi-Kämppä, J.; Bruun Hansen, H.C.; Kronvang, B. Phosphorus Retention in Riparian Buffers: Review of Their Efficiency. *J. Environ. Qual.* **2009**, *38*, 1942–1955. [CrossRef]
87. Lyons, J.B.; Görres, J.H.; Amador, J.A. Spatial and Temporal Variability of Phosphorus Retention in a Riparian Forest Soil. *J. Environ. Qual.* **1998**, *27*, 895–903. [CrossRef]
88. Young, E.O.; Ross, D.S. Total and Labile Phosphorus Concentrations as Influenced by Riparian Buffer Soil Properties. *J. Environ. Qual.* **2016**, *45*, 294–304. [CrossRef]
89. Perillo, V.; Ross, D.; Wemple, B.; Balling, C.; Lemieux, L. Stream Corridor Soil Phosphorus Availability in a Forested-Agricultural Mixed Land Use Watershed. *J. Environ. Qual.* **2019**, *48*, 185–192. [CrossRef]
90. Young, E.O.; Ross, D.S.; Alves, C.; Villars, T. Soil and Landscape Influences on Native Riparian Phosphorus Availability in Three Lake Champlain Basin Stream Corridors. *J. Soil Water Conserv.* **2012**, *67*, 1–7. [CrossRef]
91. Rosenberg, B.D.; Schroth, A.W. Coupling of Reactive Riverine Phosphorus and Iron Species during Hot Transport Moments: Impacts of Land Cover and Seasonality. *Biogeochemistry* **2017**, *132*, 103–122. [CrossRef]
92. Dosskey, M.G. Toward Quantifying Water Pollution Abatement in Response to Installing Buffers on Cropland. *Environ. Manag.* **2001**, *28*, 577–598. [CrossRef]
93. Wang, L. Effects of Watershed Best Management Practices on Habitat and Fish in Wisconsin Streams. *J. Am. Water Resour. Assoc.* **2002**, *38*, 663–680. [CrossRef]
94. Kieta, K.A.; Owens, P.N.; Lobb, D.A.; Vanrobaeys, J.A.; Flaten, D.N. Phosphorus Dynamics in Vegetated Buffer Strips in Cold Climates: A Review. *Environ. Reviews* **2018**, *26*, 255–272. [CrossRef]
95. Carlyle, G.C.; Hill, A.R. Groundwater Phosphate Dynamics in a River Riparian Zone: Effects of Hydrologic Flowpaths, Lithology and Redox Chemistry. *J. Hydrol.* **2001**, *247*, 151–168. [CrossRef]
96. Dupas, R.; Gruau, G.; Gu, S.; Humbert, G.; Jaffrézic, A.; Gascuel-Odoux, C. Groundwater Control of Biogeochemical Processes Causing Phosphorus Release from Riparian Wetlands. *Water Res.* **2015**, *84*, 307–314. [CrossRef]
97. Stutter, M.I.; Langan, S.J.; Lumsdon, D.G. Vegetated Buffer Strips Can Lead to Increased Release of Phosphorus to Waters: A Biogeochemical Assessment of the Mechanisms. *Environ. Sci. Technol.* **2009**, *43*, 1858–1863. [CrossRef]
98. Liu, J.; Macrae, M.L.; Elliott, J.A.; Baulch, H.M.; Wilson, H.F.; Kleinman, P.J.A. Impacts of Cover Crops and Crop Residues on Phosphorus Losses in Cold Climates: A Review. *J. Environ. Qual.* **2019**, *48*, 850–868. [CrossRef]
99. Costa, D.; Baulch, H.; Elliott, J.; Pomeroy, J.; Wheater, H. Modelling Nutrient Dynamics in Cold Agricultural Catchments: A Review. *Environ. Modell. Softw.* **2020**, *124*, 104586. [CrossRef]
100. Young, E.O.; Ross, D.S.; Jaynes, D.B. Editorial: Riparian Buffer Nutrient Dynamics and Water Quality. *Front. Environ. Sci.* **2019**, *7*, 76. [CrossRef]
101. Stutter, M.; Kronvang, B.; ÓhUallacháin, D.; Rozemeijer, J. Current Insights into the Effectiveness of Riparian Management, Attainment of Multiple Benefits, and Potential Technical Enhancements. *J. Environ. Qual.* **2019**, *48*, 236–247. [CrossRef]
102. Cole, L.J.; Stockan, J.; Helliwell, R. Managing Riparian Buffer Strips to Optimise Ecosystem Services: A Review. *Agric. Ecosyst. Environ.* **2020**, *296*, 106891. [CrossRef]
103. Condron, L.M.; Turner, B.L.; Cade-Menun, B.J. Chemistry and Dynamics of Soil Organic Phosphorus. In *Phosphorus: Agriculture and the Environment*; Sims, J.T., Sharpley, A.N., Eds.; ASA; CSSA; SSSA: Madison, WI, USA, 2005; pp. 87–121.
104. Young, E.O.; Ross, D.S.; Cade-Menun, B.J.; Lu, C.W. Phosphorus Speciation in Riparian Soils: A Phosphorus-31 Nuclear Magnetic Resonance Spectroscopy and Enzyme Hydrolysis Study. *Soil Sci. Soc. Am. J.* **2013**, *77*, 1636–1647. [CrossRef]
105. Griffith, K.E.; Young, E.O.; Klaiber, L.B.; Kramer, S.R. Winter Rye Cover Crop Impacts on Runoff Water Quality in a Northern New York (USA) Tile-Drained Maize Agroecosystem. *Water Air Soil Pollut.* **2020**, *231*, 1–16. [CrossRef]
106. Smeck, N.E. Phosphorus Dynamics in Soils and Landscapes. *Geoderma* **1985**, *36*, 185–199. [CrossRef]
107. Walker, T.W.; Syers, J.K. The Fate of Phosphorus during Pedogenesis. *Geoderma* **1976**, *15*, 1–19. [CrossRef]
108. Yanai, R.D. Phosphorus Budget of a 70-Year-Old Northern Hardwood Forest. *Biogeochemistry* **1992**, *17*, 1–22. [CrossRef]

109. Yanai, R.D. The Effect of Whole-Tree Harvest on Phosphorus Cycling in a Northern Hardwood Forest. *For. Ecol. Manag.* **1998**, *104*, 281–295. [CrossRef]
110. Laboski, C.A.M.; Lamb, J.A. Changes in Soil Test Phosphorus Concentration after Application of Manure or Fertilizer. *Soil Sci. Soc. Am. J.* **2003**, *67*, 544–554. [CrossRef]
111. Young, E.O.; Ross, D.S. Phosphorus Mobilization in Flooded Riparian Soils from the Lake Champlain Basin, VT, USA. *Front. Environ. Sci.* **2018**, *6*, 120. [CrossRef]
112. Zaimes, G.N.; Schultz, R.C.; Isenhart, T.M. Streambank Soil and Phosphorus Losses under Different Riparian Land-Uses in Iowa. *J. Am. Water Resour. Assoc.* **2008**, *44*, 935–947. [CrossRef]
113. Sekely, A.C.; Mulla, D.J.; Bauer, D.W. Streambank Slumping and Its Contribution to the Phosphorus and Suspended Sediment Loads to the Blue Earth River, Minnesota. *J. Soil Water Cons.* **2002**, *57*, 243–250.
114. Gupta, S.C.; Kessler, A.C.; Brown, M.K.; Zvomuya, F. Climate and Agricultural Land Use Change Impacts on Streamflow in the Upper Midwestern United States. *J. Am. Water Resour. Assoc.* **2015**, *51*, 5301–5317. [CrossRef]
115. Dosskey, M.G.; Vidon, P.; Gurwick, N.P.; Allan, C.J.; Duval, T.P.; Lowrance, R. The Role of Riparian Vegetation in Protecting and Improving Chemical Water Quality in Streams. *J. Am. Water Resour. Assoc.* **2010**, *46*, 261–277. [CrossRef]
116. Inamdar, S.; Sienkiewicz, N.; Lutgen, A.; Jiang, G.; Kan, J. Streambank Legacy Sediments in Surface Waters: Phosphorus Sources or Sinks? *Soil Syst.* **2020**, *4*, 30. [CrossRef]
117. Chardon, W.J.; Schoumans, O.F. Solubilization of Phosphorus: Concepts and Process Description of Chemical Mechanisms. In *Phosphorus Losses from Agricultural Soils: Processes at the Field-Scale*; Chardon, W.J., Schoumans, O.F., Eds.; Cost Action 832: Quantifying the Agricultural Contribution to Eutrophication; ALTERRA: Wageningen, The Netherlands, 2002; pp. 44–52.
118. McDowell, R.W.; Biggs, B.J.F.; Sharpley, A.N.; Nguyen, L. Connecting Phosphorus Loss from Agricultural Landscapes to Surface Water Quality. *Chem. Ecol.* **2004**, *20*, 1–40. [CrossRef]
119. Jordan, T.E.; Correll, D.L.; Weller, D. Nutrient Interception by a Riparian Forest Receiving Inputs from Adjacent Cropland. *J. Environ. Qual.* **1993**, *22*, 467–473. [CrossRef]
120. Osborne, L.L.; Kovacic, D.A. Riparian Vegetated Buffer Strips in Water Quality Restoration and Stream Management. *Freshwater Biol.* **1993**, *29*, 243–258. [CrossRef]
121. Clausen, J.C.; Guillard, K.; Sigmund, C.M.; Dors, K.M. Water Quality Changes from Riparian Buffer Restoration in Connecticut. *J. Environ. Qual.* **2000**, *29*, 1751–1761. [CrossRef]
122. Uusi-Kamppa, J.; Turtola, E.; Hartikainen, H.; Ylaranta, T. The Interactions of Buffer Zones and Phosphorus Runoff. In *Buffer Zones: Their Processes and Potential in Water Protection*; Haycock, N.E., Ed.; Quest Environmental: Hertfordshire, UK, 2001; pp. 43–53.
123. Roberts, W.M.; Stutter, M.I.; Haygarth, P.M. Phosphorus Retention and Remobilization in Vegetated Buffer Strips: A Review. *J. Environ. Qual.* **2012**, *41*, 389–399. [CrossRef]
124. Spruill, T.B. Statistical Evaluation of Effects of Riparian Buffers on Nitrate and Ground Water Quality. *J. Environ. Qual.* **2000**, *29*, 1523–1538. [CrossRef]
125. Young, E.O.; Briggs, R.D. Nitrogen Dynamics Among Cropland and Riparian Buffers: Soil-Landscape Influences. *J. Environ. Qual.* **2007**, *36*, 801–814. [CrossRef]
126. Gu, S.; Gruau, G.; Dupas, R.; Rumpel, C.; Crème, A.; Fovet, O.; Gascuel-Odoux, C.; Jeanneau, L.; Humbert, G.; Petitjean, P. Release of Dissolved Phosphorus from Riparian Wetlands: Evidence for Complex Interactions among Hydroclimate Variability, Topography and Soil Properties. *Sci. Total Environ.* **2017**, *598*, 421–431. [CrossRef]
127. Kumaragamage, D.; Amarawansha, E.A.G.S.; Indraratne, S.P.; Jayarathne, P.D.K.D.; Flaten, D.N.; Zvomuya, F.; Akinremi, O.O. Degree of Phosphorus Saturation as a Predictor of Redox-Induced Phosphorus Release from Flooded Soils to Floodwater. *J. Environ. Qual.* **2019**, *48*, 1817–1825. [CrossRef]
128. Hens, M.; Merckx, R. Functional Characterization of Colloidal Phosphorus Species in the Soil Solution of Sandy Soils. *Environ. Sci. Technol.* **2001**, *35*, 493–500. [CrossRef]
129. Weaver, M.M. *History of Tile Drainage*; Weaver: Waterloo, NY, USA; Valley Offset, Inc.: Deposit, NY, USA, 1964.
130. Young, E.O.; Geibel, J.G.; Ross, D.S. Influence of Controlled Drainage and Liquid Dairy Manure Application on Phosphorus Leaching from Intact Soil Cores. *J. Environ. Qual.* **2017**, *46*, 80–87. [CrossRef] [PubMed]
131. Scalenghe, R.; Edwards, A.C.; Marsan, F.A.; Barberis, E. The Effect of Reducing Conditions on the Solubility of Phosphorus in a Diverse Range of European Agricultural Soils. *Eur. J. Soil Sci.* **2002**, *53*, 439–447. [CrossRef]
132. Marjerison, R.D.; Dahlke, H.; Easton, Z.M.; Seifert, S.; Walter, M.T. A Phosphorus Index Transport Factor Based on Variable Source Area Hydrology for New York State. *J. Soil Water Conserv.* **2011**, *66*, 149–157. [CrossRef]
133. Sperotto, A.; Molina, J.L.; Torresan, S.; Critto, A.; Pulido-Velazquez, M.; Marcomini, A. A Bayesian Networks Approach for the Assessment of Climate Change Impacts on Nutrient Loading. *Environ. Sci. Policy* **2019**, *100*, 21–36. [CrossRef]
134. Chen, S.H.; Jakeman, A.J.; Norton, J.P. Artificial Intelligence Techniques: An Introduction to Their Use for Modelling Environmental Systems. *Math. Comput. Simul.* **2008**, *78*, 379–400. [CrossRef]
135. Ahmed, A.N.; Othman, F.B.; Afan, H.A.; Elsha, A. Machine Learning Methods for Better Water Quality Prediction. *J. Hydrol.* **2019**, *578*. [CrossRef]
136. Choubin, B.; Darabi, H.; Rahmati, O.; Sajedi-Hosseini, F.; Kløve, B. River Suspended Sediment Modelling Using the CART Model: A Comparative Study of Machine Learning Techniques. *Sci. Total Environ.* **2018**, *615*, 272–281. [CrossRef]

137. Chebud, Y.; Naja, G.M.; Rivero, R.G. Water Quality Monitoring Using Remote Sensing and an Artificial Neural Network. *Water Air Soil Pollut.* **2012**, *223*, 4875–4887. [CrossRef]
138. Habibiandehkordi, R.; Lobb, D.A.; Owens, P.N.; Flaten, D.N. Effectiveness of Vegetated Buffer Strips in Controlling Legacy Phosphorus Exports from Agricultural Land. *J. Environ. Qual.* **2019**, *48*, 314–321. [CrossRef]
139. Hille, S.; Graeber, D.; Kronvang, B.; Rubæk, G.H.; Onnen, N.; Molina-Navarro, E.; Baattrup-Pedersen, A.; Heckrath, G.J.; Stutter, M.I. Management Options to Reduce Phosphorus Leaching from Vegetated Buffer Strips. *J. Environ. Qual.* **2018**, *48*, 322–329. [CrossRef]
140. Penn, C.; Camberato, J. A Critical Review on Soil Chemical Processes That Control How Soil pH Affects Phosphorus Availability to Plants. *Agriculture* **2019**, *9*, 120. [CrossRef]
141. Doydora, S.; Gatiboni, L.; Grieger, K.; Hesterberg, D.; Jones, J.L.; McLamore, E.S.; Peters, R.; Sozzani, R.; Van den Broeck, L.; Duckworth, O.W. Accessing Legacy Phosphorus in Soils. *Soil Syst.* **2020**, *4*, 74. [CrossRef]

soil systems

MDPI

Article

Single and Binary Fe- and Al-hydroxides Affect Potential Phosphorus Mobilization and Transfer from Pools of Different Availability

Stella Gypser *, Elisabeth Schütze and Dirk Freese

Department of Soil Protection and Recultivation, Faculty of Environment and Natural Sciences,
Brandenburg University of Technology Cottbus-Senftenberg, Konrad-Wachsmann-Allee 6,
03046 Cottbus, Germany; Elisabeth.schuetze@b-tu.de (E.S.); dirk.freese@b-tu.de (D.F.)
* Correspondence: stella.gypser@b-tu.de

Abstract: Phosphorus (P) fixation is a global problem for soil fertility and negatively impacts agricultural productivity. This study characterizes P desorption of already fixed P by using KCl, KNO_3, histidine, and malic acid as inorganic and organic compounds, which are quite common in soil. Goethite, gibbsite, and ferrihydrite, as well as hydroxide mixtures with varying Fe- and Al-ratio were selected as model substances of crystalline and amorphous Fe- and Al-hydroxides. Especially two- and multi-component hydroxide systems are common in soils, but they have barely been included in desorption studies. Goethite showed the highest desorption in the range from 70.4 to 81.0%, followed by gibbsite with values in the range from 50.7 to 42.6%. Ferrihydrite had distinctive lower desorption in the range from 11.8 to 1.9%. Within the group of the amorphous Fe-Al-hydroxide mixtures, P desorption was lowest at the balanced mixture ratio for 1 Fe: 1 Al, increased either with increasing Fe or Al amount. Precipitation and steric effects were concluded to be important influencing factors. More P was released by crystalline Fe-hydroxides, and Al-hydroxides of varying crystallinity, but desorption using histidine and malic acid did not substantially influence P desorption compared to inorganic constituents.

Keywords: phosphorus kinetics; desorption; Fe-Al-hydroxide mixtures; histidine; malic acid

Citation: Gypser, S.; Schütze, E.;
Freese, D. Single and Binary Fe- and
Al-hydroxides Affect Potential
Phosphorus Mobilization and
Transfer from Pools of Different
Availability. *Soil Syst.* **2021**, *5*, 33.
https://doi.org/10.3390/
soilsystems5020033

Academic Editors: Donald S. Ross,
Eric O. Young and Deb Jaisi

Received: 9 March 2021
Accepted: 20 May 2021
Published: 21 May 2021

1. Introduction

P, is, on the one hand, one of the most important essential nutrients for plant growth, but on the other hand, a finite resource, limiting productivity in agriculture terrestrial ecosystems. Given the scarcity of global phosphorus reserves, and to ensure sustainable soil fertility on agricultural soils, a fundamental understanding of the mechanisms of fixation, recognized as the reduction of solubility of fertilized P in the soil [1], and mobilization of inorganic phosphorus in soils is required. Although both the inorganic and organic P pools contribute to total P availability [2,3], dissolved inorganic P is the only P fraction that can be taken up by plants and microorganisms, thus maintaining ecosystem nutrition and mineralization [4]. Dissolved P has a high affinity for adsorption to the soil matrix, which affects its bioavailability, depending on soil composition and binding motifs. In particular, adsorption, desorption, and precipitation processes on pedogenic mineral surfaces limit its availability, which is why reactions of P with selected hydroxides have been studied in detail in the past.

"Nonspecific" physisorption via electrostatic attraction provides lower binding energy and thus easy mobilization of P by ion exchange [5,6], while more "specific" chemisorption results in stronger binding at the particle surface and lower availability of P over time [7,8]. However, the most stable and long-lasting P immobilization occurs via precipitation on the particle surface, where especially amorphous Fe-hydroxides play a major role [9–11].

In addition to the fixation of inorganic P on mineral surfaces, the soil organic matter has an important influence on P adsorption and desorption. If both P and organic anions are present in the soil solution, the adsorption of P on hydroxide surfaces can be positively influenced by competition for adsorption sites, ligand exchange, and replacement of P by organic anions, dissolution of adsorbents, and changes in the surface charge of the adsorbents. Soil organic matter can also retard the crystal growth of poorly crystalline Fe- and Al-oxides and -hydroxides, which affects their specific P adsorption capacities. Moreover, organic anions can form metal-organic complexes by adsorption on metal ions (e.g., Al^{3+}, Fe^{3+}) [12–17].

P fixation is a global problem for soil fertility and negatively impacts agricultural productivity because recovery by plants in the year of application is often only 10 to 15% and P inputs from fertilizers tend to accumulate in the soil [18]. A meta-analysis of about 2000 datasets from 30 field trials from Germany and Austria showed that yield increase after P application is mainly determined by pH, soil organic matter, fertilizer type, and crop type, whereas plant-available P in the soil seems to be the most important parameter [19]. However, Syers et al. (2008) [20] showed that an irreversible immobilization of fertilized P is not supported by field studies. They divided inorganic soil P into four pools with different availability to plants based on its accessibility to plant roots and extractability using common soil analytical methods. These pools range from soil solution P characterized by immediate accessibility and availability to very low accessible, extractable, and available P, which is very strongly bonded, inaccessible, precipitated, or mineral P. Roberts and Johnston (2015) [19] have summarized that differences in P bioavailability depend on accessibility to plant roots and extractability by soil test reagents. However, when the concept of P transfer within the four pools and partial nutrient balance is considered, P recovery can exceed 70%. In particular, the distribution of P among these different pools leads to the conversion of excess P into very slowly exchangeable P that can only be partially utilized when soil P content is low [21]. These organic and inorganic P pools of different availability were generated due to the weaker bound P in surface complexes or the strongly bound precipitates. Therefore, the fundamental understanding of P binding motifs on contrasting mineral surfaces, and possible changes in binding over time [22–30] allows a more detailed characterization of soils in terms of their potential P fixation capacity, as well as short- and long-term mobilization.

Most methods for the determination of available P are based on the quantification of solubilized P using different extractants, consist of chemical equilibrium-controlled solubility, and release rates-limiting processes. They do not measure the quantity of plant-available P, but by experimentation, testing, and the application of regression equations, they allow a prediction about a soil P status related to it [31]. However, these observed results are not always applicable to different soil types or arable crops. Due to ad- and desorption processes, dissolution, or mineralization, the pool of plant-available P is strongly time-dependent [20,31]. For this, sequential extraction methods offer an inexpensive and simple approach to determine the amount of long-term mobilized P. The different extractants can be selected according to the objectives and play a role, e.g., during cultivation, in the rhizosphere, or during soil development. While no particulate speciation such as precipitated or re-adsorbed P can be provided without further investigations [1], different soil types or soil components can be investigated concerning the differentiation of potential P mobilization. Based on the empirical assignments, the P status of soils can be characterized according to a concept of different available P pools and their transformation [20,31].

Therefore, this study aims to investigate whether the crystallinity, as well as the Fe/Al ratio of hydroxides, affects the potential mobilization of adsorbed P. The results will hopefully guide as to whether the composition of soil in terms of pedogenic hydroxides affects the moderately to non-labile P reserves. For this purpose, desorption kinetics are created in a batch setup using synthetic Fe- and Al-hydroxides as well as inorganic and organic extraction agents at two concentrations.

The amount of low and very low available inorganic P that can be desorbed by increasing concentrations of organic and inorganic extractants has to be determined, and the time-depended mobilization process will be evaluated.

2. Materials and Methods

2.1. Preparation of the Fe- and Al-Hydroxides

Goethite and gibbsite were used as model substances of crystalline Fe- and Al-hydroxides. In addition to ferrihydrite, these minerals were the main model minerals of previous studies. However, two- or multi-component hydroxide systems are more common in soils, but they have barely been included so far, especially in desorption studies. Therefore, synthesized Fe- and Al-hydroxide mixtures with varying Fe- and Al-ratio represented this binary amorphous fraction appearing in soils. Poorly crystalline ferrihydrite was used as a transitional Fe-hydroxide, bridging between the initial amorphous hydroxide structure and crystalline goethite during pedogenesis.

The synthetic hydroxides investigated in this study were goethite (99%, Alfa Aesar, Haverhill, MA, USA), gibbsite (analytical grade, Merck Millipore, Merck KGaA, Darmstadt, Germany), ferrihydrite (prepared according to [32]), and mixed Fe:Al-hydroxide (prepared according to [33]).

For the preparation of 2-line-ferrihydrite, a 1 M KOH was added to 500 mL of a 0.2 M $Fe(NO_3)_3 \cdot 9\ H_2O$-solution, until a pH of 7.5 was reached. The developed precipitate was washed with ultrapure water to remove remaining salts, centrifuged for 5 min at $12,134 \times g$ (Avanti J-25 Centrifuge, Beckman Coulter, Brea, CA, USA), frozen and freeze-dried, and stored in a desiccator.

The Fe:Al-hydroxide mixtures were prepared by mixing 0.1 M $Fe(NO_3)_3 \cdot 9\ H_2O$ and 0.1 M $Al(NO_3)_3 \cdot 9\ H_2O$ in molar ratios of 1:0, 5:1. 1:1, 1:5, and 0:1, and adjusted to pH 6 with 5 M KOH. The solutions were equilibrated for 1 h. Subsequently, the precipitate was washed with ultrapure water, centrifuged for 5 min at $12,134 \times g$, dried at 60 °C, and ground into a powder. All chemicals used for the preparation were of analytical grade.

Prior to desorption experiments, 20 g of each hydroxide was adjusted to pH 6 in 50 mL ultrapure water with 0.1 M HCl or KOH, respectively, and dried at 40 °C for 5 days. P was adsorbed by adding 200 mL of a 0.3 M P solution to 17 g of the dried hydroxides. The P solution consisted of a Na_2HPO_4/KH_2PO_4 buffer solution (pH 6) with additional KH_2PO_4 to achieve the desired P concentration of 0.3 M. Subsequently, the hydroxide-P solution mixtures were shaken horizontally with 200 Motions min^{-1} for 24 h. Afterward, they were centrifuged for 15 min at $21,572 \times g$ until the supernatant was clear.

2.2. Characterization of the Fe- and Al-Hydroxides

The elemental composition of the hydroxides was verified using SEM-EDX, scanning electron microscopy (DSM 962, Zeiss, Oberkochen, Germany) with energy dispersive X-ray spectroscopy (X-Max 50 mm^2 with INCA, Oxford Instruments, Abingdon, Great Britain). The determination of the crystallization, as well as the poorly crystalline and amorphous structures was performed using X-ray diffraction (Empyrean powder diffractometer, PANalytical, Almelo, Netherlands) (for the results of gibbsite, ferrihydrite and the Fe-Al-hydroxide mixtures see [24], for the results of goethite see [23]).

Specific surface areas were determined in duplicate with an Autosorb-1 (Quantachrome, Odelzhausen, Germany) using a multi-point BET-measurement (Brunauer–Emmett–Teller) and N_2 as adsorptive. An outgas test was performed to verify the completed outgas procedure for each hydroxide. The specific surface area was substantially higher for the amorphous Fe-hydroxides and decreased with increasing crystallinity grade, as well as an increasing amount of Al for the mixtures (Table 1). The specific surface area was in the same range for the crystalline and the amorphous Al-hydroxides.

Table 1. Point of zero charge (PZC), specific surface area (SSA), and the amount of total Fe, Al, and P of crystalline and amorphous Fe- and Al-hydroxides.

Hydroxide	PZC	SSA	Fe	Al	P	
	pH	$m^2\,g^{-1}$	$mg\,g^{-1}$	$mg\,g^{-1}$	$mg\,g^{-1}$	$mg\,m^{-2}$
Goethite	8.8	17.2 ± 0.4	564.17 ± 9.74	-	7.07 ± 0.24	0.41 ± 0.01
Gibbsite	8.5	0.9 ± 0.0	-	337.39 ± 6.05	7.39 ± 0.06	8.21 ± 0.06
Ferrihydrite	7.1	251.8 ± 2.7	520.99 ± 12.27	-	30.92 ± 2.63	0.12 ± 0.01
1 Fe: 0 Al	6.0	297.3 ± 10.4	494.30 ± 25.04	-	36.50 ± 2.74	0.07 ± 0.09
5 Fe: 1 Al	7.1	203.8 ± 0.9	424.59 ± 2.51	32.10 ± 1.19	43.47 ± 5.23	0.21 ± 0.03
1 Fe: 1 Al	7.6	73.7 ± 8.0	275.44 ± 1.13	137.22 ± 0.87	31.34 ± 0.69	0.43 ± 0.01
1 Fe: 5 Al	9.8	0.8 ± 0.0	58.55 ± 0.65	162.27 ± 1.57	64.65 ± 0.64	80.82 ± 0.80
0 Fe: 1 Al	9.8	1.1 ± 0.0	-	131.44 ± 2.69	44.27 ± 0.33	40.25 ± 0.30

The point of zero charge (PZC) of each hydroxide was determined using potentiometric titration. 0.5 g of each hydroxide mixture was weighed into 100 mL PE-cups in triplicate. 30 mL of different KCL solutions (0.02, 0.2, and 2 M) were added separately to each sample. The solutions were diluted with ultrapure water to a total volume of 60 mL, leading to final concentrations of 0.01, 0.1, and 1 M KCl, respectively. The hydroxide/KCl mixtures were equilibrated for 4 days at 21 °C and shaken for 1 h d^{-1} to reach an equilibration pH value prior to the titration procedure. In the beginning, the pH of the suspensions was increased with 1 mL of 5 M KOH, followed by titration with fixed amounts of 1 M HCl. The amount of adsorbed H^+ on the hydroxide surface at each pH was determined by subtracting the titration curve of the blank KCl solutions from the titration curve of the suspension. The PZC derived from the titration curves was highest for amorphous hydroxides with a predominant Al- amount with a value of 9.8 (Table 1). The PZC decreased with increasing Fe-amount and was lowest for the pure amorphous Fe-hydroxide with a value of 6.0. Within the group of the crystalline hydroxides, gibbsite, and goethite were in a similar range with PZC values of 8.5 and 8.8, respectively. Ferrihydrite offered a slightly lower value of 7.1.

The total amount of Fe, Al, and adsorbed P was determined in duplicate by digestion of 0.02 g hydroxide in 50 mL ultrapure water with 1 mL aqua regia. The poorly crystalline ferrihydrite adsorbed initially 30.92 mg g^{-1} P, which was 4-fold higher than for goethite or gibbsite (Table 1). The amorphous Fe:Al-hydroxide mixtures had adsorbed P concentrations in the range from 31.34 to 64.65 mg g^{-1}, increasing with a predominant amount of Al. Thus, the amorphous hydroxides showed a higher P adsorption than the crystalline hydroxides. Related to the specific surface area, P adsorption values of 80.81 and 40.25 mg m^{-2} were obtained for the amorphous Al-hydroxides, and 8.21 mg m^{-2} was obtained for gibbsite. P adsorption values in the range from 0.12 to 0.43 mg m^{-2} were lower for Fe-hydroxides due to the substantially higher SSA.

2.3. Desorption Experiments

Desorption experiments were performed in a batch setup at room temperature. The investigations were carried out in quadruplicate using 50 mL PE-centrifuge tubes. Each batch contained a hydroxide-solution mixture, consisting of 0.8 g hydroxide and 40 mL reaction solution, resulting in a solid-solution ratio of 1:50.

KCl and KNO_3 were selected as inorganic compounds for desorption experiments since they are an essential part of the soil solution in agricultural soils, and components of mineral fertilizers. Histidine ($C_6H_9N_3O_2$, His) and malic acid ($C_4H_6O_5$, Mal) were chosen as organic extractants. These reaction solutions were used with concentrations of 5 mM and 50 mM, adjusted to a pH of 6 with 0.1 M KOH, HCl, or HNO_3, respectively.

At the beginning of desorption, the samples were shaken horizontally at 200 Motions min^{-1} for 24 h, afterward once a week for 1 h. For sample taking, the hydroxide-solution mixtures were centrifuged for 20 min at $21,572 \times g$. The clear supernatant was carefully decanted and filtrated using P-poor Whatman 512 1/1 folded filter papers. Afterward, 40 mL of the fresh reaction solution was added to continue desorption. Sample taking was done after 2, 6, 24, 48, 168, and 336 h desorption time. After a desorption time of 2 h, the pH of the sample solution was measured again in two randomly selected samples of each hydroxide for all treatments.

Concentrations of dissolved total P, Fe, and Al were determined by using ICP–AES (Unicam iCAP6000 Duo, Thermo Fisher Scientific, Waltham, MA, USA), total Cl was determined by using ion chromatography (Dionex DX 500 + DX 120, Thermo Fisher Scientific), total N and C were measured with a TOC-Analyzer (TOC-VCPH and TOC5000, Shimadzu, Kyoto, Japan). Repeated washing of all used materials with ultrapure water and immediate freezing of the sample solutions prevented microbial activity.

2.4. Kinetics of P Desorption

The cumulative P desorption depending on time was calculated, and different linearized kinetic models were applied to the data (Table 2). The aim was to fit the experimental data to an appropriate kinetic model and to analyze the influence of organic and inorganic solutions on desorption kinetics from contrasting Fe- and Al-hydroxides. The coefficients of determination (R^2), as well as the standard errors (S.E.) were tested using linear regression analysis to determine their applicability on the kinetics using SigmaPlot 12.0 (Systat Software Inc., San Jose, CA, USA). When not stated otherwise, the *p*-value was <0.05.

Table 2. Applied kinetic models for P desorption.

Kinetic Model	Linearized Equation	Declaration
Elovich	$Q_t = \frac{1}{\beta}\ln(\alpha\beta) + \frac{1}{\beta}\ln t$	Q_t—amount of desorbed P in mg P m^{-2} Hydroxide at time t α/a—initial P release constants in mg P m^{-2} Hydroxide min^{-1} β/b—P release rate constants in mg P m^{-2} Hydroxide min^{-1} k_p—diffusion rate constant in m s^{-1}
Exponential	$\ln Q_t = \ln a + b \ln t$	Q_0—equals value of 0 at the beginning of desorption
Parabolic	$Q_t = Q_0 + k_p t^{\frac{1}{2}}$	

3. Results

3.1. Efficiency of P Desorption

The efficiency of the desorption solutions showed in particular that the lower concentration of 5 mM led to a higher P release than the corresponding higher concentration of 50 mM for each reaction solution (Table 3). For the lower concentration of 50 mM, the efficiency of the desorption solutions ranked according to the following order: $KCl > KNO_3 > Mal > His$. This order changed for the 50 mM concentration treatment to $KCl > Mal > His > KNO_3$, where KCl also showed the highest P desorption efficiency.

Table 3. Total P desorption after 336 h by using desorption solutions KCl, KNO$_3$, histidine (His), and malic acid (Mal).

Hydroxide	Desorbed P [%]							
	KCl		KNO$_3$		His		Mal	
	5 mM	50 mM	5 mM	50 mM	5 mM	50 mM	5 mM	50 mM
Goethite	80.95 ± 0.39	72.24 ± 0.40	71.58 ± 0.51	64.72 ± 0.33	70.94 ± 0.46	65.61 ± 0.40	70.38 ± 0.32	68.91 ± 0.81
Gibbsite	46.58 ± 2.96	39.53 ± 1.85	42.60 ± 2.12	41.26 ± 2.84	50.48 ± 3.62	49.17 ± 3.56	50.74 ± 2.51	61.43 ± 3.33
Ferrihydrite	11.77 ± 0.46	4.44 ± 0.22	6.91 ± 0.23	3.72 ± 0.11	1.92 ± 0.09	1.38 ± 0.21	3.61 ± 0.12	3.01 ± 0.29
1 Fe: 0 Al	27.55 ± 0.23	15.57 ± 0.14	21.68 ± 0.17	13.52 ± 0.07	14.93 ± 0.19	10.33 ± 0.11	15.25 ± 0.13	12.14 ± 0.06
5 Fe: 1 Al	14.49 ± 0.32	5.79 ± 0.13	9.11 ± 0.23	4.40 ± 0.08	4.78 ± 0.10	2.79 ± 0.15	5.36 ± 0.10	3.87 ± 0.13
1 Fe: 1 Al	5.23 ± 0.29	2.24 ± 0.16	2.90 ± 0.20	1.95 ± 0.28	2.51 ± 0.15	2.18 ± 0.13	3.22 ± 0.27	4.38 ± 0.35
1 Fe: 5 Al	20.48 ± 1.02	12.20 ± 0.22	16.10 ± 0.09	10.80 ± 0.12	13.75 ± 0.06	10.89 ± 0.30	13.16 ± 0.32	11.86 ± 0.52
0 Fe: 1 Al	23.22 ± 0.39	14.79 ± 0.44	19.90 ± 0.54	13.69 ± 0.30	19.82 ± 0.35	16.94 ± 0.12	19.66 ± 0.44	21.33 ± 0.33
Desorbed P [mg m^{-2}]								
Goethite	0.33 ± 0.00	0.30 ± 0.00	0.29 ± 0.00	0.27 ± 0.00	0.29 ± 0.00	0.27 ± 0.00	0.29 ± 0.00	0.28 ± 0.00
Gibbsite	3.96 ± 0.25	3.36 ± 0.16	3.62 ± 0.18	3.51 ± 0.24	4.29 ± 0.31	4.18 ± 0.30	4.31 ± 0.21	5.22 ± 0.28
Ferrihydrite	0.01 ± 0.00	0.01 ± 0.00	0.01 ± 0.00	0.00 ± 0.00	0.00 ± 0.00	0.00 ± 0.00	0.00 ± 0.00	0.00 ± 0.00
1 Fe: 0 Al	0.03 ± 0.00	0.02 ± 0.00	0.03 ± 0.00	0.02 ± 0.00	0.02 ± 0.00	0.01 ± 0.00	0.02 ± 0.00	0.01 ± 0.00
5 Fe: 1 Al	0.03 ± 0.00	0.01 ± 0.00	0.02 ± 0.00	0.01 ± 0.00	0.01 ± 0.00	0.01 ± 0.00	0.01 ± 0.00	0.01 ± 0.00
1 Fe: 1 Al	0.02 ± 0.00	0.01 ± 0.00	0.01 ± 0.00	0.01 ± 0.00	0.01 ± 0.00	0.01 ± 0.00	0.01 ± 0.00	0.02 ± 0.00
1 Fe: 5 Al	16.71 ± 0.83	9.96 ± 0.18	13.13 ± 0.07	8.81 ± 0.10	11.21 ± 0.05	8.88 ± 0.25	10.73 ± 0.26	9.67 ± 0.43
0 Fe: 1 Al	9.19 ± 0.15	5.85 ± 0.17	7.87 ± 0.22	5.42 ± 0.12	7.84 ± 0.14	6.70 ± 0.05	7.78 ± 0.17	8.44 ± 0.13

For the concentration of 5 mM, goethite showed the highest P desorption in the range from 70.4 to 81.0%, followed by gibbsite with P desorption in the range from 50.7 to 42.6%. The poorly crystalline ferrihydrite had lower desorption values in the range from 11.8 to 1.9% compared to the crystalline goethite. Within the group of the amorphous Fe-Al-hydroxide mixtures, the pure hydroxides of Fe and Al had higher desorption with values in the range from 27.6 to 13.5% for 1 Fe: 0 Al, and from 23.2 to 19.7% for 0 Fe: 1 Al than the binary composites. P desorption was lowest at the balanced mixture ratio for 1 Fe: 1 Al in the range from 2.5 to 5.2%, and increased either with increasing Fe or Al amount. In total, more P was desorbed from the crystalline hydroxides than from the amorphous hydroxides.

If P desorption was related to the specific surface area, gibbsite had a higher desorption amount than goethite, because the specific surface area of goethite was substantially larger than that of gibbsite, which is why less P was ad- and desorbed from the goethite surface. Similar was observed for the hydroxide mixtures. While e.g., 1 Fe: 0 Al and 1 Fe: 5 Al had similar amounts of P desorption in %, especially for the organic treatment, the desorption amount related to the specific surface area was much lower for the pure amorphous Fe-hydroxide compared to the mixture due to the highly different surface area and hence, the amount of ad- and desorbed P per m^2.

3.2. Kinetics of P Desorption

The curves of the P desorption kinetics showed that the fast initial desorption step occurred during the first 48 h for all used reaction solutions (Figures 1 and 2). While for goethite and gibbsite, after 48 h equilibrium was nearly reached using KCl and KNO$_3$, desorption was still ongoing for ferrihydrite and the Fe-Al-hydroxide mixtures (Figure 1). Similar was observed for Mal and His, whereby it can be seen that P desorption from goethite and gibbsite continued using Mal (Figure 2).

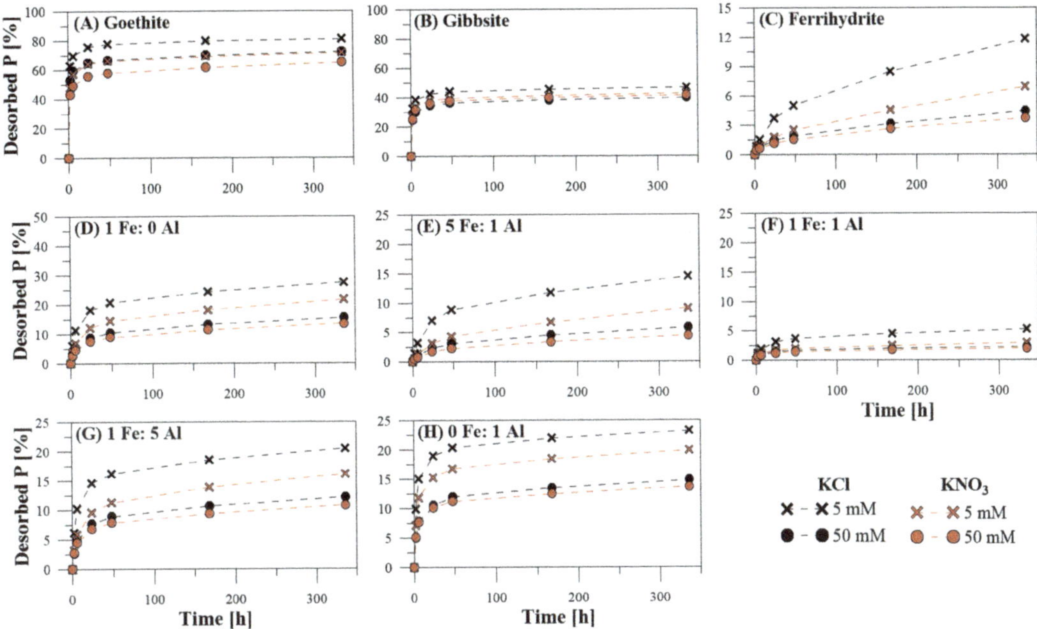

Figure 1. Kinetics of P desorption from (**A**) goethite, (**B**) gibbsite, (**C**) ferrihydrite, (**D**) 1 Fe: 0 Al, (**E**) 5 Fe: 1 Al, (**F**) 1 Fe: 1 Al, (**G**) 1 Fe: 5 Al, (**H**) 0 Fe: 1 Al with KCl and KNO$_3$ at concentrations of 5 mM and 50 mM.

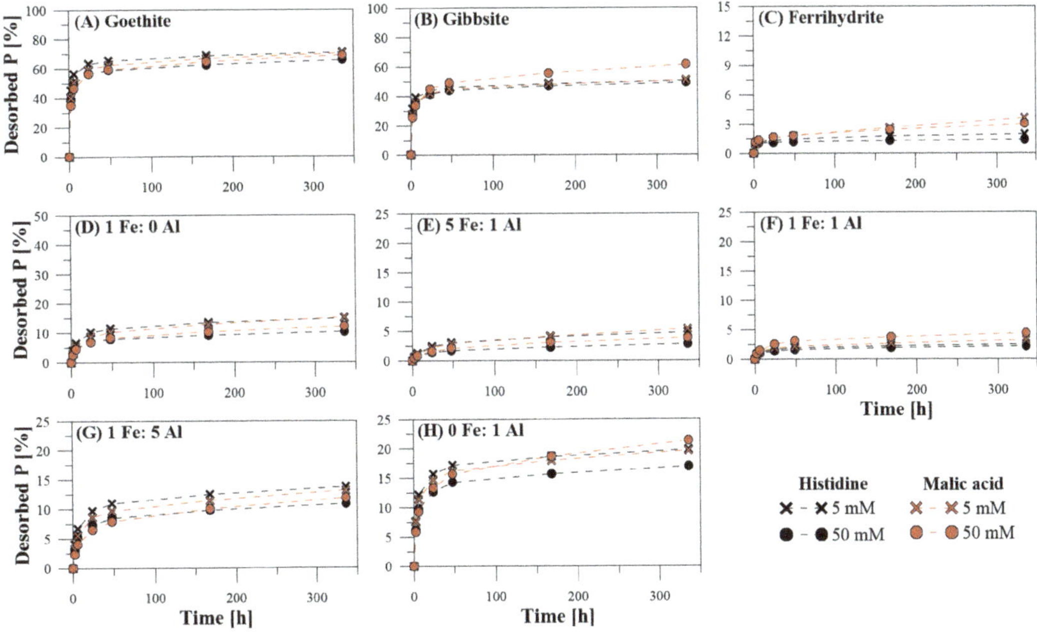

Figure 2. Kinetics of P desorption from (**A**) goethite, (**B**) gibbsite, (**C**) ferrihydrite, (**D**) 1 Fe: 0 Al, (**E**) 5 Fe: 1 Al, (**F**) 1 Fe: 1 Al, (**G**) 1 Fe: 5 Al, (**H**) 0 Fe: 1 Al with histidine and malic acid at concentrations of 5 mM and 50 mM.

The coefficients of determination for the applied kinetic models showed that desorption kinetics fitted best with the Elovich equation (mean R^2 = 0.93) (Table 4), followed by the exponential function (mean R^2 = 0.91) (Table 5), and the parabolic function (mean R^2 = 0.83) (Table 6). Only ferrihydrite had the best fit with the Parabolic equation (mean R^2 = 0.89), followed by the exponential function (mean R^2 = 0.87), and the Elovich equation (mean R^2 = 0.80). However, it is necessary to evaluate the results of the kinetic models as well as their calculated kinetic parameters (Tables 4–6) for each hydroxide separately. For goethite and gibbsite, the Elovich equation had the best fit, followed by the exponential function. In addition, the kinetic parameters obtained showed mainly a higher initial P release (α, a) and a lower P release over time (β, b). The Mal desorption treatments of goethite and the applied Elovich function showed a lower initial P release and a higher P release over time. However, noticeable were the α-values, which were overestimated compared to the actual desorption amounts. As already mentioned, for ferrihydrite the best fit was obtained by application of the exponential function. For ferrihydrite, the initial P release was lower than the release over time. By using the Elovich equation certainly, β was higher than α, where an overestimation of α for the His treatment was concluded as well. The amorphous Fe- and Al-hydroxide mixtures had the best fit with the Elovich equation, followed by the Exponential equation. For the pure amorphous Fe-hydroxide, the Fe-dominated mixtures, and the mixture with equal amounts of Fe and Al, the initial P release values α and a were lower than the release over time values β and b. The values for the mixture with predominant Al amount indicated a greater initial release of P.

Table 4. Coefficients of determination (R^2), standard error (S.E.), and calculated kinetic parameters for the Elovich equation used to describe the kinetic release of P after 336 h desorption time.

Hydroxide	Elovich	KCl		KNO$_3$		His		Mal	
		5 mM	50 mM	5 mM	50 mM	5 mM	50 mM	5 mM	50 mM
Goethite	R^2	0.93	0.97	0.94	0.98	0.93	0.94	0.96	0.97
	S.E.	0.01	0.00	0.01	0.00	0.01	0.01	0.01	0.01
	α	1.47×10^6	2.40×10^4	973.69	440.79	348.89	75.12	26.16	5.25
	β	71.32	66.86	55.46	59.67	52.10	51.00	43.47	38.44
Gibbsite	R^2	0.74	0.90	0.87	0.89	0.93	0.38	0.87	0.80
	S.E.	0.24	0.17	0.19	0.24	0.30	0.27	0.23	0.26
	α	6.62×10^4	1.62×10^3	3.03×10^3	1.17×10^3	2.84×10^3	337.64	211.40	14.21
	β	4.54	4.25	4.09	3.95	3.42	2.97	2.77	1.72
Ferrihydrite	R^2	0.92	0.87	0.85	0.80	0.77	0.84	0.89 *	0.94
	S.E.	0.00	0.00	0.00	0.00	0.00	0.00	0.00	0.00
	α	1.03×10^{-3}	4.86×10^{-4}	5.45×10^{-4}	3.91×10^{-4}	0.02	5.74	1.12×10^{-3}	3.78×10^{-3}
	β	392.88	1.12×10^3	696.87	1.33×10^3	4.58×10^3	1.02×10^4	1.69×10^3	2.40×10^3
1 Fe: 0 Al	R^2	0.99	1.00	1.00	0.99	0.99	0.99	1.00	1.00
	S.E.	0.00	0.00	0.00	0.00	0.00	0.00	0.00	0.00
	α	0.01	4.98×10^{-3}	0.01	3.90×10^{-3}	0.01	0.01	0.01	4.13×10^{-3}
	β	197.39	339.08	236.18	386.40	385.73	591.45	350.24	439.41
5 Fe: 1 Al	R^2	0.99	0.97	0.95	0.96	0.99	0.98	0.97	0.97
	S.E.	0.00	0.00	0.00	0.00	0.00	0.00	0.00	0.00
	α	3.82×10^{-3}	1.19×10^{-3}	1.47×10^{-3}	8.59×10^{-4}	1.64×10^{-3}	1.04×10^{-3}	1.22×10^{-3}	9.32×10^{-4}
	β	184.08	459.54	287.03	608.12	581.63	1.04×10^3	510.68	708.58
1 Fe: 1 Al	R^2	0.98	0.93	0.96	0.82	0.96	0.95	0.93	0.95
	S.E.	0.00	0.00	0.00	0.00	0.00	0.00	0.00	0.00
	α	0.01	0.01	0.01	0.01	0.01	0.01	0.01	0.01
	β	299.90	805.25	596.31	937.29	719.88	826.31	531.43	354.14

Table 4. *Cont.*

Hydroxide	Elovich	KCl		KNO$_3$		His		Mal	
		5 mM	50 mM	5 mM	50 mM	5 mM	50 mM	5 mM	50 mM
1 Fe: 5 Al	R^2	0.97	0.99	1.00	0.99	0.98	0.97	0.98	0.99
	S.E.	0.72	0.33	0.18	0.18	0.40	0.38	0.39	0.34
	α	14.62	3.59	3.69	3.87	8.78	6.31	5.02	2.37
	β	0.45	0.67	0.49	0.79	0.66	0.83	0.65	0.67
0 Fe: 1 Al	R^2	0.94	0.94	0.95	0.96	0.94	0.97	0.97	0.99
	S.E.	0.45	0.35	0.41	0.23	0.43	0.26	0.31	0.16
	α	52.56	7.14	16.61	10.76	23.30	13.45	18.73	4.45
	β	1.02	1.36	1.07	1.57	1.11	1.25	1.12	0.85

* $p = 0.059$.

Table 5. Coefficients of determination (R^2), standard error (S.E.), and calculated kinetic parameters for the exponential function used to describe the kinetic release of P after 336 h desorption time.

Hydroxide	Exponential	KCl		KNO$_3$		His		Mal	
		5 mM	50 mM	5 mM	50 mM	5 mM	50 mM	5 mM	50 mM
Goethite	R^2	0.91	0.96	0.91	0.97	0.89	0.90	0.92	0.92
	S.E.	0.03	0.02	0.04	0.03	0.05	0.06	0.06	0.07
	a	0.26	0.22	0.20	0.17	0.19	0.17	0.17	0.14
	b	0.05	0.06	0.07	0.08	0.08	0.09	0.10	0.12
Gibbsite	R^2	0.72	0.84	0.83	0.77	0.75	0.82	0.87	0.92
	S.E.	0.08	0.07	0.07	0.09	0.09	0.09	0.08	0.09
	a	2.80	2.12	2.34	2.17	2.72	2.39	2.38	2.09
	b	0.07	0.09	0.08	0.09	0.08	0.10	0.11	0.17
Ferrihydrite	R^2	0.99	0.99	1.00	1.00	0.95	0.38	0.95	0.82
	S.E.	0.09	0.06	0.06	0.05	0.06	0.17	0.10	0.16
	a	7.88×10^{-4}	4.78×10^{-4}	4.58×10^{-4}	3.88×10^{-4}	1.11×10^{-3}	1.09×10^{-3}	9.15×10^{-4}	1.19×10^{-3}
	b	0.51	0.41	0.49	0.42	0.13	0.07	0.25	0.18
1 Fe: 0 Al	R^2	0.92	0.96	0.96	0.97	0.93	0.93	0.96	0.96
	S.E.	0.15	0.11	0.13	0.11	0.13	0.11	0.11	0.12
	a	0.01	3.61×10^{-3}	4.44×10^{-3}	2.96×10^{-3}	4.89×10^{-3}	3.73×10^{-3}	3.61×10^{-3}	2.91×10^{-3}
	b	0.28	0.30	0.33	0.31	0.25	0.23	0.30	0.30
5 Fe: 1 Al	R^2	0.94	0.97	0.98	0.98	0.95	0.97	0.98	0.98
	S.E.	0.20	0.15	0.15	0.12	0.15	0.12	0.12	0.11
	a	2.85×10^{-3}	9.23×10^{-4}	9.98×10^{-4}	6.84×10^{-4}	1.32×10^{-3}	8.61×10^{-4}	1.03×10^{-3}	8.04×10^{-4}
	b	0.44	0.47	0.53	0.47	0.38	0.35	0.43	0.42
1 Fe: 1 Al	R^2	0.95	0.90	0.96	0.83	0.92	0.92	0.92	0.92
	S.E.	0.12	0.14	0.09	0.18	0.12	0.12	0.13	0.15
	a	4.71×10^{-3}	2.79×10^{-3}	3.14×10^{-3}	2.41×10^{-3}	3.27×10^{-3}	2.77×10^{-3}	3.32×10^{-3}	3.77×10^{-3}
	b	0.28	0.22	0.24	0.22	0.21	0.22	0.25	0.29
1 Fe: 5 Al	R^2	0.92	0.92	0.94	0.95	0.92	0.91	0.93	0.96
	S.E.	0.12	0.15	0.14	0.11	0.13	0.14	0.13	0.12
	a	5.15	2.18	2.53	2.15	3.34	2.56	2.65	1.76
	b	0.22	0.28	0.30	0.26	0.23	0.23	0.26	0.31

Table 5. Cont.

Hydroxide	Exponential	KCl		KNO_3		His		Mal	
		5 mM	50 mM	5 mM	50 mM	5 mM	50 mM	5 mM	50 mM
0 Fe: 1 Al	R^2	0.88	0.87	0.87	0.89	0.87	0.91	0.91	0.96
	S.E.	0.10	0.15	0.13	0.12	0.13	0.11	0.10	0.10
	a	4.13	1.96	2.97	2.03	3.18	2.54	3.04	2.24
	b	0.15	0.20	0.18	0.18	0.17	0.18	0.17	0.24

Table 6. Coefficients of determination (R^2), standard error (S.E.), and calculated kinetic parameters for the parabolic function used to describe the kinetic release of P after 336 h desorption time.

Hydroxide	Parabolic	KCl		KNO_3		His		Mal	
		5 mM	50 mM	5 mM	50 mM	5 mM	50 mM	5 mM	50 mM
Goethite	R^2	0.69	0.79	0.71	0.82	0.70	0.73	0.76	0.77
	S.E.	0.02	0.01	0.02	0.01	0.02	0.02	0.02	0.02
	Q_0	0.28	0.23	0.22	0.19	0.22	0.19	0.19	0.17
	k_p	3.58×10^{-3}	3.99×10^{-3}	4.67×10^{-3}	4.55×10^{-3}	4.94×10^{-3}	0.01	0.01	0.01
Gibbsite	R^2	0.55	0.67	0.64	0.61	0.59	0.65	0.71	0.81
	S.E.	0.32	0.27	0.30	0.33	0.39	0.40	0.38	0.48
	Q_0	3.09	2.41	2.65	2.48	3.09	2.78	2.79	2.59
	k_p	0.06	0.06	0.06	0.07	0.08	0.09	0.10	0.16
Ferrihydrite	R^2	0.99	0.99	0.99	1.00	0.92	0.37	0.99	0.88
	S.E.	0.00	0.00	0.00	0.00	0.00	0.00	0.00	0.00
	Q_0	2.85×10^{-4}	3.22×10^{-4}	-6.07×10^{-5}	2.26×10^{-4}	1.26×10^{-3}	1.19×10^{-3}	9.30×10^{-4}	1.33×10^{-3}
	k_p	7.83×10^{-4}	2.79×10^{-4}	4.56×10^{-4}	2.36×10^{-4}	6.44×10^{-5}	2.89×10^{-5}	1.87×10^{-4}	1.30×10^{-4}
1 Fe: 0 Al	R^2	0.84	0.91	0.91	0.92	0.83	0.84	0.91	0.91
	S.E.	0.00	0.00	0.00	0.00	0.00	0.00	0.00	0.00
	Q_0	0.01	0.01	0.01	4.12×10^{-3}	0.01	0.01	0.01	4.09×10^{-3}
	k_p	1.38×10^{-3}	8.35×10^{-4}	1.20×10^{-3}	7.38×10^{-4}	7.04×10^{-4}	4.61×10^{-4}	8.09×10^{-4}	6.43×10^{-4}
5 Fe: 1 Al	R^2	0.92	0.97	0.98	0.98	0.92	0.94	0.97	0.97
	S.E.	0.00	0.00	0.00	0.00	0.00	0.00	0.00	0.00
	Q_0	4.44×10^{-3}	1.06×10^{-3}	6.39×10^{-4}	6.87×10^{-4}	1.95×10^{-3}	1.18×10^{-3}	1.20×10^{-3}	9.50×10^{-4}
	k_p	1.56×10^{-3}	6.45×10^{-4}	1.05×10^{-3}	4.93×10^{-4}	4.90×10^{-4}	2.78×10^{-4}	5.82×10^{-4}	4.19×10^{-4}
1 Fe: 1 Al	R^2	0.88	0.84	0.91	0.76	0.84	0.85	0.87	0.86
	S.E.	0.00	0.00	0.00	0.00	0.00	0.00	0.00	0.00
	Q_0	0.01	3.65×10^{-3}	3.95×10^{-3}	3.11×10^{-3}	4.29×10^{-3}	3.60×10^{-3}	4.31×10^{-3}	0.01
	k_p	9.38×10^{-4}	3.49×10^{-4}	4.86×10^{-4}	3.04×10^{-4}	3.86×10^{-4}	3.39×10^{-4}	5.41×10^{-4}	7.99×10^{-4}
1 Fe: 5 Al	R^2	0.80	0.86	0.88	0.88	0.81	0.82	0.85	0.91
	S.E.	1.87	1.05	1.28	0.83	1.25	0.96	1.12	0.85
	Q_0	7.07	3.17	3.70	2.97	4.65	3.53	3.74	2.46
	k_p	0.60	0.41	0.57	0.35	0.41	0.33	0.42	0.43
0 Fe: 1 Al	R^2	0.71	0.78	0.74	0.79	0.72	0.77	0.79	0.88
	S.E.	1.00	0.66	0.91	0.55	0.91	0.72	0.77	0.74
	Q_0	5.25	2.63	3.96	2.63	4.17	3.33	3.89	2.99
	k_p	0.25	0.20	0.25	0.17	0.23	0.21	0.24	0.33

3.3. Dissolved Elemental Composition during P Desorption

With regard to possible release mechanisms of P from the investigated hydroxides, the concentrations of dissolved total Fe, Al, Cl, N, as well as C were measured in the reaction solution during and after desorption experiments. Since anion exchange is the dominating mechanism during P desorption, an increase of Cl^- and NO_3^- could be observable. The concentrations of total Cl and N showed despite some variation a decreasing trend. The correlation of the change of dissolved total Cl and N with the concentration of dissolved P in the sample solution showed no clear relationship (Figure 3C,D).

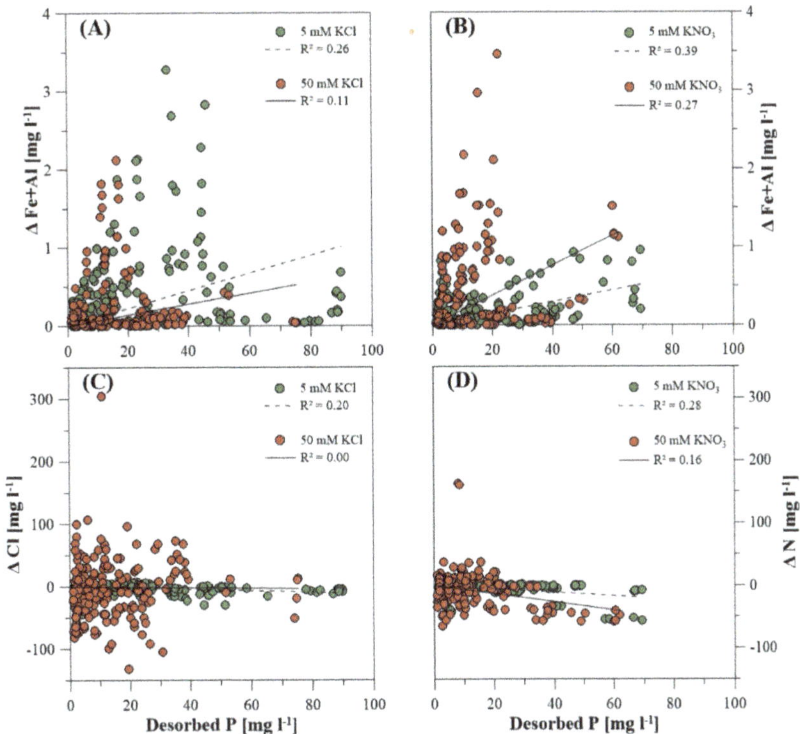

Figure 3. Correlation between desorbed P and the change of (**A**) dissolved total Fe + Al for KCl, (**B**) dissolved total Fe + Al for KNO_3, (**C**) dissolved total Cl for KCl, and (**D**) dissolved total N for KNO_3 in the sample solution for all hydroxides and each desorption time step.

The concentration of dissolved total Fe and Al (separately Fe or Al for the sole hydroxides, sum of Fe and Al for the hydroxide-mixtures) showed enrichment in the sample solution during P desorption, however, the values had a high variation and no distinct correlation with the amount of desorbed P for the inorganic treatment (Figure 3A,B). While the KCl treatment with the lower 5 mM concentration had higher concentrations of dissolved total Fe and Al, the opposite was observed for KNO_3. The same missing correlations of Fe, Al, and P were observed for the organic desorption treatments, whereas the concentration of dissolved Fe and Al increased in the sample solution as well. The Mal treatment showed higher concentrations of dissolved total Fe and Al than His (Figure 4A,B).

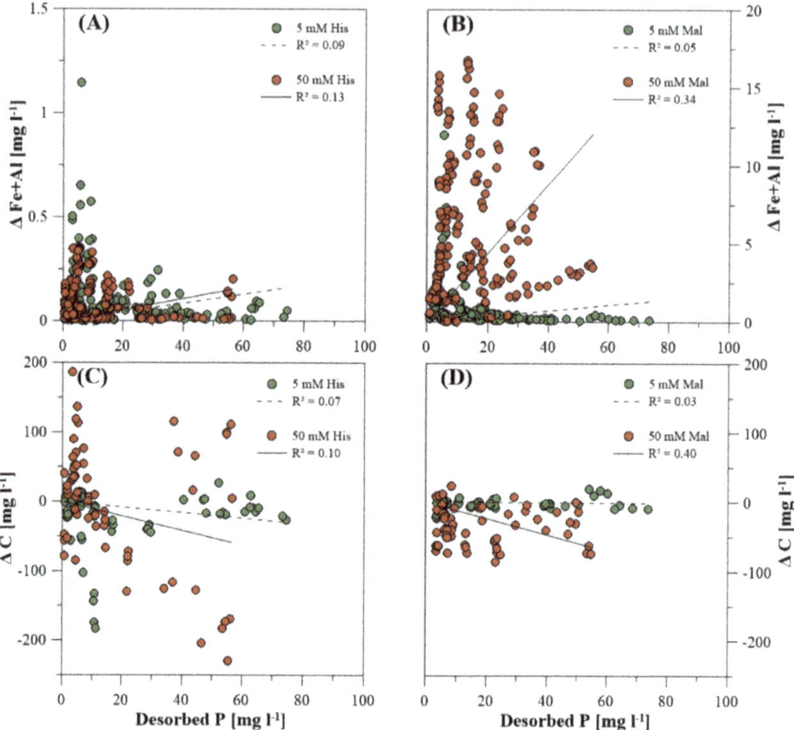

Figure 4. Correlation between desorbed P and the change of (**A**) dissolved total Fe + Al for histidine, (**B**) dissolved total Fe + Al for malic acid, (**C**) dissolved total C for histidine, and (**D**) dissolved total C for malic acid in the sample solution for all hydroxides and each desorption time step.

In addition, for desorption using organic constituents, total C was measured in the sample solution, and the difference from the initial total C concentration (240 mg L^{-1} for Mal and 3400 mg L^{-1} for His, respectively) to total C in the sample solution was calculated. Again, no clear relationship to the amount of desorbed P was observed, whereas the C concentration tended to decrease (Figure 4C,D).

3.4. Solution pH

The pH of the investigated hydroxides and the reaction solutions was adjusted to 6 prior to the experiments. After a 2 h desorption time, the pH of the sample solutions increased for nearly all treatments and hydroxides (Table 7). Only goethite, ferrihydrite, 5 Fe: 1 Al, and 1 Fe: 1 Al had a decreasing or constant pH during desorption using 50 mM KCl. The same was observed for goethite and ferrihydrite during desorption using 50 mM His. No correlation was detected for the change of solution pH and the amount of desorbed P (not shown).

Table 7. Change of H^+ concentrations in the sample solutions after 2 h desorption time.

Hydroxide	$\Delta\ c(H^+)$ $(10^{-7}$ mol $L^{-1})$							
	KCl		KNO$_3$		His		Mal	
	5 mM	50 mM	5 mM	50 mM	5 mM	50 mM	5 mM	50 mM
Goethite	-0.79 ± 0.00	1.18 ± 0.07	-3.36 ± 0.04	-2.91 ± 0.00	-3.44 ± 0.04	0.45 ± 0.64	-2.43 ± 0.00	-3.38 ± 0.46
Gibbsite	-2.25 ± 0.04	-0.48 ± 0.01	-4.24 ± 0.01	-4.11 ± 0.03	-3.84 ± 0.26	0.00 ± 0.24	-2.73 ± 0.04	-4.44 ± 0.39
Ferrihydrite	-1.87 ± 0.11	0.26 ± 0.30	-3.47 ± 0.16	-2.85 ± 0.20	-4.30 ± 0.24	-0.57 ± 0.11	-3.27 ± 0.01	-3.61 ± 0.78
1 Fe: 0 Al	-2.30 ± 0.03	-0.44 ± 0.01	-4.18 ± 0.02	-4.06 ± 0.01	-5.71 ± 0.01	-1.72 ± 0.09	-3.05 ± 0.01	-7.08 ± 0.22
5 Fe: 1 Al	-2.07 ± 0.00	0.17 ± 0.21	-3.58 ± 0.07	-3.39 ± 0.08	-5.39 ± 0.03	-1.52 ± 0.19	-2.89 ± 0.01	-5.34 ± 0.17
1 Fe: 1 Al	-1.96 ± 0.11	0.00 ± 0.09	-3.40 ± 0.15	-2.75 ± 0.38	-4.77 ± 0.16	-1.17 ± 0.30	-2.88 ± 0.02	-3.86 ± 0.43
1 Fe: 5 Al	-2.44 ± 0.01	-0.57 ± 0.06	-4.36 ± 0.06	-4.31 ± 0.02	-5.88 ± 0.02	-2.51 ± 0.32	-3.18 ± 0.04	-6.53 ± 0.19
0 Fe: 1 Al	-2.44 ± 0.01	-0.63 ± 0.04	-4.46 ± 0.00	-4.33 ± 0.03	-5.78 ± 0.18	-3.78 ± 0.24	-3.21 ± 0.01	-8.88 ± 0.51

4. Discussion

4.1. Influence of Crystallinity and Fe/Al Content on P Sorption

During the preparation of the desorption experiments, it was already shown that the different hydroxides had different reactive surface areas and hence, individual P adsorption capacities. It is well known that P adsorption on Fe- and Al-hydroxides occurs via inner-sphere complex formation [15,25,26,28,29,34], but also by surface precipitation [24,29,35,36], at which the crystallinity grade of the hydroxides played an important role. For goethite, several studies described the formation of either monodentate inner-sphere complexes [36,37] or bidentate complexes [23,38] with a minor fraction of monodentate complexes [23] in the intermediate pH range. Li and Stanforth (2000) [10] observed a more negative surface charge of goethite due to the replacement of surface OH groups by protonated and more acidic P anions, led to further decreasing P adsorption. For gibbsite, the formation of simultaneously existent monodentate and bidentate surface complexes with hydrogen-bonding to outer-sphere complexes was concluded [28,29]. In addition, the formed inner-sphere complex was described as a precursor for Al-P precipitation [28,29,35], decreasing with increasing pH due to the increased solubility of Al-phosphates at pH values around 6 [35].

Studies on poorly crystalline ferrihydrite divided the process of P adsorption into the formation of a monodentate [15,26] or bidentate inner-sphere complexes [22,25], the migration of P to surface sorption sites of decreasing accessibility within the particles [39], and with longer equilibration time also the formation of stable Fe-P precipitate [24,36]. For the amorphous Fe-hydroxide, the preferred formation of bidentate surface complexes as well as the formation of Fe-P precipitate with increasing P concentration and equilibration time was described, whereas for the amorphous Al-hydroxides prevalent monodentate inner-sphere complexes were reasoned. In the hydroxide mixtures, the Fe content is particularly contributed to a stable P fixation by precipitation reactions [24], whereby also P bindings via inner-sphere complexes were formed [40].

Summarized, the higher accessibility of both surface and structural binding sites of amorphous hydroxides led to a higher amount of adsorbed and precipitated P compared to well crystalline hydroxides, underlines the important role of amorphous Al and Fe fractions for the release of labile P in soils [41]. The more rigid, but poorly crystalline character allowed the migration of P into mineral particles, which also enabled a stable and effective P adsorption related to the specific surface area [24]. The decrease of initially P adsorption on the pure amorphous Al-hydroxide compared to the hydroxide mixtures with predominant Al content can, thus, also be attributed to a transitional phase with greater crystallization observed in the sample [24,34], which was already indicated by the XRD measurements in the present study. It is therefore possible that the release of adsorbed P from the inner mineral particle surface by means of anion exchange, in particular of more complex organic anions, is sterically inhibited by the initial crystallization.

As a consequence, the different binding motifs also have an impact on P mobilization. "Nonspecific" physisorption via electrostatic attraction provides lower binding energy and thus easy mobilization of P via ion exchange [5,6] while more "specific" chemisorption results in stronger binding at the particle surface and lower availability of P over time [7,8]. However, the most stable and long-lasting P fixation occurs via precipitation on the particle surface [9,10]. The preferred formation of Fe-P precipitates during adsorption on ferrihydrite and the amorphous hydroxides with predominant Fe amount led to a more stable P binding, and hence, lower desorption capacities than the crystalline hydroxides. With increasing Al content, the influence of the surface complexes also increased, which is why the desorption capacity increased, in this study independently from the type of extracting agent.

4.2. Effect of PZC and Electrolytes on P Adsorption

In addition to crystallinity or Fe/Al ratio, surface properties of the hydroxides also play a major role regarding exchange reactions. Besides the specific reactive surface area, a varying PZC influences sorption processes. The reported PZC from literature for goethite varies from 6.4 to 9.7 [10,42,43], are in the range from 7.5 to 11.3 for gibbsite [44–47], and between 7.6 and 8.0 for ferrihydrite [46,48]. Sujana et al. (2009) [33] reported PZC values of the amorphous Fe- and Al-hydroxide mixtures in a range from 4.8 to 6.1, increasing with an increasing amount of Al. In the present study, the measured PZC values of the amorphous Fe- and Al-hydroxides were significantly higher and in a range from 6.0 to 9.8, also increasing with increasing Al-amount. As the experimental pH was set to a value of 6, the positive net charge and thus a positive electric potential below the PZC values led to a charge surplus, and hence, stronger adsorption of anions. If the pH of the surrounding solution will be increased, the positive potential decreases and becomes negative at pH values above the PZC, constraining further specific adsorption.

But simultaneously with P adsorption, also sorption of protons (H^+) takes place [49]. Along with the P adsorption process, a higher surface charge was produced by diffusion of protons from and to the hydroxide surface [43,50]. Similar was observed for background electrolyte solutions such as NaCl, $NaNO_3$, or KNO_3. The presence and concentration of electrolytes in the reaction solution can lead to a decrease of the positive electric potential and hence, a weaker P adsorption at low pH (<4) [42,51]. Certainly, the adsorption of electrolytes or P will be mutually affected. Higher adsorption of cations can be supported by a higher surface coverage of the hydroxide with negatively charged P [51]. Although the PZC changes during P adsorption, the amount of bound P can be affected, depending on the surrounding pH. In combination with the different binding motifs, depending on the crystallinity and the Fe-amount of the hydroxides, the later release of P can be influenced.

4.3. Desorption Kinetics

Similar to the observations for P adsorption kinetics, desorption showed a biphasic behavior with a first rapid and a second slower stage, which was described in previous studies for Fe- and Al-hydroxides by applying organic and inorganic reagents [11,26,52]. A great P release within the first 24 h of desorption time was observed for all investigated hydroxides, independent of their degree of crystallization or the Fe/Al ratio. However, compared to the crystalline hydroxides, ferrihydrite, and the amorphous Fe- and Al-hydroxide mixtures showed in general lower release rates and a continuous P mobilization. This was shown in both the P desorption kinetic curves and the calculated kinetic parameters. This typical time-dependent trend can be attributed to the different P binding mechanisms of the crystalline and amorphous hydroxides [11,24,53] and therefore, an easier release of P from weaker outer-sphere bindings [54] and nonspecific adsorbed P from low-affinity sites, followed by a slower release of specific adsorbed P from high-affinity sites as well as diffusion of structural bound P [55]. Thus, during the adsorption process, related to the specific surface, comparatively more P was bound to goethite or gibbsite than to ferrihydrite, but this P can be released again in the short-term. Meanwhile, the binding to amorphous hydroxides mainly contributes to a long-term release and a distinct ongoing mobilization over time. The better fit of the Elovich equation suggested that P was desorbed by chemisorption reactions, which was corroborated by previous studies using soil [53,56,57]. For ferrihydrite, in particular, the better applicability of the exponential function suggested that P mobilization is slow at first and stronger with time. This would be in good agreement with an inner-particulate P binding, which precedes the actual desorption with a diffusion phase and the migration of inner-particle bound P.

But independently of the desorption reactions, the high calculated values of both α (goethite and gibbsite) and β (ferrihydrite and amorphous hydroxide mixtures with predominant Fe) do not seem realistic compared to the measured values. For example, a cumulative P desorption of 3.44 ± 0.22 mg g^{-1} after 336 h desorption time was measured for ferrihydrite. This equals 0.01 ± 0.00 mg m^{-2}, while the calculated P release constant over time β for the KCl treatment in the Elovich model amounted 392.88 mg min^{-1} m^{-2}. Since an empirical model like the Elovich equation describes processes in an ideal system, an interpretation of the values can be difficult or misleading. An application to less ideal or even natural systems such as soils can therefore be problematic [58]. An increase or decrease of the fit parameters can display a change of reaction rates, whereas the slope of the function depends more on the reaction conditions than on their characterization. Therefore, it is possible to overestimate the initial or mid-term release due to either sharp or weak curvature of the P desorption kinetics [58].

4.4. Inorganic Extracting Agents

A further aspect of P release is the mechanism of action and, thus, the efficiency of both organic and inorganic extraction agents. Concerning inorganic anions, ion exchange reactions are the main mechanisms during nutrient mobilization in general and P desorption in particular [1,59]. While hydrated monovalent ions such as K$^+$, Cl$^-$, or NO$_3$$^-$ usually form weak non-spherical complexes on oppositely charged surfaces, P can be attached via ligand exchange to the hydroxide surfaces [6]. However, the anion exchange of adsorbed P by Cl$^-$ and NO$_3$$^-$ had still a clear effect on the crystalline hydroxides goethite and gibbsite (from 40 to 81% desorption capacity), but its effectiveness was lower for ferrihydrite and the amorphous Fe- and Al-hydroxides.

The successively decreasing desorption and the incomplete release of P, which varies depending on the hydroxide, can be explained, among other factors, by the change of anion exchange sites as a result of P adsorption, which was partly irreversible with respect to adsorption of NO$_3$$^-$ and Cl$^-$ [34]. During P desorption using KCl and KNO$_3$, the concentration of total Cl and N decreased with increasing P release, whereby the effect was more pronounced for KNO$_3$. This gives an indication of anion exchange reactions, where the anion concentration can vary greatly depending on the extracting agent and investigated hydroxide. However, in the absence of a clear trend of the anion concentration change in the reaction solution, mainly equilibrium reactions between the solid hydroxide surface and the reaction solution took place.

4.5. Organic Extracting Agents

If both P and organic anions were present, the adsorption of P can already be affected by the competition for adsorption sites, dissolution of adsorbents, change of the adsorbents surface charge, the formation of new adsorption sites by formation of metal-organic complexes through adsorption of metal ions (e.g., Al^{3+}, Fe^{3+}), as well as the retardation of crystal growth of poorly crystalline Fe- and Al-oxides and hydroxides [14–16,60,61]. If P was already adsorbed and hence, fixed, the further release can be controlled by the dissolution of low soluble minerals, ligand exchange and the replacement P by organic anions, as well as the formation of metal-organic complexes and thus the blocking of adsorption sites [12,17,62–64].

The measurement of total C showed no clear trend supporting concentration changes of organic anions during desorption using His and Mal, even though a stronger decrease was correlated for the 50 mM treatments. It can also be assumed that, in addition to anion exchange, equilibrium reactions took place. If the changes of the Fe and Al concentrations in the reaction solution were considered, in particular, more Fe and Al dissolved from the amorphous hydroxides with increasing desorption. Due to the higher Fe and Al concentrations at the beginning of desorption as well as the low release from the crystalline hydroxides, dissolution of the hydroxides by organic reactants, in particular, was assumed to be improbable. Moreover, the use of the organic reagents His and Mal showed a distinctive higher release from crystalline than from amorphous hydroxides. Certainly, the amount of released P was similar between organic and inorganic extracting agents, and, therefore, a clear beneficial influence of the organic compounds was not detected.

Basak (2019) [13] reported an average amount of P released from rock phosphates by organic acids in the range from 0.015 to 83.5% after 6 d reaction time, increasing with an increasing concentration of the acids. The effectiveness of the used acids followed the order: oxalic acid > citric acid > tartaric acid > formic acid > malic acid > succinic acid > acetic acid. However, Basak (2019) [13] also described a decreasing P release for the increase of the organic acid concentration from 0.3 to 0.5 M. Xu et al. (2004) [65] demonstrated that citric acid has the highest capacity to solubilize P from rock and iron phosphates after 24 h reaction time and that an increase in the concentration of organic acids enhanced significantly P solubilization (except oxalic acid). For P release from Fe phosphates, the effectiveness of the organic acids followed the order: citric acid > oxalic acid > malonic acid > tartaric acid > malic acid > acetic acid (not complete). Especially for citric acid, the amount of released P increased from 13.7 to 67.0 mg g^{-1} with an increasing acid concentration from 0.001 to 0.01 M. Wang et al. (2015) [53] studied P release from acidic, neutral, and calcareous soils using low molecular weight organic acids. They reported a high efficiency of oxalic acid (0.6 to 3.2 mg kg^{-1}); followed by citric acid (0.6 to 2.8 mg kg^{-1}) on cumulative released organic P, regardless form the soil type. Concerning inorganic P fractions, oxalic acid was more effective on calcareous soils, while citric acid had the highest amounts of P release from neutral and acidic soils.

While the enhanced release of inorganic P was likely due to accelerating desorption and dissolution processes, the acid strength contributed to the release of organic P. Gypser et al. (2019) [11] showed a clear lower effect of inorganic constituents than organic acids on P release from Fe- and Al-hydroxides using a concentration of 0.01 M at pH 6 over 1344 h reaction time. While P desorption using CaCl$_2$ and CaSO$_4$ amounted between 0.0 and 57.4%, humic acid showed a desorption capacity in the range from 0.3 to 87.2%. Citric acid had the highest P release in the range from 6.7 to 90.5%. Moreover, desorption increased with increasing crystallinity grade and Al content of the hydroxides. However, the reaction time is a crucial factor in P release, also with respect to P that was strongly bound on amorphous hydroxides. Thus, desorption reactions were still detected after an experimental reaction time of 8 weeks [11]. Taghipour and Jalali (2013) [66] also reported a lower efficiency of malic acid in comparison to citric and oxalic acid for calcareous soils. In particular, the chemical structure, type, and location of the functional groups of the ligands of organic acids influence the efficiency of P mobilization, where di- and tri-carboxylic acids were more effective during mobilization than mono-carboxylic acids [63]. In terms of carboxylation, the P release capacity of Mal (di-carboxylic) is expected to be higher than that of His (mono-carboxylic). In the present study, the release capacities of both compounds were equal for most hydroxides or only slightly higher for Mal. Therefore, the effect related to the number of carboxylic groups may have been relativized by at least one other mechanism.

Although the above-mentioned previous studies have shown that organic compounds support P release, this effect could not be observed in comparison with the inorganic compounds. One possible reason can be organic molecules acting as P adsorbing surfaces in some circumstances [67], and form loosely surface-bound complexes with already released P. Especially His has its isoelectric point in a neutral pH range (7.47 [68]) and can act both as a proton donor as well as an acceptor. In addition, it has a simple aromatic ring and is therefore considered more stable than Mal [69]. This is relevant, considering that P could be adsorbed on His and thus remained in soils, but can be released over a longer time. Another reason, which is essentially also related to the complex structure of organic molecules, is the formation of a "physical barrier" on the mineral surface and hence, limiting P desorption [65,70].

As mentioned above, several processes can be considered for the release of initially adsorbed P by organic constituents. These mechanisms can take place separately or in combination. A mechanism that can take place both when P is adsorbed and when P and organic anions occur simultaneously in the solution, is the dissolution of the adsorbent, which is responsible for a very effective P release. In particular, the dissolution reveals a clear relationship between released P and major components of the adsorbent [71]. In the present study, the dissolved concentrations of Fe and Al in comparison to the initial Fe and Al contents of the hydroxides (see Table 2) were too low to indicate the dissolution of the hydroxides during P release. Furthermore, C_{Total} fluctuated around the initial concentration of C_{Total} during the experiments with 5 mM His and Mal, while C_{Total} decreased using 50 mM His and Mal. The concentrations of Fe and Al slightly increased at the beginning of P desorption and fluctuated around zero during ongoing P mobilization, which rather indicated an equilibrium reaction between solid and liquid phase than the formation of metal-organic complexes in the reaction solution. Thus, ligand exchange and the replacement of P by organic anions were concluded to be the predominant mechanisms of P release by His and Mal. At the same time, the sorption of anions can induce adsorption of H^+ and thus, explain the increasing pH in the reaction solution.

Initially, it was expected that a higher concentration of the respective reactant would also increase P release. In the present study, the opposite was observed. A possible explanation could be an increasing P co-adsorption of already desorbed P, and the formation of outer-sphere complexes by electrostatic interaction. Due to the addition of negative charge by adsorption of organic anions, the electrostatic repulsive force decreased and induced H^+ adsorption as mentioned above. This more positive surface charge influenced the sorption behavior of P on the mineral surface. The similar was observed for Ca^{2+} [42,72], and Na^+ [30] as background electrolytes. Duputel et al. (2013) [73] reported a decrease of available P for citrate concentrations below 20 µM due to large adsorption of citrate, enhancing Ca^{2+} adsorption and facilitate P binding through Ca-bridging. Hence, it can be assumed that a purely additive effect is invalid or limited to a small range [34].

5. Conclusions

The Fe- and Al-hydroxides showed different capacities to retain inorganic P, depending on the crystallinity of the hydroxides and, thus, the specific P binding motifs, govern the extent and strength of desorption.

The poorly crystalline and amorphous hydroxides especially contribute to a stable fixation of P. In addition, the proportion of Fe and Al plays a considerable role. Precipitation of poorly soluble Fe-phosphates inhibits or prevents effective short- to medium-term P mobilization from amorphous Fe-hydroxides. Al-hydroxides adsorb more P related to the specific reactive surface area than Fe-hydroxides, but they also show a substantially greater P release over time. Certainly, the release of adsorbed P from the inner mineral particle surface through complex organic anions can be sterically inhibited by an initial crystallization process.

In the present study, desorption using His and Mal might not be expected to substantially influence P desorption compared to selected inorganic constituents of the soil solution. An increase in concentration tends to have a detrimental effect on P release as well. It was suggested that organic molecules act as P adsorbing surfaces, which is relevant, considering that P could be adsorbed on His, but can be released over a longer time. Another reason, which is essentially also related to the complex structure of organic molecules, is the formation of a "physical barrier" on the mineral surface and, hence, limiting P desorption.

The assumption that the majority of P fertilization is stably bound to components of the soil and thus permanently unavailable to plants can therefore not be supported. The implication of these results for a sustainable P management in agricultural soils is that the consideration of the reactions of the P ions with soil particles should be included in balancing of potential plant-available P. There will be a transfer of inorganic P from very low/low available P pools (amorphous hydroxides) and low/readily available P pools (crystalline Fe-hydroxides, Al-hydroxides), but with time as one decisive factor. Therefore, a combination of extraction methods should be chosen for a comprehensive characterization of the current soil P status, and a prediction of the potential recovery of P reserves. Thus, at least the questions of the right rate and time of fertilizer application can be taken into account to supply plants with previously unused P reserves, and to reduce fertilizer or to use them more efficiently.

However, P adsorption and desorption will vary greatly in natural systems compared to purely artificial systems, as various physico–chemical properties significantly limit the assumption of pure additive effects. For an accurate calculation, a transfer from the lab to the field and the consideration of further factors, such as soil organic matter, pH, or crop type is necessary. The purpose should be the establishment of advanced methods and protocols to evaluate these implications and to link soil research with agronomic implementations.

Author Contributions: Conceptualization, E.S., S.G. and D.F.; methodology, E.S., S.G. and D.F.; investigation, E.S. and S.G.; characterization and chemical analysis, batch experiments, E.S. and S.G.; batch experiments, E.S.; kinetic modeling, S.G.; validation, S.G.; resources, S.G. and D.F.; writing—original draft preparation, S.G.; writing—review and editing, S.G. and D.F.; visualization, S.G.; supervision, D.F.; project administration, D.F.; funding acquisition, D.F. and S.G. All authors have read and agreed to the published version of the manuscript.

Funding: This research was funded by the German Federal Ministry of Education and Research (BMBF) in the BonaRes project InnoSoilPhos (No. 031B509C).

Institutional Review Board Statement: Not applicable.

Informed Consent Statement: Not applicable.

Conflicts of Interest: The authors declare no conflict of interest. The funders had no role in the design of the study; in the collection, analyses, or interpretation of data; in the writing of the manuscript, or in the decision to publish the results.

References

1. Kruse, J.; Abraham, M.; Amelung, W.; Baum, C.; Bol, R.; Kühn, O.; Lewandowski, H.; Niederberger, J.; Oelmann, Y.; Rüger, C.; et al. Innovative methods in soil phosphorus research: A review. *J. Plant Nutr. Soil Sci.* **2015**, *178*, 43–88. [CrossRef] [PubMed]
2. Celi, L.; Prati, M.; Magnacca, G.; Santoro, V.; Martin, M. Role of crystalline iron oxides on stabilization of inositol phosphates in soil. *Geoderma* **2020**, *374*, 114442. [CrossRef]
3. Jiang, X.; Bol, R.; Willbold, S.; Vereecken, H.; Klumpp, E. Speciation and distribution of P associated with Fe and Al oxides in aggregate-sized fraction of an arable soil. *Biogeosciences* **2015**, *12*, 6443–6452. [CrossRef]
4. Weihrauch, C. *Phosphor-Dynamiken in Böden. Grundlagen, Konzepte und Untersuchungen zur räumlichen Verteilung des Nährstoffs*; Springer Spektrum: Wiesbaden, Germany, 2018.
5. Weng, Y.; Vekeman, J.; Zhang, H.; Chou, L.; Elskens, M.; Tielens, F. Unravelling phosphate adsorption on hydrous ferric oxide surfaces at the molecular level. *Chemosphere* **2020**, *261*, 127776. [CrossRef] [PubMed]
6. Blume, H.-P.; Brümmer, G.W.; Horn, R.; Kandeler, E.; Kögel-Knabner, I.; Kretzschmar, R.; Stahr, K.; Wilke, B.-M. *Scheffer/Schachtschabel Lehrbuch der Bodenkunde*, 16th ed.; Springer Spektrum: Berlin/Heidelberg, Germany, 2016.
7. Mnthambala, F.; Maida, J.; Lowole, M.W.; Kabambe, V.H. Phosphorus sorption and external phosphorus requirements of ultisols and oxisols in Malawi. *J. Soil Sci. Environ. Manag.* **2015**, *6*, 35–41.
8. Reddy, K.R.; DeLaune, R.D. *Biogeochemistry of Wetlands: Science and Applications*; CRC Press Taylor & Francis Group: Boca Raton, FL, USA; London, UK; New York, NY, USA, 2008.
9. Lang, F.; Bauhus, J.; Frossard, E.; George, E.; Kaiser, K.; Kaupenjohann, M.; Krüger, J.; Matzner, E.; Polle, A.; Prietzel, J.; et al. Phosphorus in forest ecosystems: New insights from an ecosystem nutrition perspective. *J. Plant Nutr. Soil Sci.* **2016**, *179*, 129–135. [CrossRef]
10. Li, L.; Stanforth, R. Distinguishing adsorption and surface precipitation of phosphate on goethite (alpha-FeOOH). *J. Colloid Interf. Sci.* **2000**, *230*, 12–21. [CrossRef]

11. Gypser, S.; Schütze, E.; Freese, D. Crystallization of single and binary iron- and aluminum hydroxides affect phosphorus desorption. *J. Plant Nutr. Soil Sci.* **2019**, *182*, 741–750. [CrossRef]

12. Bais, H.P.; Weir, T.L.; Perry, L.G.; Gilroy, S.; Vivanco, J.M. The role of root exudates in rhizosphere interactions with plants and other organisms. *Annu. Rev. Plant Biol.* **2006**, *57*, 233–266. [CrossRef]

13. Basak, B.B. Phosphorus release by low molecular weight organic acids from low-grade indian rock phosphate. *Waste Biomass Valor.* **2019**, *10*, 3225–3233. [CrossRef]

14. Bhatti, J.S.; Comerford, N.B.; Johnston, C.T. Influence of oxalate and soil organic matter on sorption and desorption of phosphate onto a spodic horizon. *Soil Sci. Soc. Am. J.* **1998**, *62*, 1089–1095. [CrossRef]

15. Borggaard, O.K.; Raben-Lange, B.; Gimsing, A.L.; Strobel, B.W. Influence of humic substances on phosphate adsorption by aluminium and iron oxides. *Geoderma* **2005**, *127*, 270–279. [CrossRef]

16. Hinsinger, P. Bioavailability of soil inorganic P in the rhizosphere as affected by root-induced chemical changes: A review. *Plant Soil* **2001**, *237*, 173–195. [CrossRef]

17. Johnson, S.E.; Loeppert, R.H. Role of organic acids in phosphate mobilization from iron oxide. *Soil Sci. Soc. Am. J.* **2006**, *70*, 222–234. [CrossRef]

18. Roberts, T.L.; Johnston, A.E. Phosphorus use efficiency and management in agriculture. *Resour. Conserv. Recycl.* **2015**, *105*, 275–281. [CrossRef]

19. Buczko, U.; van Laak, M.; Eichler-Löbermann, B.; Gans, W.; Merbach, I.; Panten, K.; Peiter, E.; Reitz, T.; Spiegel, H.; von Tucher, S. Re-evaluation of the yield response to phosphorus fertilization based on meta-analyses of long-term field experiments. *Ambio* **2018**, *47*, 50–61. [CrossRef] [PubMed]

20. Syers, J.K.; Johnston, A.E.; Curtin, D. *Efficiency of Soil and Fertilizer Phosphorus Use: Reconciling Changing Concepts of Soil Phosphorus Behaviour with Agronomic Information*; FAO Fertilizer and Plant Nutrition No. 18; Food and Agriculture Organization of the United Nations: Rome, Italy, 2008.

21. van der Bom, F.; McLaren, T.I.; Doolette, A.L.; Magid, J.; Frossard, E.; Oberson, A.; Jensen, L.S. Influence of long-term phosphorus fertilisation history on the availability and chemical nature of soil phosphorus. *Geoderma* **2019**, *355*, 113909. [CrossRef]

22. Antelo, J.; Fiol, S.; Pérez, C.; Mariño, S.; Arce, F.; Gondar, D.; López, R. Analysis of phosphate adsorption onto ferrihydrite using the CD-MUSIC model. *J. Colloid Interf. Sci.* **2010**, *347*, 112–119. [CrossRef]

23. Ahmed, A.A.; Gypser, S.; Leinweber, P.; Freese, D.; Kühn, O. Infrared spectroscopic characterization of phosphate binding at the goethite-water interface. *Phys. Chem. Chem. Phys.* **2019**, *21*, 4421–4434. [CrossRef]

24. Gypser, S.; Hirsch, F.; Schleicher, A.M.; Freese, D. Impact of crystalline and amorphous iron- and aluminum hydroxides on mechanisms of phosphate adsorption and desorption. *J. Environ. Sci.* **2018**, *70*, 175–189. [CrossRef]

25. Khare, N.; Martin, J.D.; Hesterberg, D. Phosphate bonding configuration on ferrihydrite based on molecular orbital calculations and XANES fingerprinting. *Geochim. Cosmochim. Acta* **2007**, *71*, 4405–4415. [CrossRef]

26. Krumina, L.; Kenney, J.P.; Loring, J.S.; Persson, P. Desorption mechanisms of phosphate from ferrihydrite and goethite surfaces. *Chem. Geol.* **2016**, *427*, 54–64. [CrossRef]

27. Kubicki, J.D.; Kwon, K.D.; Paul, K.W.; Sparks, D.L. Surface complex structures modelled with quantum chemical calculations: Carbonate, phosphate, sulphate, arsenate and arsenite. *Eur. J. Soil Sci.* **2007**, *58*, 932–944. [CrossRef]

28. van Emmerik, T.J.; Sandström, D.E.; Antzutkin, O.N.; Angove, M.J.; Johnson, B.B. 31P solid-state nuclear magnetic resonance study of the sorption of phosphate onto gibbsite and kaolinite. *Langmuir* **2007**, *23*, 3205–3213. [CrossRef] [PubMed]

29. Zheng, T.-T.; Sun, Z.-X.; Yang, X.-F.; Holmgren, A. Sorption of phosphate onto mesoporous γ-alumina studied with in-situ ATR-FTIR spectroscopy. *Chem. Cent. J.* **2012**, *6*, 26. [CrossRef]

30. Arai, Y.; Sparks, D.L. ATR–FTIR spectroscopic investigation on phosphate adsorption mechanisms at the ferrihydrite–water interface. *J. Colloid Interf. Sci.* **2001**, *241*, 317–326. [CrossRef]

31. Tiessen, H.; Moir, J.O. Characterization of available P by sequential extraction. In *Soil Sampling and Methods of Analysis*; Carter, M.R., Gregorich, E.G., Eds.; CRC Press: Boca Raton, FL, USA, 2007; ISBN 9780429126222.

32. Schwertmann, U.; Cornell, R.M. *Iron Oxides in the Laboratory: Preparation and Characterization*; Wiley-VCH: Hoboken, NJ, USA, 2008.

33. Sujana, M.G.; Soma, G.; Vasumathi, N.; Anand, S. Studies on fluoride adsorption capacities of amorphous Fe/Al mixed hydroxides from aqueous solutions. *J. Fluor. Chem.* **2009**, *130*, 749–754. [CrossRef]

34. Harvey, O.R.; Rhue, R.D. Kinetics and energetics of phosphate sorption in a multi-component Al(III)-Fe(III) hydr(oxide) sorbent system. *J. Colloid Interf. Sci.* **2008**, *322*, 384–393. [CrossRef]

35. Johnson, B.B.; Ivanov, A.V.; Antzutkin, O.N.; Forsling, W. 31P nuclear magnetic resonance study of the adsorption of phosphate and phenyl phosphates on γ-Al$_2$O$_3$. *Langmuir* **2002**, *18*, 1104–1111. [CrossRef]

36. Persson, P.; Nilsson, N.; Sjöberg, S. Structure and bonding of orthophosphate ions at the iron oxide-aqueous interface. *J. Colloid Interf. Sci.* **1996**, *177*, 263–275. [CrossRef]

37. Loring, J.S.; Sandström, M.H.; Norén, K.; Persson, P. Rethinking arsenate coordination at the surface of goethite. *Chemistry* **2009**, *15*, 5063–5072. [CrossRef]

38. Luengo, C.; Brigante, M.; Antelo, J.; Avena, M. Kinetics of phosphate adsorption on goethite: Comparing batch adsorption and ATR-IR measurements. *J. Colloid Interf. Sci.* **2006**, *300*, 511–518. [CrossRef] [PubMed]

39. Torrent, J. Fast and slow phosphate sorption by goethite-rich natural materials. *Clay. Clay. Miner.* **1992**, *40*, 14–21. [CrossRef]

40. Lǚ, J.; Liu, H.; Liu, R.; Zhao, X.; Sun, L.; Qu, J. Adsorptive removal of phosphate by a nanostructured Fe–Al–Mn trimetal oxide adsorbent. *Powder Technol.* **2013**, *233*, 146–154. [CrossRef]
41. Arai, Y.; Livi, K.J. Underassessed phosphorus fixation mechanisms in soil sand fraction. *Geoderma* **2013**, *192*, 422–429. [CrossRef]
42. Antelo, J.; Avena, M.; Fiol, S.; López, R.; Arce, F. Effects of pH and ionic strength on the adsorption of phosphate and arsenate at the goethite-water interface. *J. Colloid Interf. Sci.* **2005**, *285*, 476–486. [CrossRef]
43. Strauss, R.; Brümmer, G.W.; Barrow, N.J. Effects of crystallinity of goethite: II. Rates of sorption and desorption of phosphate. *Eur. J. Soil Sci.* **1997**, *48*, 101–114. [CrossRef]
44. Adekola, F.; Fédoroff, M.; Geckeis, H.; Kupcik, T.; Lefèvre, G.; Lützenkirchen, J.; Plaschke, M.; Preocanin, T.; Rabung, T.; Schild, D. Characterization of acid-base properties of two gibbsite samples in the context of literature results. *J. Colloid Interf. Sci.* **2011**, *354*, 306–317. [CrossRef]
45. Jodin, M.-C.; Gaboriaud, F.; Humbert, B. Limitations of potentiometric studies to determine the surface charge of gibbsite gamma-Al(OH)$_3$ particles. *J. Colloid Interf. Sci.* **2005**, *287*, 581–591. [CrossRef]
46. Kosmulski, M. Compilation of PZC and IEP of sparingly soluble metal oxides and hydroxides from literature. *Adv. Colloid Interf. Sci.* **2009**, *152*, 14–25. [CrossRef]
47. Liu, J.; Wang, X.; Lin, C.-L.; Miller, J.D. Significance of particle aggregation in the reverse flotation of kaolinite from bauxite ore. *Miner. Eng.* **2015**, *78*, 58–65. [CrossRef]
48. Cornell, R.M.; Schwertmann, U. *The Iron Oxides. Structure, Properties, Reactions, Occurences and Uses*, 2nd ed.; Wiley-VCH: Weinheim, Germany, 2003.
49. Hiemstra, T.; van Riemsdijk, W.H. A surface structural approach to ion adsorption: The charge distribution (CD) model. *J. Colloid Interf. Sci.* **1996**, *179*, 488–508. [CrossRef]
50. Ahmed, A.A.; Gypser, S.; Freese, D.; Leinweber, P.; Kühn, O. Molecular level picture of the interplay between pH and phosphate binding at the goethite-water interface. *Phys. Chem. Chem. Phys.* **2020**, *22*, 26509–26524. [CrossRef] [PubMed]
51. Rietra, R.P.; Hiemstra, T.; van Riemsdijk, W.H. Interaction between calcium and phosphate adsorption on goethite. *Environ. Sci. Technol.* **2001**, *35*, 3369–3374. [CrossRef]
52. Wang, X.; Phillips, B.; Boily, J.-F.; Hu, Y.; Hu, Z.; Yang, P.; Feng, X.; Xu, W.; Zhu, M. Phosphate sorption speciation and precipitation mechanisms on amorphous aluminum hydroxide. *Soil Syst.* **2019**, *3*, 20. [CrossRef]
53. Wang, Y.; Chen, X.; Whalen, J.K.; Cao, Y.; Quan, Z.; Lu, C.; Shi, Y. Kinetics of inorganic and organic phosphorus release influenced by low molecular weight organic acids in calcareous, neutral and acidic soils. *J. Plant Nutr. Soil Sci.* **2015**, *178*, 555–566. [CrossRef]
54. Onishi, B.S.D.; dos Reis Ferreira, C.S.; Urbano, A.; Santos, M.J. Modified hydrotalcite for phosphorus slow-release: Kinetic and sorption-desorption processes in clayey and sandy soils from North of Paraná state (Brazil). *Appl. Clay Sci.* **2020**, *197*, 105759. [CrossRef]
55. Wang, X.; Liu, F.; Tan, W.; Li, W.; Feng, X.; Sparks, D.L. Characteristics of phosphate adsorption-desorption onto ferrihydrite. *Soil Sci.* **2013**, *178*, 1–11. [CrossRef]
56. Shariatmadari, H.; Shirvani, M.; Jafari, A. Phosphorus release kinetics and availability in calcareous soils of selected arid and semiarid toposequences. *Geoderma* **2006**, *132*, 261–272. [CrossRef]
57. Steffens, D. Phosphorus release kinetics and extractable phosphorus after long-term fertilization. *Soil Sci. Soc. Am. J.* **1994**, *58*, 1702. [CrossRef]
58. Lammers, A. *Phosphatformen und Phosphatfreisetzung in hochgedüngten Böden Europas*; Dissertation, Utz: München, Germany, 1997.
59. Arai, Y.; Sparks, D.L. Phosphate reaction dynamics in soils and soil components: A multiscale approach. *Advances Agron.* **2007**, *94*, 135–179.
60. Gerke, J. Phosphate adsorption by humic/Fe-oxide mixtures aged at pH 4 and 7 and by poorly ordered Fe-oxide. *Geoderma* **1993**, *59*, 279–288. [CrossRef]
61. Borggaard, O.K.; Jørgensen, S.S.; Moberg, J.P.; Raben-Lange, B. Influence of organic matter on phosphate adsorption by aluminium and iron oxides in sandy soils. *Eur. J. Soil Sci.* **1990**, *41*, 443–449. [CrossRef]
62. Jones, D.L. Organic acids in the rhizosphere—A critical review. *Plant Soil* **1998**, *205*, 25–44. [CrossRef]
63. Kpomblekou-A, K.; Tabatabai, M.A. Effect of low-molecular weight organic acids on phosphorus release and phytoavailabilty of phosphorus in phosphate rocks added to soils. *Agric. Ecosyst. Environ.* **2003**, *100*, 275–284. [CrossRef]
64. Chen, C.R.; Condron, L.M.; Davis, M.R.; Sherlock, R.R. Effects of plant species on microbial biomass phosphorus and phosphatase activity in a range of grassland soils. *Biol. Fertil. Soils* **2004**, *40*, 313–322. [CrossRef]
65. Xu, R.; Zhu, Y.; Chittleborough, D. Phosphorus release from phosphate rock and iron phosphate by low-molecular-weight organic acids. *J. Environ. Sci.* **2004**, *16*, 5–8.
66. Taghipour, M.; Jalali, M. Effect of low-molecular-weight organic acids on kinetics release and fractionation of phosphorus in some calcareous soils of western Iran. *Environ. Monit. Assess.* **2013**, *185*, 5471–5482. [CrossRef]
67. Zhu, J.; Li, M.; Whelan, M. Phosphorus activators contribute to legacy phosphorus availability in agricultural soils: A review. *Sci. Total Environ.* **2018**, *612*, 522–537. [CrossRef]
68. Vinu, A.; Hossain, K.Z.; Satish Kumar, G.; Ariga, K. Adsorption of l-histidine over mesoporous carbon molecular sieves. *Carbon* **2006**, *44*, 530–536. [CrossRef]
69. Lathrop, E.C. The organic nitrogen compounds of soils and fertilizers. *J. Frankl. Inst.* **1917**, 303–321. [CrossRef]

70. Mimmo, T.; Ghizzi, M.; Marzadori, C.; Gessa, C.E. Organic acid extraction from rhizosphere soil: Effect of field-moist, dried and frozen samples. *Plant Soil* **2008**, *312*, 175–184. [CrossRef]
71. Schütze, E.; Gypser, S.; Freese, D. Kinetics of phosphorus release from vivianite, hydroxyapatite, and bone char influenced by organic and inorganic compounds. *Soil Syst.* **2020**, *4*, 15. [CrossRef]
72. Talebi Atouei, M.; Rahnemaie, R.; Goli Kalanpa, E.; Davoodi, M.H. Competitive adsorption of magnesium and calcium with phosphate at the goethite water interface: Kinetics, equilibrium and CD-MUSIC modeling. *Chem. Geol.* **2016**, *437*, 19–29. [CrossRef]
73. Duputel, M.; van Hoye, F.; Toucet, J.; Gérard, F. Citrate adsorption can decrease soluble phosphate concentration in soil: Experimental and modeling evidence. *Appl. Geochem.* **2013**, *39*, 85–92. [CrossRef]

MDPI

St. Alban-Anlage 66

4052 Basel

Switzerland

Tel. +41 61 683 77 34

Fax +41 61 302 89 18

www.mdpi.com

Soil Systems Editorial Office

E-mail: soilsystems@mdpi.com

www.mdpi.com/journal/soilsystems